T0351084

Undergraduate Texts in Mathematics

Editors
S. Axler
F. W. Gehring
K. A. Ribet

Springer
New York
Berlin
Heidelberg
Barcelona
Budapest
Hong Kong
London
Milan
Paris
Tokyo

I. M. Singer
J. A. Thorpe

Lecture Notes on Elementary Topology and Geometry

With 109 Illustrations

Springer

I.M. Singer
Department of Mathematics
Massachusetts Institute of Technology
Cambridge, Massachusetts 02139
USA

J.A. Thorpe
Provost
Queens College CUNY
Flushing, New York 11367
USA

AMS Subject Classifications: 50-01, 53-01, 54-01

Library of Congress Cataloging in Publication Data

Singer, Isadore Manuel, 1924–
    Lecture notes on elementary topology and geometry.

    (Undergraduate texts in mathematics)
    Reprint of the ed. published by Scott, Foresman,
Glenview, Ill.
    Bibliography: p. 230
    Includes index.
    1. Topology. 2. Algebraic topology.
3. Geometry, Differential. I. Thorpe, John A., joint author. II. Title.
[QA611.S498   1976]   514   76-26137

Printed and bound by Braun-Brumfield, Inc., Ann Arbor, MI.
Printed in the United States of America.

9 8 7 6

ISBN 0-387-90202-3 Springer-Verlag New York Berlin Heidelberg
ISBN 3-540-90202-3 Springer-Verlag Berlin Heidelberg New York SPIN 10681565

Preface

At the present time, the average undergraduate mathematics major finds mathematics heavily compartmentalized. After the calculus, he takes a course in analysis and a course in algebra. Depending upon his interests (or those of his department), he takes courses in special topics. If he is exposed to topology, it is usually straightforward point set topology; if he is exposed to geometry, it is usually classical differential geometry. The exciting revelations that there is some unity in mathematics, that fields overlap, that techniques of one field have applications in another, are denied the undergraduate. He must wait until he is well into graduate work to see interconnections, presumably because earlier he doesn't know enough.

These notes are an attempt to break up this compartmentalization, at least in topology-geometry. What the student has learned in algebra and advanced calculus are used to prove some fairly deep results relating geometry, topology, and group theory. (De Rham's theorem, the Gauss-Bonnet theorem for surfaces, the functorial relation of fundamental group to covering space, and surfaces of constant curvature as homogeneous spaces are the most noteworthy examples.)

In the first two chapters the bare essentials of elementary point set topology are set forth with some hint of the subject's application to functional analysis. Chapters 3 and 4 treat fundamental groups, covering spaces, and simplicial complexes. For this approach the authors are indebted to E. Spanier. After some preliminaries in Chapter 5 concerning the theory of manifolds, the De Rham theorem (Chapter 6) is proven as in H. Whitney's *Geometric Integration Theory*. In the two final chapters on Riemannian geometry, the authors follow E. Cartan and S. S. Chern. (In order to avoid Lie group theory in the last two chapters, only oriented 2-dimensional manifolds are treated.)

These notes have been used at M.I.T. for a one-year course in topology and geometry, with prerequisites of at least one semester of modern algebra and one semester of advanced calculus "done right." The class consisted of about seventy students, mostly seniors. The ideas for such a course originated in one of the author's tour of duty for the Committee on the Undergraduate Program in Mathematics of the Mathematical Association of America. A program along these lines, but more ambitious, can be found in the CUPM pamphlet "Pregraduate Preparation of Research Mathematicians" (1963). (See Outline III on surface theory, pp. 68–70.) The authors believe, however, that in lecturing to a large class without a textbook, the material in these notes was about as much as could be covered in a year.

Contents

Contents

Some point set topology 1

1.1 Naive set theory

We shall accept as primitive (undefined) the concepts of a *set* (collection, family) of objects and the concept of an object *belonging to* a set.

We merely remark that, given a set S and an object x, one can determine if the object belongs to (is an element of) the set, written $x \in S$, or if it does not belong to the set, written $x \notin S$.

Definition. Let A and B be sets. A is a *subset* of B, written $A \subset B$, if $x \in A$ implies $x \in B$. A is *equal* to B, written $A = B$, if $A \subset B$ and $B \subset A$.

Notation. The *empty set*, that is, the set with no objects in it, is denoted by $\varnothing$.

Remark
(1) $\varnothing \subset A$ for all sets A.
(2) The empty set $\varnothing$ is unique; that is, any two empty sets are equal. For if $\varnothing_1$ and $\varnothing_2$ are two empty sets, $\varnothing_1 \subset \varnothing_2$ and $\varnothing_2 \subset \varnothing_1$.
(3) $A \subset A$ for all sets A.

Definition. Let A and B be sets. The *union* $A \cup B$ of A and B is the set of all x such that $x \in A$ or $x \in B$, written

$$A \cup B = [x; x \in A \text{ or } x \in B].$$

The *intersection* $A \cap B$ of A and B is defined by

$$A \cap B = [x; x \in A \text{ and } x \in B].$$

Similarly, if $\mathcal{S}$ is a set (collection) of sets, the union and intersection of all the sets in $\mathcal{S}$ are defined respectively by

$$\bigcup_{S \in \mathcal{S}} S = [x; x \in S \text{ for some } S \in \mathcal{S}],$$

$$\bigcap_{S \in \mathcal{S}} S = [x; x \in S \text{ for every } S \in \mathcal{S}].$$

If $A \subset B$, the *complement* of A in B, denoted A' or $B - A$, is defined by

$$A' = [x \in B; x \notin A].$$

Theorem 1. *Let A, B, C, and S be sets. Then*

(1) $A \cup B = B \cup A$.
(2) $A \cap B = B \cap A$.
(3) $(A \cup B) \cup C = A \cup (B \cup C)$.
(4) $(A \cap B) \cap C = A \cap (B \cap C)$.
(5) $A \cup (B \cap C) = (A \cup B) \cap (A \cup C)$.
(6) $A \cap (B \cup C) = (A \cap B) \cup (A \cap C)$.
(7) *If $A \subset S$ and $B \subset S$ then,* $(A \cup B)' = A' \cap B'$.
(8) *If $A \subset S$ and $B \subset S$, then* $(A \cap B)' = A' \cup B'$.
(9) *If $\mathcal{S}_1$ and $\mathcal{S}_2$ are two sets (collections) of sets, then*

$$\left(\bigcup_{S \in \mathcal{S}_1} S \right) \cup \left(\bigcup_{S \in \mathcal{S}_2} S \right) = \bigcup_{S \in \mathcal{S}_1 \cup \mathcal{S}_2} S$$

and

$$\left(\bigcap_{S \in \mathcal{S}_1} S \right) \cap \left(\bigcap_{S \in \mathcal{S}_2} S \right) = \bigcap_{S \in \mathcal{S}_1 \cup \mathcal{S}_2} S.$$

(10) *For $\mathcal{S}_1$ and $\mathcal{S}_2$ as in (9),*

$$\left(\bigcup_{S_1 \in \mathcal{S}_1} S_1 \right) \cap \left(\bigcup_{S_2 \in \mathcal{S}_2} S_2 \right) = \bigcup_{\substack{S_1 \in \mathcal{S}_1 \\ S_2 \in \mathcal{S}_2}} (S_1 \cap S_2).$$

PROOF. The proof of this theorem is left to the student.

Definition. Let A and B be sets. The *Cartesian product* $A \times B$ of A and B is the set of ordered pairs

$$A \times B = [(a, b); a \in A, b \in B].$$

A *relation* between A and B is a subset R of $A \times B$. a and b are said to be *R-related* if $(a, b) \in R$.

EXAMPLE. Let $A = B =$ the set of real numbers. Then $A \times B$ is the plane. The order relation $x < y$ is a relation between A and B. This relation is the shaded set of points in Figure 1.1.

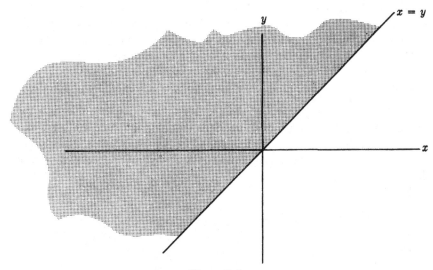

Figure 1.1

Definition. A relation $R \subset A \times A$ is a *partial ordering* if

(1) $(s_1, s_2) \in R$ and $(s_2, s_3) \in R \Rightarrow (s_1, s_3) \in R$ and
(2) $(s_1, s_2) \in R$ and $(s_2, s_1) \in R \Rightarrow s_1 = s_2$.

A relation R is a *simple ordering* if it is a partial ordering, and, in addition:

(3) either $(s_1, s_2) \in R$ or $(s_2, s_1) \in R$ for every pair $s_1, s_2 \in S$.

The order relation for $S =$ real numbers is an example of a simple ordering. In general, we say that S is partially ordered (simply ordered) by R.

Definition. Let A and B be sets. A *function f* mapping A to B, denoted $f: A \to B$, is a relation ($f \subset A \times B$) between A and B satisfying the following properties.

(1) If $a \in A$, then there exists $b \in B$ such that $(a, b) \in f$.
(2) If $(a, b) \in f$ and $(a, b_1) \in f$, then $b = b_1$.

Property (1) says that the function f is defined everywhere on A. Property (2) says that f is a "single-valued" function.

Notation. Let $f: A \to B$. By $f(a) = b$ we mean $(a, b) \in f$.

3

Definition. Let $f: A \to B$. f is *surjective* (onto) if for each $b \in B$ there exists $a \in A$ such that $f(a) = b$. If f is surjective, we write $f(A) = B$. f is *injective* (one-to-one) if $f(a) = f(a_1) \Rightarrow a = a_1$. If f is both surjective and injective, we say f is a *one-to-one correspondence* between A and B.

Definition. A set A is *countable* if there exists a one-to-one correspondence between the set of all integers and A. A set A is *finite* if for some positive integer n there exists a one-to-one correspondence between the set $\{1, \ldots, n\}$ and A, in which case we say A has n elements.

Theorem 2. *If $A = \{a_1, \ldots, a_n\}$ is a finite set of n elements, then the set of all subsets of A has 2^n elements.*

PROOF. Consider the set F of all functions mapping A to the set {YES, NO} consisting of the two elements YES and NO. F has 2^n elements. The set $\mathscr{S}$ of all subsets of A is in one-to-one correspondence with F. For let $f: \mathscr{S} \to F$ be defined as follows.

For $B \in \mathscr{S}$, that is, for $B \subset A$, $f(B)$ is that element of F (that is,

$$f(B): A \to [\text{YES, NO}])$$

given by

$$f(B)(x) = \begin{cases} \text{YES if } x \in B, \\ \text{NO} \ \text{ if } x \notin B. \end{cases}$$

f is injective because if $f(B) = f(C)$, then $f(B)(x) = f(C)(x)$ for all $x \in A$. Thus $f(B)(x) = \text{YES}$ if and only if $f(C)(x) = \text{YES}$; that is, $x \in B$ if and only if $x \in C$. Thus $B = C$. f is surjective because every function $g: A \to [\text{YES, NO}]$ determines a $B \subset A$ by

$$B = [x; g(x) = \text{YES}]$$

and $f(B) = g$. $\qquad \square$

Notation. Motivated by this proposition, we denote by 2^A the set of all subsets of A. Given two sets A and B, B^A denotes the set of all functions $A \to B$.

Definition. Let $f \in B^A$. The *inverse* f^{-1} of f is the function $2^B \to 2^A$ defined by

$$f^{-1}(B_1) = [a \in A; f(a) \in B_1] \qquad (B_1 \subset B).$$

$f^{-1}(B_1)$ is called the *inverse image* of B_1. Note that

$$f^{-1} \in (2^A)^{2^B}.$$

Notation. Let W be a set, and let $\mathscr{S}$ be a collection of sets. We say $\mathscr{S}$ is *indexed* by W if there is given a surjective function $\varphi: W \to \mathcal{S}$. For $w \in W$, we denote $\varphi(w)$ by S_w and denote the *indexing* of $\mathcal{S}$ by W as $\{S_w\}_{w \in W}$.

Definition. Let $\{S_w\}_{w \in W}$ be indexed by W. The *product* of the sets $\{S_w\}_{w \in W}$ is the set

$$\prod_{w \in W} S_w = \left[f: W \to \bigcup_{w \in W} S_w; f(w) \in S_w \text{ for all } w \in W \right].$$

If the set W is not finite, this product is called an *infinite product*. Note that this notion of the product of sets extends the notion of the product of two sets $S_1 \times S_2$. For let $W = \{1, 2\}$, let $\mathcal{S} = \{S_1, S_2\}$, and let $\varphi: W \to \mathcal{S}$ by $\varphi(j) = S_j$, $j = 1, 2$. Then $S_1 \times S_2 = [(s_1, s_2), s_j \in S_j]$, and $\prod_{w \in W} S_w = [f: \{1, 2\} \to S_1 \cup S_2; f(j) \in S_j]$, which can be identified with $[(f(1), f(2)); f(j) \in S_j]$, which can be identified with $S_1 \times S_2$.

Remark. $\prod_{w \in W} S_w$ is a set of functions. One might ask whether there exist any such functions; that is, is $\prod_{w \in W} S_w \neq \varnothing$? In other words, given infinitely many nonempty sets, is it possible to make a choice of one element from each set? It can be shown in axiomatic set theory that this question cannot be answered by appealing to the usual axioms of set theory. We accept the affirmative answer here as an axiom.

Axiom of choice. *Let $\{S_w\}_{w \in W}$ be sets indexed by W. Assume $S_w \neq \varnothing$ for all $w \in W$. Then*

$$\prod_{w \in W} S_w \neq \varnothing.$$

The axiom of choice is equivalent to several other axioms, one of which is the following.

Maximum principle. *If S is partially ordered by R, and T is a simply ordered subset, then there exists a set M such that the following statements are valid.*

(1) *$T \subset M \subset S$.*

(2) *M is simply ordered by R.*

(3) *If $M \subset N \subset S$, and N is simply ordered by R, then $M = N$; that is, M is a maximal simply ordered subset containing T.*

1.2 Topological spaces

Definition. A *metric space* is a set S together with a function $\rho: S \times S \to$ the nonnegative real numbers, such that for each $s_1, s_2, s_3 \in S$:

(1) $\rho(s_1, s_2) = 0$ if and only if $s_1 = s_2$.

(2) $\rho(s_1, s_2) = \rho(s_2, s_1)$.

(3) $\rho(s_1, s_3) \leq \rho(s_1, s_2) + \rho(s_2, s_3)$.

The function ρ is called a *metric* on S.

Given a point s_0 in a metric space S and a real number a, the *ball of radius a about* s_0 is defined to be the set

$$B_{s_0}(a) = [s \in S; \rho(s, s_0) < a].$$

EXAMPLE. Let S be the plane, that is, the product of the set of the real numbers with itself. We define three metrics on S as follows.

For $P_1 = (x_1, y_1)$ and $P_2 = (x_2, y_2)$ two points in S,

$$\rho_1(P_1, P_2) = \sqrt{(x_2 - x_1)^2 + (y_2 - y_1)^2},$$
$$\rho_2(P_1, P_2) = \max\{|x_2 - x_1|, |y_2 - y_1|\},$$
$$\rho_3(P_1, P_2) = |x_2 - x_1| + |y_2 - y_1|.$$

The ball of radius a about the point $0 = (0, 0)$ relative to each of these metrics is indicated by the shaded areas in Figure 1.2. Note that a ball does not necessarily have a circular, or even a smooth, boundary.

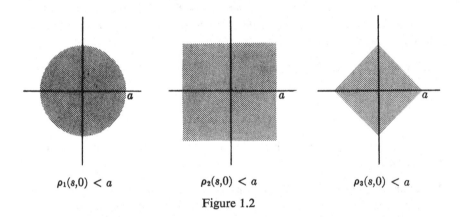

$$\rho_1(s,0) < a \qquad \rho_2(s,0) < a \qquad \rho_3(s,0) < a$$

Figure 1.2

Remark. The three metrics defined above provide the plane with three distinct structures as a metric space. Yet for studying certain properties of these spaces, these metrics are equivalent. Thus, if we want to know, for example, whether 0 is a limit point of a set $T \subset S$, we ask whether there is a sequence of points in T which converges to 0; that is, whether a sequence $\{s_n\}$ of points in T exists such that given any $\varepsilon > 0$, there exists an N such that

$$\rho(s_n, 0) < \varepsilon$$

for all $n > N$. It is not difficult to see that the answer to this question is independent of whichever of the above metrics we use for ρ; that is, given $\varepsilon > 0$, there exists such an N using ρ_1 if and only if there exists such an N using ρ_2, etc. The answer does not depend on the shape of the ball of radius ε, but only on its "fatness" or "openness." For this reason among others, it is

convenient to gather together those properties of a metric space that are
essential for describing "openness" and to use such properties to define a
more abstract structure, a *topological structure*, in which we can still talk
about limit points and in which the three metric structures on the plane
described above will give the same "open sets."

Definition. A *topological space* is a set S together with a collection $\mathcal{U}$ of subsets of S (that is, $\mathcal{U}$ is a subset of 2^S) satisfying the following conditions:

(1a) $\varnothing \in \mathcal{U}, S \in \mathcal{U}$.

(2a) If $U_1, \ldots, U_n \in \mathcal{U}$ then $\bigcap_{i=1}^{n} U_i \in \mathcal{U}$.

(3a) Arbitrary unions of elements in $\mathcal{U}$ lie in $\mathcal{U}$; that is, if $\tilde{\mathcal{U}} \subset \mathcal{U}$, then
$\bigcup_{U \in \tilde{\mathcal{U}}} U \in \mathcal{U}$.

The elements of $\mathcal{U}$ are called *open sets* in S. The collection $\mathcal{U}$ is called a
topology on S.

Remark. We shall often suppress the $\mathcal{U}$ and simply refer to S as a topological space.

Definition. Let $(S, \mathcal{U})$ be a topological space. A set $A \subset S$ is *closed* if it is the
complement of an open set, that is, if $A' \in \mathcal{U}$.

Remark. By taking complements in conditions (1a), (2a), and (3a) above,
one sees that the collection $\mathscr{C}$ of closed sets satisfies the following conditions.

(1b) $\varnothing \in \mathscr{C}, S \in \mathscr{C}$.

(2b) If $A_1, \ldots, A_n \in \mathscr{C}$, then $\bigcup_{i=1}^{n} A_i \in \mathscr{C}$.

(3b) Arbitrary intersections of elements in $\mathscr{C}$ lie in $\mathscr{C}$.

Remark. A topology can be described by specifying the collection of closed
sets equally as well as specifying the collection of open sets.

Definition. Let $(S, \mathcal{U})$ be a topological space. Let $A \subset S$. A point $s \in S$ is a
limit point of A if for each $U \in \mathcal{U}$ such that $s \in U$,

$$(U - \{s\}) \cap A \neq \varnothing.$$

Definition. The *closure* of a set $A \subset S$, denoted by $\bar{A}$, is the set

$$\bar{A} = A \cup [s \in S; s \text{ is a limit point of } A].$$

Theorem 1. *The closure $\bar{A}$ of a set A is closed.*

PROOF. We must show that $\bar{A}'$ is open. For this it suffices to show that for
each $s \in \bar{A}'$ there exists an open set U_s with $s \in U_s \subset \bar{A}'$. Then $s \in U_s$ for each s

7

implies $\bar{A}' \subset \bigcup_{s\in\bar{A}'} U_s$ and $U_s \subset \bar{A}'$ for each s implies $\bigcup_{s\in\bar{A}'} U_s \subset \bar{A}'$. Thus $\bar{A}' = \bigcup_{s\in\bar{A}'} U_s$ is a union of open sets and hence is open.

Now let $s \in \bar{A}'$. Then s is not a limit point of A, so there exists an open set U_s such that $s \in U_s$ and $(U_s - \{s\}) \cap A = \varnothing$. Furthermore, $s \notin A$ because $s \notin \bar{A}$ and hence, in fact, $U_s \cap A = \varnothing$. Since each element of U_s is contained in an open set, namely U_s itself, whose intersection with A is $\varnothing$, it follows that U_s contains no limit points of A and $U_s \cap \bar{A} = \varnothing$; that is, $U_s \subset \bar{A}'$. $\qquad\square$

Theorem 2. *A set A is closed if and only if $A = \bar{A}$.*

PROOF. Assume A is closed. Then A' is open. If $s \notin A$, then A' is an open set containing s such that $(A' - \{s\}) \cap A = \varnothing$. Thus s is not a limit point of A. Hence all limit points of A lie in A; that is, $A = \bar{A}$.

Conversely, if $A = \bar{A}$, then A is closed by the previous theorem. $\qquad\square$

Definition. A set $\mathscr{B} \subset 2^S$ is a *basis* for a topology on S if the following conditions are satisfied:

(1c) $\varnothing \in \mathscr{B}$.
(2c) $\bigcup_{B\in\mathscr{B}} B = S$.
(3c) If B_1 and $B_2 \in \mathscr{B}$, then $B_1 \cap B_2 = \bigcup_{B\in\tilde{\mathscr{B}}} B$ for some subset $\tilde{\mathscr{B}} \subset \mathscr{B}$.

Theorem 3. *Let S be a set and $\mathscr{B}$ be a basis for a topology on S. Let*

$$\mathscr{U}_\mathscr{B} = [U \in 2^S; U \text{ is a union of elements of } \mathscr{B}].$$

Then $\mathscr{U}_\mathscr{B}$ is a topology on S, the topology generated *by $\mathscr{B}$.*

PROOF. We must verify that $\mathscr{U}_\mathscr{B}$ satisfies the three open-set axioms for a topology on S.

By (1c) and (2c) in the definition of a basis, both $\varnothing$ and $S \in \mathscr{U}_\mathscr{B}$ so that condition (1a) in the definition of a topological space is satisfied.

Suppose $\mathscr{V} \subset \mathscr{U}_\mathscr{B}$. Then $\bigcup_{V\in\mathscr{V}} V = \bigcup_{V\in\mathscr{V}} (\bigcup_{B\in\mathscr{B}_V} B) = \bigcup_{B\in\tilde{\mathscr{B}}} B$ where $\mathscr{B}_V \subset \mathscr{B}$ for each $V \in \mathscr{V}$ and $\tilde{\mathscr{B}} = \bigcup_{V\in\mathscr{V}} \mathscr{B}_V \subset \mathscr{B}$. Hence condition (3a) holds.

We prove condition (2a) by induction. We assume that the intersection of k sets in $\mathscr{U}_\mathscr{B}$ lies in $\mathscr{U}_\mathscr{B}$. (For $k = 1$, the statement is automatically true.) Suppose then $U_1, \ldots, U_{k+1} \in \mathscr{U}_\mathscr{B}$. By the inductive hypothesis, $U_1 \cap \cdots \cap U_k \in \mathscr{U}_\mathscr{B}$; that is, there exists a subset $\mathscr{B}_1 \subset \mathscr{B}$ such that $U_1 \cap \cdots \cap U_k = \bigcup_{B_1\in\mathscr{B}_1} B_1$. Since $U_{k+1} \in \mathscr{U}_\mathscr{B}$, there exists a subset $\mathscr{B}_2 \subset \mathscr{B}$ such that $U_{k+1} = \bigcup_{B_2\in\mathscr{B}_2} B_2$. Hence

$$U_1 \cap \cdots \cap U_{k+1} = \left(\bigcup_{B_1\in\mathscr{B}_1} B_1\right) \cap \left(\bigcup_{B_2\in\mathscr{B}_2} B_2\right) = \bigcup_{\substack{B_1\in\mathscr{B}_1 \\ B_2\in\mathscr{B}_2}} (B_1 \cap B_2).$$

But by condition (3c) in the definition of a basis, $B_1 \cap B_2 \in \mathscr{U}_\mathscr{B}$.

Hence $U_1 \cap \cdots \cap U_{k+1} \in \mathscr{U}_\mathscr{B}$. $\qquad\square$

Theorem 4. *Let (S, ρ) be a metric space. Let*

$$\mathscr{B} = [B_s(a); s \in S \text{ and } a \text{ is a nonnegative real number}].$$

Then $\mathscr{B}$ is a basis for a topology on S.

PROOF

(1) $B_s(0) = \varnothing$ for any $s \in S$, so $\varnothing \in \mathscr{B}$.
(2) For any $a > 0$, $S = \bigcup_{s \in S} B_s(a)$, so $S = \bigcup_{B \in \mathscr{B}} B$.
(3) Let $s_1, s_2 \in S$, let $a_1, a_2 > 0$, and let $T = B_{s_1}(a_1) \cap B_{s_2}(a_2)$. We may assume $T \neq \varnothing$.

To show that T is a union of elements of $\mathscr{B}$, it suffices to show that for each $s \in T$ there exists $a_s > 0$ such that $B_s(a_s) \subset T$. For then $T \subset \bigcup_{s \in T} B_s(a_s) \subset T$. The first inclusion follows because $s \in B_s(a_s)$ for each $s \in T$, and the second because $B_s(a_s) \subset T$ for each s. Thus $T = \bigcup_{s \in T} B_s(a_s)$ is a union of elements of $\mathscr{B}$.

Now for $s \in T$, let $b_j = \rho(s, s_j)$ for $j = 1, 2$. Then $b_j < a_j$ since $s \in B_{s_j}(a_j)$. Let $a_s = \min\{a_1 - b_1, a_2 - b_2\}$. Then $a_s > 0$, and we claim that $B_s(a_s) \subset T$. For suppose $t \in B_s(a_s)$. Then

$$\rho(t, s_j) \leq \rho(t, s) + \rho(s, s_j) < a_s + b_j \leq a_j - b_j + b_j = a_j;$$

so $t \in B_{s_j}(a_j)$, $j = 1, 2$. $\qquad\square$

Corollary. *A metric space has the structure of a topological space in which the open sets are unions of balls.*

Definition. Let S be a set, and let $\mathscr{B}_1$ and $\mathscr{B}_2$ be bases for topologies on S. $\mathscr{B}_1$ and $\mathscr{B}_2$ are *equivalent* if they generate the same topology; that is, if $\mathscr{U}_{\mathscr{B}_1} = \mathscr{U}_{\mathscr{B}_2}$.

Theorem 5. *Let S be a set, and let $\mathscr{B}_1$ and $\mathscr{B}_2$ be bases for topologies on S. Then $\mathscr{B}_1$ and $\mathscr{B}_2$ are equivalent if and only if*

(1) *for each $s \in S$ and $B_1 \in \mathscr{B}_1$ with $s \in B_1$, there exists $B_2 \in \mathscr{B}_2$ such that $s \in B_2 \subset B_1$, and*
(2) *for each $s \in S$ and $B_2 \in \mathscr{B}_2$ with $s \in B_2$, there exists $B_1 \in \mathscr{B}_1$ such that $s \in B_1 \subset B_2$.*

PROOF. Suppose $\mathscr{B}_1$ and $\mathscr{B}_2$ are equivalent, and let $s \in B_1 \in \mathscr{B}_1$. Then $B_1 \in \mathscr{U}_{\mathscr{B}_1} = \mathscr{U}_{\mathscr{B}_2}$, so $B_1 = \bigcup_{B_2 \in \tilde{\mathscr{B}}_2} B_2$ for some subset $\tilde{\mathscr{B}}_2 \subset \mathscr{B}_2$. Hence $s \in B_2 \subset B_1$ for some $B_2 \in \tilde{\mathscr{B}}_2 \subset \mathscr{B}_2$. Thus (1) is proved, and (2) is proved similarly.

Conversely, suppose (1) and (2) are satisfied. We first show that $\mathscr{U}_{\mathscr{B}_1} \subset \mathscr{U}_{\mathscr{B}_2}$. Let $B \in \mathscr{B}_1$. By (1) for each $s \in B$ there exists $B_s \in \mathscr{B}_2$ such that $s \in B_s \subset B$. Now $B \subset \bigcup_{s \in B} B_s \subset B$, so $B = \bigcup_{s \in B} B_s \in \mathscr{U}_{\mathscr{B}_2}$. Thus $\mathscr{U}_{\mathscr{B}_1} \subset \mathscr{U}_{\mathscr{B}_2}$. Similarly, using (2), $\mathscr{U}_{\mathscr{B}_2} \subset \mathscr{U}_{\mathscr{B}_1}$, and so $\mathscr{U}_{\mathscr{B}_1} = \mathscr{U}_{\mathscr{B}_2}$. $\qquad\square$

Corollary. *The three metrics* ρ_1, ρ_2, ρ_3 *on the plane S described earlier all determine the same topology on S.*

PROOF. Conditions (1) and (2) of Theorem 5 are clearly satisfied. ☐

Remark. It is *not* true, however, that all metrics on a given space give the same topology. For example, consider the space R of real numbers with the following two metrics.

$$\rho_1(r_1, r_2) = |r_1 - r_2|$$
$$\rho_2(r_1, r_2) = \begin{cases} 0 & (r_1 = r_2) \\ 1 & (r_1 \neq r_2) \end{cases}$$

The open sets in the topology defined by ρ_1 are the usual open sets (generated by open intervals) in R, whereas the collection of open sets defined by ρ_2 is the set 2^R of all subsets of R. For, in fact, relative to ρ_2, $B_r(\frac{1}{2}) = \{r\}$ for each $r \in R$, so each "point" is an open set; and, hence, so is each union of "points," that is, each subset of R.

Definition. If $(S, \mathcal{U})$ is a topological space and $\mathcal{U} = 2^S$, then S is said to have the *discrete topology*.

Theorem 6. *Let* $(S, \mathcal{U})$ *be a topological space and let* $A \subset S$. *Let*

$$\mathcal{U}_A = [A \cap U; U \in \mathcal{U}].$$

Then $\mathcal{U}_A$ *is a topology on* A, *called the* relative topology *on* A.

PROOF. This is a consequence of the following facts.

(1) $\varnothing \cap A = \varnothing$, $S \cap A = A$.
(2) $\bigcap_{i=1}^n (U_i \cap A) = (\bigcap_{i=1}^n U_i) \cap A$.
(3) $\bigcup_{\tilde{U} \in \tilde{\mathcal{U}}} (\tilde{U} \cap A) = (\bigcup_{\tilde{U} \in \tilde{\mathcal{U}}} \tilde{U}) \cap A$. ☐

Remark. When dealing with subspaces, one must be careful to specify which topology is being used at any given time. Thus, for example, if U is the open interval $(1, 3) \subset R$ and A is the closed interval $[2, 4] \subset R$, then $U \cap A$ is open in A, but not open in R. However, if A is either open or closed, we do have the following theorem.

Theorem 7. *Let* S *be a topological space and let* $A \subset S$. *If* A *is open in* S, *then every open set in* A *is open in* S. *If* A *is closed in* S, *then every closed set in* A *is closed in* S.

PROOF. If A is open in S and B is open in A, then $B = U \cap A$ for some open set U in S; so B is open in S since both U and A are open in S.

If A is closed in S, and B is closed in A, then $A - B$ is open in A; so $A - B = A \cap U$ for some open set U in S. Thus

$$B = A - (A - B) = A - A \cap U = A \cap U'.$$

Since A and U' are both closed in S, so is $A \cap U' = B$. ☐

Theorem 8. *Let S be a topological space and let $Q \subset S$ have the relative topology. Let P be a subset of Q. Then the relative topology $\mathcal{U}_1$ on P regarded as a subset of Q is the same as the relative topology $\mathcal{U}_2$ on P regarded as a subset of S.*

PROOF. Let $A \subset P$. Then

$$A \in \mathcal{U}_1 \Leftrightarrow A = P \cap U \text{ for some open set } U \subset Q$$
$$\Leftrightarrow A = P \cap (Q \cap \tilde{U}) \text{ for some open set } \tilde{U} \subset S$$
$$\Leftrightarrow A = P \cap \tilde{U} \text{ for some open set } \tilde{U} \subset S \quad (\text{since } P \subset Q)$$
$$\Leftrightarrow A \in \mathcal{U}_2. \qquad \square$$

1.3 Connected and compact spaces

Definition. A topological space S is *connected* if the only sets which are both open and closed are $\varnothing$ and S.

Theorem 1. *A topological space S is connected if and only if it is not the union of two disjoint nonempty open sets.*

PROOF. Assume S is connected. Suppose $S = V_1 \cup V_2$ for open sets V_1 and V_2 with $V_1 \cap V_2 = \varnothing$. Then $V_1 = V_2'$, so V_1 is closed as well as open. Since S is connected, either $V_1 = \varnothing$ or $V_1 = S$. If $V_1 = S$, then $V_2 = \varnothing$; so in both cases either V_1 or V_2 must be empty.

Conversely, suppose S is not the union of two disjoint non-empty sets. Let $V \subset S$ be both open and closed. Then V' is also both open and closed, and S is the union of the disjoint open sets V and V'. Thus either $V = \varnothing$ or $V' = \varnothing$; that is, either $V = \varnothing$ or $V = S$, so S is connected. $\qquad \square$

EXAMPLES. It is shown in real analysis that the following spaces are connected:

(1) The space R of real numbers,
(2) Any interval in R,
(3) Real n-space R^n, and
(4) Any ball or cube in R^n.

Remark. A subset of a topological space is said to be connected if it is connected in the relative topology.

Theorem 2. *Let S be a topological space, and let T_0 and $\{T_w\}_{w \in W}$ be connected subsets of S. Assume $T_0 \cap T_w \neq \varnothing$ for each $w \in W$. Then $T_0 \cup (\bigcup_{w \in W} T_w)$ is connected.*

Remark. This theorem can be used to prove that R^n is connected, given the fact that lines in R^n are connected. For, in fact, let $T_0 = \{0\}$ and $\{T_w\}_{w \in W}$ denote the set of all lines through 0. (The indexing set W can be taken to be the unit sphere in R^n.) Then $R^n = T_0 \cup (\bigcup_{w \in W} T_w)$.

PROOF. Let $T = T_0 \cup (\bigcup_{w \in W} T_w)$. Suppose $T = V_1 \cup V_2$ for some disjoint sets V_1 and V_2 open in T. Then for each w, $V_1 \cap T_w$ and $V_2 \cap T_w$ are disjoint open sets in T_w, and their union is T_w. Since T_w is connected, either $V_1 \cap T_w = \varnothing$ or $V_2 \cap T_w = \varnothing$. Similarly, either $V_1 \cap T_0 = \varnothing$ or $V_2 \cap T_0 = \varnothing$; say $V_2 \cap T_0 = \varnothing$. Then $V_1 \cap T_0 = T_0$; that is, $T_0 \subset V_1$. Therefore, since $T_0 \cap T_w \neq \varnothing$, $V_1 \cap T_w \neq \varnothing$ for each $w \in W$. Thus, from above, $V_2 \cap T_w = \varnothing$ for all $w \in W$ and $V_1 \cap T_w = T_w$; that is, $T_w \subset V_1$ for all $w \in W$. Therefore, $V_1 = T_0 \cup (\bigcup_{w \in W} T_w)$, and $V_2 = \varnothing$. $\qquad\square$

EXAMPLE. $R^n - \{0\}$ is connected for $n > 1$. Prove that it is.

Definitions. Let S be a set. A collection $\mathscr{V} \subset 2^S$ is a *covering* of S if $\bigcup_{V \in \mathscr{V}} V = S$. If S is a topological space and each $V \in \mathscr{V}$ is an open set, $\mathscr{V}$ is called an *open covering* of S.

A topological space S is *compact* if every open covering has a finite subcovering; that is, if for every open covering $\mathscr{V}$, there exist a finite number of sets, say $V_1, \ldots, V_k \in \mathscr{V}$ for some k, such that $S = \bigcup_{j=1}^{k} V_j$.

EXAMPLE. It is shown in real analysis that

(1) the compact subsets of R^n are the closed bounded subsets of R^n (Heine-Borel theorem),
(2) R^n is not compact, and
(3) an open interval in R^1 is not compact.

Definition. A topological space is said to have the *finite intersection property* (abbreviated f.i.p.) if every collection $\mathscr{F}$ of closed sets with property (α) also has property (β):

(α) $F_1 \cap \cdots \cap F_k \neq \varnothing$ for each finite subcollection $\{F_1, \ldots, F_k\} \subset \mathscr{F}$.
(β) $\bigcap_{F \in \mathscr{F}} F \neq \varnothing$.

Theorem 3. *Let S be a topological space. Then S is compact if and only if S has the f.i.p.*

PROOF. There is a one-to-one correspondence between collections $\mathscr{F}$ of closed sets in S and collections $\mathscr{V}$ of open sets in S given by complementation; that is, $\mathscr{V} \leftrightarrow \mathscr{F}$ if and only if $\mathscr{V} = [F'; F \in \mathscr{F}]$ or, equivalently, $\mathscr{F} = [V'; V \in \mathscr{V}]$. Now if $\mathscr{V}$ corresponds to $\mathscr{F}$, then

$$\bigcap_{F \in \mathscr{F}} F = \varnothing \Leftrightarrow \bigcup_{V \in \mathscr{V}} V = S.$$

That is, $\mathscr{V}$ is an open covering if and only if $\mathscr{F}$ does not satisfy property (β); and $\mathscr{V}$ has a finite subcovering if and only if $\mathscr{F}$ does not satisfy property (α). $\qquad\square$

Theorem 4. *Every closed subset of a compact space is compact in its relative topology.*

PROOF. Let A be a closed subset of a compact space S. We show that A has the f.i.p. Let $\mathcal{F}$ be any collection of closed subsets of A satisfying property (α). Since A is closed and each $F \in \mathcal{F}$ is closed in A, each $F \in \mathcal{F}$ is closed in S. Thus $\mathcal{F}$ is a collection of closed sets in S satisfying (α). Since S is compact, $\bigcap_{F \in \mathcal{F}} F \neq \varnothing$. But $\bigcap_{F \in \mathcal{F}} F \subset A$, so A has the f.i.p. $\qquad\square$

Theorem 5. *Let S be a compact topological space. Then every infinite subset of S has a limit point.*

PROOF. We show that if $A \subset S$ has no limit point, then A is finite. The proof is in three steps.

Step (1). A is discrete; that is, its relative topology is the discrete topology. For suppose $a \in A$. Since a is not a limit point of A, there exists an open set $U_a \subset S$ containing a such that $(U_a - \{a\}) \cap A = \varnothing$; that is, $U_a \cap A = \{a\}$. Thus each $\{a\}$ is an open set in the relative topology of A and hence A is discrete.

Step (2). A is compact. For in fact, since A has no limit points, $A = \bar{A}$, and hence A is closed. But, by Theorem 4, this implies A is compact.

Step (3). A is finite. For let $U_a = \{a\}$ for each $a \in A$. Then, by Step 1, $\{U_a\}_{a \in A}$ is an open covering of A. By Step (2), A is compact, so there exists a finite subcovering $\{U_{a_1}, \ldots, U_{a_k}\}$.

Thus $A = \bigcup_{j=1}^{k} U_{a_j} = \{a_1, \ldots, a_k\}$ is finite. $\qquad\square$

1.4 Continuous functions

Definition. Let S and T be topological spaces. A function $f: S \to T$ is *continuous* if the inverse images of open sets are open; that is, for each open set $U \subset T$, $f^{-1}(U)$ is open in S.

Theorem 1. *Let S and T be topological spaces. Let $\mathcal{B}_S$ and $\mathcal{B}_T$ be bases for the topologies on S and T respectively. Then $f: S \to T$ is continuous if and only if for each $s \in S$ and each $V \in \mathcal{B}_T$ with $f(s) \in V$, there exists a $U \in \mathcal{B}_S$ such that $s \in U$ and $f(U) \subset V$.*

Remark. For metric spaces, Theorem 1 shows that our definition of continuity is equivalent to the usual ε, δ definition of continuity.

PROOF. Assume f is continuous. If $s \in S$ and $f(s) \in V \in \mathcal{B}_T$, then $f^{-1}(V)$ is open in S and so is a union of elements of $\mathcal{B}_S$. Since $s \in f^{-1}(V)$, s must belong to one of these basis elements. Call it U. Then $U \subset f^{-1}(V)$, so $f(U) \subset V$.

Conversely, suppose for each such s and V there exists such a U. Let $\tilde{V}$ be open in T. We must show that $f^{-1}(\tilde{V})$ is open in S. Let $s \in f^{-1}(\tilde{V})$. Then $f(s) \in \tilde{V}$, so $f(s)$ lies in some basis element $V_s \in \mathcal{B}_T$ with $V_s \subset \tilde{V}$. Therefore, there exists a basis element $U_s \in \mathcal{B}_S$ such that $s \in U_s$ and $f(U_s) \subset V_s \subset \tilde{V}$; that is, $U_s \subset f^{-1}(\tilde{V})$. Then $f^{-1}(\tilde{V}) = \bigcup_{s \in f^{-1}(\tilde{V})} U_s$; so $f^{-1}(\tilde{V})$ is open in S. $\qquad\square$

Theorem 2. *Let R, S, T be topological spaces. If $g: R \to S$ and $f: S \to T$ are both continuous, then $f \circ g: R \to T$ is continuous.*

PROOF. Let V be open in T. Since f is continuous, $f^{-1}(V)$ is open in S. Hence, since g is continuous, $g^{-1}(f^{-1}(V))$ is open in R. But $(f \circ g)^{-1} = g^{-1} \circ f^{-1}$, so $(f \circ g)^{-1}(V)$ is open in R, and $f \circ g$ is continuous. □

Theorem 3. *Let S and T be topological spaces. Let $f: S \to T$ be continuous and surjective. If S is connected, then so is T.*

PROOF. Suppose V_1 and V_2 are disjoint open sets in T with $V_1 \cup V_2 = T$. Then $f^{-1}(V_1)$ and $f^{-1}(V_2)$ are disjoint open sets in S with $f^{-1}(V_1) \cup f^{-1}(V_2) = S$. Since S is connected, either $f^{-1}(V_1)$ or $f^{-1}(V_2)$ must be empty. But since f is surjective, $f^{-1}(V_j) = \varnothing$ for some j implies $V_j = \varnothing$. Thus T is connected. □

Corollary. *If $f: S \to T$ is continuous, and if S is connected, then $f(S)$ is connected.*

EXAMPLE 1. Let $f: [a, b] \to R^1$ be a continuous real-valued function defined on the closed interval from a to b. If $f(x_1) > 0$ for some $x_1 \in [a, b]$, and $f(x_2) < 0$ for some $x_2 \in [a, b]$, then $f(x_0) = 0$ for some $x_0 \in [a, b]$.

PROOF. Let $T = f([a, b])$. Then T is connected by the corollary since $[a, b]$ is connected. But if $0 \notin T$, then $T = (R_- \cap T) \cup (R_+ \cap T)$ is the union of nonempty disjoint open sets. Here R_+ is the set of positive real numbers, and R_- the set of negative real numbers. Thus $0 \in T$; that is, $0 = f(x_0)$ for some x_0. □

EXAMPLE 2. Let S^n be the unit sphere in R^{n+1}; that is,

$$S^n = \left[(x_1, \ldots, x_{n+1}) \in R^{n+1}; \sum_{j=1}^{n+1} x_j^2 = 1 \right].$$

Then S^n is connected for $n > 0$.

PROOF. From Section 1.3 we know that $R^{n+1} - \{0\}$ is connected. Let

$$f: R^{n+1} - \{0\} \to S^n$$

be defined by

$$f(x_1, \ldots, x_{n+1}) = \left(\frac{x_1}{\sqrt{\sum x_j^2}}, \frac{x_2}{\sqrt{\sum x_j^2}}, \ldots, \frac{x_{n+1}}{\sqrt{\sum x_j^2}} \right).$$

Then f is continuous and surjective; so S^n is connected. □

14

EXAMPLE 3. Let $GL(n, R)$ and $GL(n, C)$ denote respectively the sets of non-singular $n \times n$ matrices with real and complex entries. Then by stringing out the rows of each matrix in a line, $GL(n, R)$ may be regarded as a subset of R^{n^2}, and hence is a topological space as a subset of R^{n^2} with the relative topology. Similarly $GL(n, C) \subset C^{n^2}$ is a topological space. (Note that complex n^2-space C^{n^2} has the topology of the product of R^2 with itself n^2 times.) Now $GL(n, R)$ is *not* connected. For let

$$\Delta: R^{n^2} \to R^1 - \{0\}$$

be the determinant function. Δ is continuous and surjective. Since $R^1 - \{0\}$ is not connected, $GL(n, R)$ is not connected.

Note that this argument fails for $GL(n, C)$ since $\Delta: GL(n, C) \to C - \{0\}$, and $C - \{0\} = R^2 - \{0\}$ is connected. In fact, as we shall see later, $GL(n, C)$ is connected.

Theorem 4. *Let S and T be topological spaces, and let $f: S \to T$ be continuous and surjective. If S is compact, then so is T.*

PROOF. Let $\mathscr{V}$ be an open covering of T. Then $[f^{-1}(V); V \in \mathscr{V}]$ is an open covering of S. Since S is compact, there exists a finite subcovering

$$\{f^{-1}(V_1), \ldots, f^{-1}(V_k)\}.$$

Since f is surjective, $f(f^{-1}(V_j)) \doteq V_j$ for $j = 1, \ldots, k$, and $\bigcup_{j=1}^{k} V_j = T$ because $\bigcup_{j=1}^{k} f^{-1}(V_j) = S$. Thus T is compact. $\square$

Corollary. *If $f: S \to T$ is continuous and S is compact, then $f(S)$ is compact.*

Corollary 2. *Let $f: S \to R^1$ be continuous. If S is compact, then f assumes its maximum and its minimum.*

PROOF. By Theorem 4, $f(S)$ is compact, and therefore a closed bounded set in R^1. The maximum of f is the l.u.b. of $f(S)$. It is a limit point of $f(S)$, hence is in $f(S)$ because $f(S)$ is closed. Similarly the minimum of f is in $f(S)$. $\square$

Definition. Let S and T be topological spaces and $f: S \to T$. f is a *homeomorphism* if f is a one-to-one correspondence and both f and f^{-1} are continuous.

Remark 1. Note that although f^{-1} has been defined as a map $2^T \to 2^S$, it may, in the case where f is a one-to-one correspondence, be regarded as a map $T \to S$ by identifying the one-point set $\{t\}$ with t and $f^{-1}(\{t\})$ with s, where $f(s) = t$.

Remark 2. The fact that, for a homeomorphism f, both f and f^{-1} are continuous means that f not only maps points of S to points of T in a one-to-one manner, but f also maps open sets of S to open sets of T in a one-to-one manner. This means that S and T are topologically the same; that is, any topological property enjoyed by S is also enjoyed by T and conversely. Thus if $f: S \to T$ is a homeomorphism, then S is connected if and only if T is connected; and S is compact if and only if T is compact.

1.5 Product spaces

Definition. Let T be a set and let $\mathcal{U}$ and $\mathcal{V}$ be two topologies on T. The topology $\mathcal{U}$ is said to be *weaker* than the topology $\mathcal{V}$ if $\mathcal{U} \subset \mathcal{V}$. Or equivalently, if the identity map

$$I: (T, \mathcal{V}) \to (T, \mathcal{U})$$

is continuous.

Theorem 1. *Let S be a set, and let $\mathcal{W} \subset 2^S$. Then there exists a weakest topology $\mathcal{U}$ on S such that $\mathcal{W} \subset \mathcal{U}$.*

PROOF. Let

$$\mathcal{A} = [\mathcal{V}; \mathcal{W} \subset \mathcal{V} \subset 2^S, \mathcal{V} \text{ a topology on } S].$$

$\mathcal{A} \neq \varnothing$ because $2^S \in \mathcal{A}$. Let $\mathcal{U} = \bigcap_{\mathcal{V} \in \mathcal{A}} \mathcal{V}$. Then $\mathcal{W} \subset \mathcal{U}$. Moreover, $\mathcal{U}$ is a topology on S; that is, $\mathcal{U}$ satisfies the open set axioms. For, in fact,

(1) $\varnothing, S \in \mathcal{U}$ because $\varnothing, S \in \mathcal{V}$ for each $\mathcal{V} \in \mathcal{A}$.
(2) If $U_1, \ldots, U_k \in \mathcal{U}$, then $U_1, \ldots, U_k \in \mathcal{V}$ for each $\mathcal{V} \in \mathcal{A}$ so $\bigcap_{j=1}^{k} U_j \in \mathcal{V}$ for each $\mathcal{V} \in \mathcal{A}$ and hence $\bigcap_{j=1}^{k} U_j \in \mathcal{U}$.
(3) If $U_w \in \mathcal{U}$ for each $w \in W$, W some indexing set, then $U_w \in \mathcal{V}$ for each $w \in W$ and each $\mathcal{V} \in \mathcal{A}$ so $\bigcup_{w \in W} U_w \in \mathcal{V}$ for each $\mathcal{V} \in \mathcal{A}$ and hence $\bigcup_{w \in W} U_w \in \mathcal{U}$.

Thus $\mathcal{U}$ is a topology on S. $\mathcal{U}$ is the weakest topology containing W because if $\tilde{\mathcal{U}}$ is any other such topology, then $\tilde{\mathcal{U}} \in \mathcal{A}$ and hence $\mathcal{U} = \bigcap_{\mathcal{V} \in \mathcal{A}} \mathcal{V} \subset \tilde{\mathcal{U}}$. $\qquad\square$

Remark. A basis for the weakest topology on S containing $\mathcal{W}$ is

$$\mathcal{B} = [U_1 \cap \cdots \cap U_k; U_j \in \mathcal{W} \text{ for } j = 1, \ldots, k,$$
$$\text{where } k \text{ is any positive integer}].$$

(Throw in $\varnothing$ and S.)

Exercise. Prove the above remark.

Remark. We can regard $\mathcal{U}$ in two distinct ways: as the intersection of all topologies on S containing $\mathcal{W}$ and as the topology "generated" by $\mathcal{W}$. This is analogous to the situation in linear algebra where, given a subset W of a vector space V, the smallest subspace of V containing W may be regarded either as the intersection of all subspaces containing W or as the linear space spanned by W.

Theorem 2. *Let W be an indexing set and let $\{T_w\}_{w \in W}$ be topological spaces. Let S be a set, and let $\{f_w\}_{w \in W}$ be a collection of functions, $f_w : S \to T_w$ for each $w \in W$. Then there exists a weakest topology on S such that $f_w : S \to T_w$ is continuous for each $w \in W$.*

PROOF. Let $\mathscr{W} = [f_w{}^{-1}(V_w); V_w \text{ open in } T_w, w \in W]$. Let $\mathscr{U}$ be the weakest topology such that $\mathscr{W} \subset \mathscr{U}$. Clearly, any topology on S such that each f_w is continuous must contain $\mathscr{W}$. $\mathscr{U}$ is the weakest such topology. $\square$

Remark. The set

$$\mathscr{B} = \left[\bigcap_{w \in W_1} f_w{}^{-1}(V_w); W_1 \text{ is a finite subset of } W, \right.$$
$$\left. V_w \text{ open in } T_w \text{ for each } w \in W_1 \right]$$

is a basis for the topology $\mathscr{U}$ of Theorem 2.

Definition. Let $\{S_w\}_{w \in W}$ be a collection of topological spaces. Let $P = \prod_{w \in W} S_w$ (Cartesian product of sets). Let $\pi_w : P \to S_w$ be defined for each $w \in W$ by

$$\pi_w(f) = f(w), \qquad (f \in P).$$

(Recall that a point $f \in \prod_{w \in W} S_w$ is a function

$$f : W \to \bigcup_{w \in W} S_w$$

such that $f(w) \in S_w$ for each $w \in W$.) The *product topology* on P is the weakest topology in which each π_w is continuous. This topology exists by Theorem 2.

Remark. The function $\pi_w (w \in W)$ is projection onto the factor S_w. A basis for the product topology is

$$\mathscr{B} = \left[\bigcap_{w \in W_1} \pi_w{}^{-1}(U_w); W_1 \text{ a finite subset of } W \text{ and } U_w \right.$$
$$\left. \text{an open set in } S_w \text{ for each } w \in W_1 \right].$$

Let us take a closer look at these basic open sets. First, for each $w_0 \in W$ and each open set $U_{w_0} \subset S_{w_0}$, $\pi_{w_0}{}^{-1}(U_{w_0})$ is the "cylinder" $\prod_{w \in W} T_w$, where

$$T_w = \begin{cases} S_w & (w \neq w_0) \\ U_{w_0} & (w = w_0). \end{cases}$$

For, in fact, $f \in \pi_{w_0}{}^{-1}(U_{w_0}) \Leftrightarrow \pi_{w_0}(f) \in U_{w_0} \Leftrightarrow f(w_0) \in U_{w_0} \Leftrightarrow f(w) \in T_w$ for each $w \in W \Leftrightarrow f \in \prod_{w \in W} T_w$. Taking finite intersections, it follows easily that

$$\bigcap_{w \in W_1} \pi_w{}^{-1}(U_w) = \prod_{w \in W} Q_w$$

17

where

$$Q_w = \begin{cases} S_w & (w \notin W_1) \\ U_w & (w \in W_1). \end{cases}$$

Thus the basis $\mathscr{B}$ for the topology on P consists of products of open sets, one in each factor S_w, with all but finitely many of these being the whole space S_w.

EXAMPLE 1. Let I denote the open interval $(0, 1)$. Then $I \times I$ is the open unit square in R^2. The projection maps are $\pi_j \colon I \times I \to I (j = 1, 2)$, defined by

$$\pi_1(a, b) = a, \qquad \pi_2(a, b) = b.$$

For U_1, an open interval in I, the cylinder $\pi_1^{-1}(U_1)$ is the shaded vertical strip in Figure 1.3. For U_2, an open interval in I, $\pi_2^{-1}(U_2)$ is the shaded

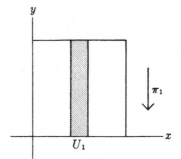

Figure 1.3

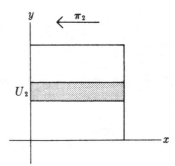

Figure 1.4

horizontal strip in Figure 1.4. Thus $\pi_1^{-1}(U_1) \cap \pi_2^{-1}(U_2)$ is the open square in Figure 1.5. Since the open intervals form a basis for the topology on I, it follows that the open squares form a basis for the product topology on $I \times I$. That is, the product topology on $I \times I$ is the same as the topology induced on $I \times I$ by the metric space topology on R^2.

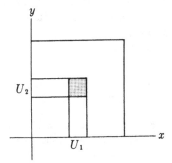

Figure 1.5

EXAMPLE 2. The circle S^1 is a subset of R^2, so it is a topological space in the induced topology. The product $S^1 \times S^1$ of the circle with itself is a *torus* (the surface of a tire or doughnut). In the product topology on the torus, a basis for the open sets is given by "rectangular patches," as in Figure 1.6. The student should find it instructive to convince himself that the product topology on $S^1 \times S^1$ is the same as the induced topology on the torus considered as a subset of R^3.

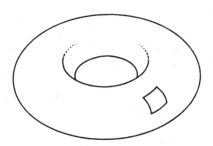

Figure 1.6

EXAMPLE 3. Let R_+ denote the positive real numbers, and let S^n denote the unit sphere in R^{n+1}. Then there exists a one-to-one correspondence

$$\varphi: R^{n+1} - \{0\} \to R_+ \times S^n$$

given by

$$\varphi(x_1, \ldots, x_{n+1}) = \left(\|x\|, \left(\frac{x_1}{\|x\|}, \ldots, \frac{x_{n+1}}{\|x\|} \right) \right)$$

where $\|x\| = (\sum_{j=1}^{n+1} x_j^2)^{1/2}$. φ is in fact a homeomorphism between $R^{n+1} - \{0\}$ in the induced topology and $R_+ \times S^n$ in the product topology. Here R_+ and S^n are topologized as subsets of R^1 and R^{n+1} respectively.

19

EXAMPLE 4. Let $J^+ = [n; n$ is a positive integer] and let $I_n = [0, 1/n]$. Then $P = \prod_{n \in J^+} I_n$ is a topological space in the product topology. We introduce a metric on the point set of P as follows. If $\bar{x} = (x_1, \ldots, x_n, \ldots)$ and $\bar{y} = (y_1, \ldots, y_n, \ldots)$ are in P, then

$$\rho(\bar{x}, \bar{y}) = \left(\sum_{n=1}^{\infty} (x_n - y_n)^2 \right)^{1/2}.$$

Exercise. Show that the series converges so that ρ is well-defined. Show that ρ is a metric on P and the metric topology on P is the same as the product topology.

1.6 The Tychonoff theorem

One of the most important properties of product spaces is given by the following theorem. In fact, this theorem is one strong reason why the product topology we have described in Section 1.5 is a good topology for infinite products (better, for example, than one where products of infinitely many proper open subsets of the factors are allowed as basic open sets).

Theorem 1 (Tychonoff). *The product of compact spaces is compact.*

PROOF. The proof is deferred to the end of Section 1.6.

Remark. Let us first consider an application of Theorem 1. Recall from analysis that there are many nice properties enjoyed by continuous real-valued functions on compact spaces which do not hold when the domain is not compact. For example, every continuous real-valued function on a compact set assumes its maximum. Thus it is reasonable to inquire when a continuous function, defined on some noncompact set S, can be extended to a continuous function defined on some compact set $C \supset S$.

Definition. If $f: S \xrightarrow{\text{cont}} T$ and if $S \subset C$, then a *continuous extension* g of f is a continuous map $g: C \to T$, such that $g(s) = f(s)$, $s \in S$; that is, $g|_S = f$.

EXAMPLE 1. Let $S = (0, 1]$, the half-open interval. Let $f: S \to R^1$ be defined by $f(x) = \sin x$ for $x \in (0, 1]$. Although S is not compact, it is clear that f can be extended to a continuous function on the compact set $[0, 1]$, the closed interval, by setting $f(0) = 0$.

EXAMPLE 2. Let $S = (0, 1]$ as before. Now let $f: S \to R^1$ be defined by

$$f(x) = \sin 1/x \qquad \text{for } x \in S.$$

Once again, f is a continuous function on the noncompact set S, but now it is clear that f cannot be extended to a continuous function on the compact set

[0, 1]; that is, we cannot, by adding one point, obtain a compact space on which f can be continuously extended. But it turns out that if we "add" sufficiently many points, we can obtain a compact space P; a continuous function $\varphi: S \to P$ (φ will in fact turn out to be a homeomorphism onto $\varphi(S) \subset P$); and a continuous function $\tilde{f}: P \to R^1$, such that

(1) $\varphi(S)$ is dense in P; that is, $\overline{\varphi(S)} = P$,
(2) $\tilde{f} \circ \varphi = f$.

Thus we replace S by a homeomorphic copy $\varphi(S)$, carry f over to $\varphi(S)$ through the homeomorphism φ^{-1}, and then we can extend to a continuous function on a compact space $P \supset \varphi(S)$.

We construct P as follows. Let $I = [-1, 1]$, and let $P_1 = [0, 1] \times I$, the closed square.

Let $\varphi: S \to P_1$ be defined by

$$\varphi(x) = (x, \sin 1/x) \qquad (x \in (0, 1]).$$

(Note that if $\tilde{f}: [0, 1] \times I \to R^1$ is defined by

$$\tilde{f}(x, y) = y,$$

then $\tilde{f} \circ \varphi = f$.) Now P_1 is compact, but $\overline{\varphi(S)} \neq P_1$. However, if we set

$$P = \overline{\varphi(S)} = \text{the closure of } \varphi(S) \text{ in } P_1,$$

then $\varphi: S \to P$. Now P, φ, and the restriction of $\tilde{f}$ to P have the required properties.

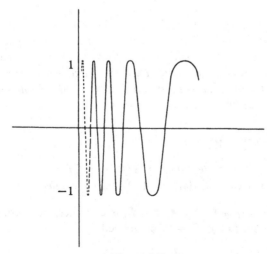

Figure 1.7

In summary, we have replaced $(0, 1]$ by the graph $\varphi(S)$ of sin $1/x$ (a homeomorphic copy of $(0, 1]$ (see Figure 1.7). We have carried the function f over to $\varphi(S)$ through the homeomorphism φ^{-1}. Then we have extended our function to the compact set

$$(\{0\} \times [-1, 1]) \cup ([(x, \sin 1/x); x \in (0, 1]).$$

Another way to look at the result of this construction is the following. Let us take the set $\varphi(S)$ and identify it with S through the homeomorphism φ. Then the set P is identified with the set (Figure 1.8)

$$\tilde{P} = (\{0\} \times [-1, 1]) \cup ((0, 1] \times \{0\}).$$

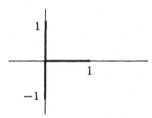

Figure 1.8

The function $\tilde{f}$ is now a continuous function on $\tilde{P}$ that agrees with f on $(0, 1] \times \{0\}$. However, note that the topology on $\tilde{P}$ is *not* the induced topology from R^2; it is the topology of P carried over through the identification $S \rightarrow \varphi(S)$. Thus, for example, each point of $\{0\} \times [-1, 1]$ is a limit point of the set $(0, 1] \times \{0\}$.

Example 2 illustrates the main points in the statement and proof of the following theorem, which is a consequence of the Tychonoff theorem.

Theorem 2. *Let W be a collection of continuous real-valued (or complex-valued) functions on a topological space S with the property that each $w \in W$ is a bounded function; that is, $w(S) \subset I_w$, where I_w is, for each $w \in W$, some closed bounded interval (or ball). Then there exists a compact space P; a continuous function $\varphi: S \rightarrow P$; and a family $\mathcal{F}$ of continuous real-valued (complex-valued) functions on P such that*

(1) $\overline{\varphi(S)} = P$
(2) $[f \circ \varphi; f \in \mathcal{F}] = W.$

If, in addition, W separates points of S (that is, if, for each $s_1 \neq s_2 \in S$, there exists a $w \in W$ such that $w(s_1) \neq w(s_2)$), then φ is injective.

PROOF. Let $P_1 = \prod_{w \in W} I_w$. Since each I_w is compact, P_1 is compact by the Tychonoff theorem. Let $\varphi: S \rightarrow P_1$ be defined by

$$\varphi(s)(w) = w(s).$$

Then φ is continuous. To prove this it suffices to show that inverse images of basic open sets are open, because each open set is a union of such sets. Suppose U is a basic open set in P_1. Then

$$U = \bigcap_{w \in W_1} \pi_w^{-1}(V_w)$$

for some finite subset $W_1 \subset W$, where V_w is open in I_w for each $w \in W_1$. Thus

$$
\begin{aligned}
\varphi^{-1}(U) &= \left[s \in S; \varphi(s) \in \bigcap_{w \in W_1} \pi_w^{-1}(V_w) \right] \\
&= [s \in S; \varphi(s)(w) \in V_w \text{ for each } w \in W_1] \\
&= [s \in S; w(s) \in V_w \text{ for each } w \in W_1] \\
&= \bigcap_{w \in W_1} w^{-1}(V_w).
\end{aligned}
$$

Thus, since each $w \in W$ is continuous, $\varphi^{-1}(U)$ is a finite intersection of open sets. Hence $\varphi^{-1}(U)$ is open, and φ is continuous.

Now let $P = \overline{\varphi(S)}$. P is compact, for it is a closed subset of a compact space. Then $\varphi: S \to P$ is continuous, where P has the relative topology in P_1, because $\varphi: S \to P_1$ is continuous. Moreover, from the definition of φ, it is clear that

$$\pi_w \circ \varphi(s) = w(s)$$

for each $s \in S$ and each $w \in W$. Thus, if we define

$$\mathscr{F} = [\pi_w|_P; w \in W],$$

then

$$[f \circ \varphi; f \in \mathscr{F}] = W.$$

It remains only to prove that if W separates points, then φ is injective. But if, for each $s_1 \neq s_2 \in S$, there exists $w \in W$ such that $w(s_1) \neq w(s_2)$, then

$$\varphi(s_1)(w) = w(s_1) \neq w(s_2) = \varphi(s_2)(w)$$

for that w, so $\varphi(s_1) \neq \varphi(s_2)$. $\qquad \square$

Remark. In Example 2 above, $W = \{f_1, f_2\}$ where $f_1(x) = \sin 1/x$ and $f_2(x) = x$. The function f_2 was added to obtain separation of points.

PROOF OF THEOREM 1. Let $\mathscr{F}$ be a collection of closed sets in $P = \prod_{w \in W} S_w$, where S_w is compact for each $w \in W$. Suppose $\mathscr{F}$ has the following property:

(α) Intersections of finitely many elements of $\mathscr{F}$ are always nonempty.

To prove that P is compact it suffices, by Theorem 3, Section 1.3, to show that $\bigcap_{F \in \mathscr{F}} F \neq \varnothing$ for each such $\mathscr{F}$.

23

Now even though each collection

$$\mathscr{F}_w = [\overline{\pi_w(F)}; F \in \mathscr{F}]$$

is a collection of closed sets in the compact space S_w with property (α) (and hence each $\bigcap_{F \in \mathscr{F}} \overline{\pi_w(F)} \neq \varnothing$)—one cannot hope to find a point in $\bigcap_{F \in \mathscr{F}} F$ merely by selecting a point x such that $\pi_w(x) \in \bigcap_{F \in \mathscr{F}} \overline{\pi_w(F)}$ for each $w \in W$. For example, if P is a closed square in R^2, let $\mathscr{F}$ be the following collection of pairs of closed balls about two given points $P_1 = (x_1, y_1)$ and $P_2 = (x_2, y_2) \in P$:

$$\mathscr{F} = [(\overline{B_{P_1}(r)} \cup \overline{B_{P_2}(r)}) \cap P; r \in R^1].$$

Then the point $P_3 = (x_1, y_2)$ has the property that

$$\pi_1(P_3) \in \bigcap_{F \in \mathscr{F}} \overline{\pi_1(F)}$$

and

$$\pi_2(P_3) \in \bigcap_{F \in \mathscr{F}} \overline{\pi_2(F)}$$

but yet $P_3 \notin \bigcap_{F \in \mathscr{F}} F$.

To remedy this situation we shall enlarge $\mathscr{F}$ by adding enough sets to eliminate "poor choices." Let $\Lambda = [\mathscr{G}; \mathscr{G}$ is a collection of subsets of P with property (α) and $\mathscr{F} \subset \mathscr{G}]$. Note that $\mathscr{F} \in \Lambda$. Furthermore, Λ is partially ordered by inclusion; that is, a partial ordering relation R in $\Lambda \times \Lambda$ is given by

$$(\mathscr{G}_1, \mathscr{G}_2) \in R \Leftrightarrow \mathscr{G}_1 \subset \mathscr{G}_2.$$

Now $\{\mathscr{F}\} \subset \Lambda$ is a simply ordered subset of Λ consisting of one point, since $\mathscr{F} \subset \mathscr{F}$. By the maximum principle (see Section 1.1), there exists a maximal simply ordered $\Lambda_1 \subset \Lambda$ such that $\{\mathscr{F}\} \subset \Lambda_1$. Let $\mathscr{H} = \bigcup_{\mathscr{G} \in \Lambda_1} \mathscr{G}$. Then $\mathscr{F} \subset \mathscr{H}$. We claim $\mathscr{H} \in \Lambda_1$. Note first that $\mathscr{H} \in \Lambda$; that is, $\mathscr{H}$ has property (α). For, if $H_1, \ldots, H_k \in \mathscr{H}$, then $H_j \in \mathscr{G}_j$ for some $\mathscr{G}_j \in \Lambda_1$ ($j = 1, \ldots, k$). Since Λ_1 is simply ordered, there exists a j_0 such that $\mathscr{G}_j \subset \mathscr{G}_{j_0}$ for all $j = 1, \ldots, k$. Therefore, all $H_j \in \mathscr{G}_{j_0}$ and, since $\mathscr{G}_{j_0}$ has property (α), $\bigcap_{j=1}^k H_j \neq \varnothing$. Thus $\mathscr{H} \in \Lambda$.

But, in fact, $\mathscr{H} \in \Lambda_1$. For if $\mathscr{H} \notin \Lambda_1$, consider the set $\Lambda_2 = \Lambda_1 \cup \{\mathscr{H}\}$. Then Λ_2 is a simply ordered subset of Λ containing $\{\mathscr{F}\}$ and $\Lambda_1 \subsetneq \Lambda_2$. This contradicts the maximality of Λ_1.

Note that $\mathscr{H}$ has the following property:

(β) If $K \subset P$ and $K \cap H \neq \varnothing$ for all $H \in \mathscr{H}$, then $K \in \mathscr{H}$.

For otherwise $\mathscr{K} = \mathscr{H} \cup \{K\}$ satisfies property (α); hence $\mathscr{K} \in \Lambda$. Thus $\Lambda_1 \cup \{\mathscr{K}\}$ is a simply ordered subset of Λ ($\mathscr{K} \supset \mathscr{G}$ for each $\mathscr{G} \in \Lambda_1$) containing Λ_1, which contradicts the maximality of Λ_1.

Now, for each $w \in W$, let

$$\mathcal{H}_w = [\overline{\pi_w(H)}; H \in \mathcal{H}].$$

Then $\mathcal{H}_w$ is, for each $w \in W$, a collection of closed sets in S_w, satisfying property (α) because $\mathcal{H}$ satisfies property (α). By compactness of S_w,

$$T_w = \bigcap_{H \in \mathcal{H}} \overline{\pi_w(H)} \neq \varnothing$$

for all $w \in W$. By the axiom of choice (see Section 1.1), $\prod_{w \in W} T_w \neq \varnothing$; that is, there exists $f \in P$ such that $f(w) \in \bigcap_{H \in \mathcal{H}} \overline{\pi_w(H)}$ for each $w \in W$.

It remains only to show that $f \in \bigcap_{F \in \mathcal{F}} F$. To prove this it suffices to show that if V is an open set in P containing f, then $V \in \mathcal{H}$. For if $V \cap H \neq \varnothing$ for all $H \in \mathcal{H}$, then in particular $V \cap F \neq \varnothing$ for all $F \in \mathcal{F}$. That this is true for each open V containing f implies that $f \in \bar{F}$; that is, $f \in F$ since F is closed for each $F \in \mathcal{F}$. Thus $f \in \bigcap_{F \in \mathcal{F}} F$, proving the theorem.

So suppose V is an open set containing f. Since V is a union of basis elements, $f \in \tilde{V} \subset V$ for some basis element $\tilde{V}$. Since $\tilde{V}$ is a basis element,

$$\tilde{V} = \bigcap_{w \in W_1} \pi_w^{-1}(V_w)$$

for some finite subset $W_1 \subset W$, where V_w is, for each $w \in W_1$, an open set in S_w. Since $f \in \tilde{V}$, $f(w) = \pi_w(f) \in V_w$ for each $w \in W_1$. Furthermore, $f(w) \in \overline{\pi_w(H)}$ for each $w \in W_1$ and each $H \in \mathcal{H}$, by our choice of f; that is, for each $w \in W_1$ and $H \in \mathcal{H}$, either $f(w) \in \pi_w(H)$ or $f(w)$ is a limit point of $\pi_w(H)$. In either case, it follows that $V_w \cap \pi_w(H) \neq \varnothing$ for each $w \in W_1$, where $H \in \mathcal{H}$. Therefore, $\pi_w^{-1}(V_w) \cap H \neq \varnothing$ for all $w \in W_1$, $H \in \mathcal{H}$. By property (β), $\pi_w^{-1}(V_w) \in \mathcal{H}$ for each $w \in W_1$.

Now by property (α) for any $H \in \mathcal{H}$,

$$V \cap H \supset \left(\bigcap_{w \in W_1} \pi_w^{-1}(V_w) \right) \cap H \neq \varnothing.$$

Hence, again by property (β), $V \in \mathcal{H}$. $\qquad\square$

2

More point set topology

2.1 Separation axioms

In Chapter 1 we dealt with topological concepts where it was not particularly important whether open sets were plentiful or scarce. However, the topological spaces usually encountered in applications have plenty of open sets, almost always enough to "separate" points, and often enough to "separate" closed sets.

Definitions. Let S be a topological space.

S is a T_0 *space* if, given any pair of distinct points, $s_1, s_2 \in S$, there exists an open set U in S containing one of these but not the other.

S is a T_1 *space* if, whenever $s_1 \neq s_2$, there exists an open set U_1 such that $s_1 \in U_1$ but $s_2 \notin U_1$, and there exists an open set U_2 such that $s_2 \in U_2$ but $s_1 \notin U_2$.

S is a T_2 *space*, or *Hausdorff space*, if, whenever $s_1 \neq s_2$, there exist open sets U_j with $s_j \in U_j$ $(j = 1, 2)$ such that $U_1 \cap U_2 = \varnothing$; that is, U_1 and U_2 are disjoint.

Remark. Note that every T_2 space is a T_1 space and every T_1 space is a T_0 space. An example of a T_0 space which is not a T_1 space is given by

$$S = \{s_1, s_2\} = \text{a set consisting of two points.}$$

The topology on S is $\mathscr{U} = \{\varnothing, S, \{s_1\}\}$.

Exercise. Find a T_1 space which is not T_2.

Remark. Any metric space is a Hausdorff space. For if $s_1 \neq s_2$, let

$$a = \rho(s_1, s_2) > 0,$$

and take $U_j = B_{s_j}(a/2),\ (j = 1, 2)$.

Theorem 1. *S is a T_1 space if and only if each point of S is closed as a subset of S.*

PROOF. Suppose S is T_1. Let $F = \{s_0\}$ and consider F'. Since S is T_1, there exists about each point $s \in F'$ an open set U_s such that $s_0 \notin U_s$, that is, such that $U_s \subset F'$. Thus $F' = \bigcup_{s \in F'} U_s$ is open in S, and hence F is closed.

Conversely, suppose points of S are closed. Let $U_1 = \{s_2\}'$ and let $U_2 = \{s_1\}'$. Then if $s_1 \neq s_2$, $s_j \in U_j$ and U_j are open. $\qquad\square$

Theorem 2. *Every compact subset of a Hausdorff space is closed.*

PROOF. Let C be a compact subset of S. To show that C' is open, and hence C is closed, it suffices to show that for each $s \notin C$ there exists an open set U_s containing s with $U_s \cap C = \varnothing$. Since S is Hausdorff, there exists for each $c \in C$ disjoint open sets U_c and V_c such that $s \in U_c$, $c \in V_c$. The collection $\{V_c \cap C\}_{c \in C}$ is then an open covering of C. (These are open sets in the relative topology.) By compactness of C there exists a finite subcovering; that is, there exists a finite subset $C_1 \subset C$ such that $\bigcup_{c \in C_1} (V_c \cap C) = C$. Thus $C \subset \bigcup_{c \in C_1} V_c$. Let $U_s = \bigcap_{c \in C_1} U_c$. Then U_s is open, $s \in U_s$, and $U_s \cap C \subset U_s \cap (\bigcup_{c \in C_1} V_c) = \varnothing$. This last equality holds because if $t \in U_s \cap (\bigcup_{c \in C_1} V_c)$, then $t \in U_c$ for all $c \in C_1$ and $t \in V_{c_0}$ for some $c_0 \in C_1$; hence $t \in U_{c_0} \cap V_{c_0} = \varnothing$, which is impossible. $\qquad\square$

Remark. We have actually shown, in the course of proving Theorem 2, that if C is a compact subset of a Hausdorff space S and if $s \notin C$, then there exist disjoint open sets U_1 and U_2 such that $s \in U_1$ and $C \subset U_2$. We can, in fact, prove more.

Theorem 3. *Let C_1 and C_2 be disjoint compact subsets of a Hausdorff space S. Then there exist disjoint open sets U_j such that $C_j \subset U_j$ $(j = 1, 2)$.*

PROOF. For each $s \in C_1$ there exist, according to the remark above, disjoint open sets U_s and V_s such that $s \in U_s$ and $C_2 \subset V_s$. The collection $\{U_s \cap C_1\}_{s \in C_1}$ is an open covering of C_1. Since C_1 is compact, there exists a finite subset $D_1 \subset C_1$ such that $\{U_s \cap C_1\}_{s \in D_1}$ actually cover C_1; that is, $C_1 \subset \bigcup_{s \in D_1} U_s$. Let $U_1 = \bigcup_{s \in D_1} U_s$ and $U_2 = \bigcap_{s \in D_1} V_s$. Then $C_2 \subset U_2$, and $U_1 \cap U_2 = \varnothing$ since each $U_s \cap V_s = \varnothing$. $\qquad\square$

Theorem 4. *Let S be a compact space, and let T be Hausdorff. Then any continuous one-to-one correspondence $f: S \to T$ is a homeomorphism.*

PROOF. We must show that $f^{-1}: T \to S$ is continuous; that is, if U is open in S, then $f(U) = (f^{-1})^{-1}(U)$ must be open in T. By considering complements, this is the same as proving that if F is closed in S, then $f(F)$ is closed in T. But by Theorem 4, Section 1.3, F is compact. Thus, by Theorem 4, Section 1.4, $f(F)$ is compact. Hence $f(F)$ is closed by Theorem 2 above. $\qquad\square$

EXAMPLES. That both assumptions are necessary in Theorem 4 is illustrated by the following two examples.

(1) $S = R^1$ with the discrete topology (S is not compact). $T = R^1$ with the usual topology.
(2) $S = \{s_1, s_2\}$ with the discrete topology (S is compact). $T = \{s_1, s_2\}$ with topology $\mathscr{U} = \{\varnothing, S, \{s_1\}\}$ (T is not Hausdorff).

$f: S \to T =$ the identity map in both examples.

Remark. Theorem 3 above shows that in a Hausdorff space there are enough open sets to separate compact sets. Sometimes we need to be able to separate closed sets. This requires a stronger axiom.

Definitions. A topological space S is *regular* (or T_3) if

(1) S is T_1, and
(2) for every closed set F in S and each $s \notin F$, there exist disjoint open sets U_1 and U_2 such that $s \in U_1$ and $F \subset U_2$.

A topological space S is *normal* (or T_4) if

(1) S is T_1, and
(2) for every pair of disjoint closed sets F_j in S, there exist disjoint open sets U_j such that $F_j \subset U_j$ ($j = 1, 2$).

Remark. Notice that every T_k space ($k = 0, 1, \ldots, 4$) is a T_j space for each $j < k$.

EXAMPLE. We shall construct a space which is Hausdorff, but not regular. Let $S = [x = (x_1, x_2) \in R^2; x_2 \geq 0]$; that is, S is the "closed" upper half-plane. We define a topology on S by giving it a basis $\mathscr{B}$ as follows. Let R^1 denote the x_1 axis in R^2. Let $S_+ = S - R^1$, so S_+ is the "open" upper half-plane. Let

$$\mathscr{B}_1 = [B_x(r); x \in S_+ \text{ and } r < x_2, \text{ where } x = (x_1, x_2)]$$

and

$$\mathscr{B}_2 = [(B_x(r) \cap S_+) \cup \{x\}; x \in R^1],$$

where the balls $B_x(r)$ are defined relative to the usual metric on R^2. Now let $\mathscr{B} = \mathscr{B}_1 \cup \mathscr{B}_2$. Verification that $\mathscr{B}$ is a basis for a topology on S is left to the student. The basis elements in $\mathscr{B}$ are illustrated in Figure 2.1.

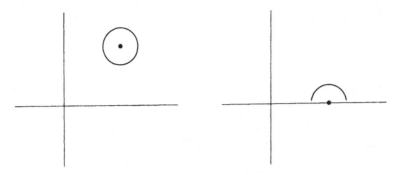

Figure 2.1

Note that the topology on S is not the topology induced from the usual topology on R^2. Now it is easy to check that S is T_2. Hence points are closed. Furthermore, the set $R^1 - \{0\}$ is closed in S since its complement $S_+ \cup \{0\}$ is clearly open. However, the closed set $R^1 - \{0\}$ and the point 0 cannot be separated by open sets: if U_1 is open and $0 \in U_1$, then some basis element $(B_0(r) \cap S_+) \cup \{0\}$ is contained in U_1 (for some real number r). Thus any open set containing $(r/2, 0)$ must intersect U_1. Hence S is not regular. (We shall later give an example of a regular space which is not normal.)

Remark. Every compact Hausdorff space is normal. This is a consequence of Theorem 3 by the fact that closed subsets of a compact space are compact.

Remark. The open sets U_1 and U_2 whose existence is asserted in the definition of a regular space (or of a normal space) might conceivably have the property that their boundaries intersect; that is, $\overline{U}_1 \cap \overline{U}_2$ might be nonempty. According to the following two theorems (Theorems 5 and 6), it is possible to choose our sets U_j more judiciously.

Theorem 5. *Let S be a regular topological space. Suppose F is closed in S and $s \notin F$. Then there exist open sets U_1 and U_2 in S such that $s \in U_1$, $F \subset U_2$, and $\overline{U}_1 \cap \overline{U}_2 = \varnothing$.*

PROOF. By the definition of regularity, there exist disjoint open sets V_1 and V_2 with $s \in V_1$ and $F \subset V_2$. Consider the set V_1'. It is closed in S, and $s \notin V_1'$. Therefore, since S is regular, there exist disjoint open sets W_1 and W_2 such that $s \in W_1$ and $V_1' \subset W_2$. Let $U_1 = W_1$ and $U_2 = V_2$. Then $s \in U_1$, and $F \subset U_2$. Moreover, $\overline{U}_1 \cap \overline{U}_2 = \varnothing$ because

$$\overline{U}_1 = \overline{W}_1 \subset \overline{W_2'} = W_2' \subset V_1$$

and

$$\overline{U}_2 = \overline{V}_2 \subset \overline{V_1'} = V_1'$$

so that

$$\overline{U}_1 \cap \overline{U}_2 \subset V_1 \cap V_1' = \varnothing. \qquad \square$$

29

Theorem 6. *Let S be a normal topological space. Suppose F_1 and F_2 are closed subsets of S. Then there exist open sets U_1 and U_2 in S such that $F_1 \subset U_1$, $F_2 \subset U_2$, and $\bar{U}_1 \cap \bar{U}_2 = \varnothing$.*

PROOF. Similar to the proof of Theorem 5.

Theorem 7. *Let (S, ρ) be a metric space. Then S, with the associated metric topology, is normal.*

PROOF. We have already seen that S is Hausdorff, hence T_1. For C any subset of S, we consider the function $d_C \colon S \to R^1$ defined by

$$d_C(s) = \operatorname*{g.l.b.}_{c \in C} \{\rho(s, c)\},$$

where g.l.b. denotes greatest lower bound. Now, for F_1 and F_2 disjoint closed sets in S, let $d_1 = d_{F_1}$ and $d_2 = d_{F_2}$, and let

$$U_1 = [s \in S; d_1(s) - d_2(s) < 0],$$
$$U_2 = [s \in S; d_1(s) - d_2(s) > 0].$$

Then $U_1 \cap U_2 = \varnothing$. We shall show that $F_1 \subset U_1$, $F_2 \subset U_2$, and that U_1 and U_2 are open, thereby proving that S is normal.

First, suppose $s \in F_1$. Then $d_1(s) = 0$. So to prove that $s \in U_1$, we must show that $d_2(s) > 0$. But if $d_2(s) = 0$, then for every $\varepsilon > 0$ there exists $f_2 \in F_2$ such that $f_2 \in B_s(\varepsilon)$. Hence $s \in \bar{F}_2 = F_2$, contradicting the fact that $F_1 \cap F_2 = \varnothing$.

Thus $F_1 \subset U_1$. Similarly, $F_2 \subset U_2$. It now remains only to show that U_1 and U_2 are open.

To show that U_1 is open, it suffices to show that each $s_1 \in U_1$ is contained in a ball which is contained in U_1. (The proof for U_2 is identical.) Let $a = d_2(s_1) - d_1(s_1)$. Then $a > 0$. We prove that $B_{s_1}(a/3) \subset U_1$. In fact, let $s \in B_{s_1}(a/3)$. Then $\rho(s, s_1) < a/3$, and we must show that $d_1(s) - d_2(s) < 0$. But

$$d_1(s) - d_2(s) = [d_1(s) - d_1(s_1)] + [d_1(s_1) - d_2(s_1)] + [d_2(s_1) - d_2(s)]$$
$$= [d_1(s) - d_1(s_1)] - a + [d_2(s_1) - d_2(s)].$$

Now, for each $f_1 \in F_1$,

$$\rho(s, f_1) < \rho(s, s_1) + \rho(s_1, f_1) < \frac{a}{3} + \rho(s_1, f_1).$$

Taking the g.l.b. over all $f_1 \in F_1$,

$$d_1(s) \le \frac{a}{3} + d_1(s_1)$$

or

$$d_1(s) - d_1(s_1) \le \frac{a}{3}.$$

Similarly, for each $f_2 \in F_2$,

$$\rho(s_1, f_2) \leq \rho(s_1, s) + \rho(s, f_2) < \frac{a}{3} + \rho(s, f_2)$$

so

$$d_2(s_1) \leq \frac{a}{3} + d_2(s)$$

or

$$d_2(s_1) - d_2(s) \leq \frac{a}{3}.$$

Thus

$$d_1(s) - d_2(s) \leq \frac{a}{3} - a + \frac{a}{3} < 0,$$

so $s \in U_1$; that is, $B_{s_1}(a/3) \subset U_1$, and U_1 is open, as claimed. $\qquad\square$

2.2 Separation by continuous functions

If S is a topological space in which the only open sets are $\varnothing$ and S, then one sees easily that the only continuous real-valued functions on S are the constant functions. On the other hand, if the space is such that, for every pair s_1, s_2 of distinct points in S, there exists a continuous real-valued function on S with $f(s_1) \neq f(s_2)$, then S is Hausdorff. (Take U_1 and U_2 disjoint open sets in R^1 with $f(s_1) \in U_1$ and $f(s_2) \in U_2$; then $s_j \in f^{-1}(U_j)$ $(j = 1, 2)$, and $f^{-1}(U_1) \cap f^{-1}(U_2) = \varnothing$.) Thus the existence of continuous real-valued functions on a space is related to the separation axiom satisfied by the space. Theorem 1 and the following remark characterize normality in these terms.

Theorem 1 (Urysohn's lemma). *Let A_0 and A_1 be disjoint closed subsets of a normal space S. Then there exists a continuous function $f: S \to [0, 1]$ such that $f(A_0) = 0$ (that is, $f(s) = 0$ for all $s \in A_0$) and $f(A_1) = 1$.*

Remark. Conversely, if a T_1 space S has the property that for each pair A_0, A_1 of disjoint closed sets there exists such a function, then S is normal. For let $U_0 = f^{-1}([0, \frac{1}{2}))$ and $U_1 = f^{-1}((\frac{1}{2}, 1])$. Then $A_0 \subset U_0$, $A_1 \subset U_1$, and $U_0 \cap U_1 = \varnothing$.

PROOF. Given A_0 and A_1, we construct f by first constructing approximations U_r to the sets $f^{-1}([0, r))$ for all $r = k/2^n$ $(k = 1, \ldots, 2^n; n = 1, 2, \ldots)$.

First, by normality, there exist disjoint open sets $U_{1/2}$ and $V_{1/2}$ such that $A_0 \subset U_{1/2}$ and $A_1 \subset V_{1/2}$. Then we have

$$A_0 \subset U_{1/2} \subset V'_{1/2} \subset A'_1.$$

Second, consider the disjoint closed sets A_0 and $U'_{1/2}$. By normality, there exist disjoint open sets $U_{1/4}$ and $V_{1/4}$ such that $A_0 \subset U_{1/4}$ and $U'_{1/2} \subset V_{1/4}$. The sets $V'_{1/2}$ and A_1 are disjoint closed sets, so, by normality again, there exist disjoint open sets $U_{3/4}$ and $V_{3/4}$ such that $V'_{1/2} \subset U_{3/4}$ and $A_1 \subset V_{3/4}$. Now we have

$$A_0 \subset U_{1/4} \subset V'_{1/4} \subset U_{1/2} \subset V'_{1/2} \subset U_{3/4} \subset V'_{3/4} \subset A'_1.$$

Continuing by induction, we obtain open sets U_r and V_r, defined for each dyadic $r = k/2^n$, with this property: for each n,

$$A_0 \subset U_{1/2^n} \subset V'_{1/2^n} \subset U_{2/2^n} \subset V'_{2/2^n} \subset \cdots \subset U_{(2^n-1)/2^n} \subset V'_{(2^n-1)/2^n} \subset A'_1.$$

In particular, for each pair r_1 and r_2 of dyadics with $r_1 < r_2$,

$$U_{r_1} \subset V'_{r_1} \subset U_{r_2} \subset V'_{r_2}.$$

Now we define $f\colon S \to [0, 1]$ by

$$f(s) = \begin{cases} \text{g.l.b. } [r; s \in U_r] & \text{if } s \in \bigcup_r U_r \\ 1 & \text{if } s \notin \bigcup_r U_r. \end{cases}$$

Since $A_0 \subset U_r$ for all r, $f(A_0) = 0$. Similarly, since $U_r \subset A'_1$ for all r, $U_r \cap A_1 = \varnothing$ for all r and $f(A_1) = 1$. Thus to complete the proof we need only verify that f is continuous. To do so, it suffices to show that for some basis $\mathscr{B}$ of the open sets in $[0, 1]$, $f^{-1}(U)$ is open for each $U \in \mathscr{B}$. Let us take as basis:

$$\mathscr{B} = [[0, a), (c, d), (b, 1]; a, b, c, d \text{ are irrational}].$$

Now, since $(c, d) = [0, d) \cap (c, 1]$, and hence $f^{-1}(c, d) = f^{-1}([0, d)) \cap f^{-1}((c, 1])$, it suffices to show that $f^{-1}([0, d))$ and $f^{-1}((c, 1])$ are open for all irrationals c and d.

But $f^{-1}([0, d))$ is open because $f^{-1}([0, d)) = \bigcup_{r<d} U_r$. For, if $s \in U_{r_0}$ for some $r_0 < d$, then $f(s) \leq r_0 < d$, so $s \in f^{-1}([0, d))$. Conversely, if $s \in f^{-1}([0, d))$, then $f(s) < d$, so there exists a dyadic $r_0 < d$ such that $s \in U_{r_0} \subset \bigcup_{r<d} U_r$.

Similarly, $f^{-1}((c, 1])$ is open because $f^{-1}((c, 1]) = \bigcup_{r>c} V_r$. For if $s \in V_{r_0}$ for some $r_0 > c$, then $s \notin V'_{r_0}$, so $s \notin U_{r_0}$, and $f(s) \geq r_0 > c$. Conversely, if $s \in f^{-1}((c, 1])$, then $s \notin U_{r_0}$ for some $r_0 > c$. Since the dyadics are dense in $[0, 1]$, there exists a dyadic r_1 with $c < r_1 < r_0$. Now $s \notin U_{r_0}$ implies $s \notin V'_{r_1} \subset U_{r_0}$; that is, $s \in V_{r_1} \subset \bigcup_{r>c} V_r$. $\qquad\square$

Corollary. *Let A_0 and A_1 be disjoint closed subsets of a normal space S. Then there exists a continuous function $g\colon S \to [a, b]$ with $g(A_0) = a$ and $g(A_1) = b$.*

PROOF. Let $f: S \to [0, 1]$ be the function given by Theorem 1. Define $g = h \circ f$ where $h: [0, 1] \to [a, b]$ is defined by

$$h(x) = a + (b - a)x \qquad (x \in [0, 1]). \qquad \square$$

Remark. Theorem 1 may be rephrased in terms of continuous extensions as follows.

Let S be a normal space and let A_0 and A_1 be closed subsets of S. Let $S_1 = A_0 \cup A_1$, and let $f: S_1 \to [0, 1]$ be defined by $f(A_0) = 0$, $f(A_1) = 1$. Then there exists a continuous extension g of f to all of S.

Note that A_0 and A_1 are each both open and closed in S_1: they are open because S is normal. Hence the function $f: S_1 \to [0, 1]$ is continuous.

Remark. Theorem 2 of Section 1.6 was also an extension theorem. Essentially it says that any family of continuous bounded real-valued or complex-valued functions on a topological space S can be extended to a family of continuous functions on a compact space containing S.

EXAMPLE. One should not expect that, given any $S_1 \subset S$ and $f: S_1 \to T$, there exists a continuous extension of f to S. For example, let S be the disc

$$S = [(x, y) \in R^2; x^2 + y^2 \le 1],$$

and let S_1 be the circle

$$S_1 = [(x, y) \in R^2; x^2 + y^2 = 1].$$

Let $T = S_1$, and let $f: S_1 \to T$ be the identity map. Then there exists no extension g to f to all of S. (This fact will later be proved rigorously when we have more machinery at our disposal.)

Intuitively, we can see why no extension exists: If g were a continuous extension, consider the behaviour of g on the concentric circles S_r of radius r ($0 < r \le 1$). Since g is continuous, nearby points are mapped into nearby points. Thus, any point s of S_r, for r close to 1, must be mapped close to the point s/r on S_1. Letting s move around S_r, we see that $g(s)$ must move around S_1, and g must map S_r onto S_1. Now, as r gets smaller, the same argument still shows that g must map S_r onto S_1 for all $r > 0$. But, for small enough r, the center 0 is close to every point on S_r. Hence $g(0)$ must be close to every point in the image of S_r; that is, $g(0)$ must be close to every point of S_1. Of course, this is impossible.

Remark. The question *When does a function defined on a subset of topological space admit an extension to the whole space?* is one of the fundamental questions of topology. It is this question which motivates much of the machinery developed in algebraic topology. However, for real-valued functions on normal spaces, we can already prove the following.

Theorem 2 (Tietze extension theorem). *Let S_1 be a closed subset of a normal space S, and let $f: S_1 \to [-1, 1]$ be continuous. Then f has a continuous extension g to all of S.*

PROOF. Let $A_1 = f^{-1}([\frac{1}{3}, 1])$ and $A_{-1} = f^{-1}([-1, -\frac{1}{3}])$. Then A_1 and A_{-1} are disjoint closed subsets of S_1. Since S_1 is closed, A_1 and A_{-1} are closed in S. By Urysohn's lemma, there exists a continuous function $f_1: S \to [-\frac{1}{3}, \frac{1}{3}]$ such that $f_1(A_{\pm 1}) = \pm \frac{1}{3}$. (We are actually using here the corollary to Theorem 1 above.) Then, for all $s \in S_1$,

$$|f(s) - f_1(s)| \le \tfrac{2}{3}.$$

Next, consider the function $f - f_1: S_1 \to [-\frac{2}{3}, \frac{2}{3}]$. Let $A_2 = (f - f_1)^{-1}([\frac{2}{9}, \frac{2}{3}])$ and $A_{-2} = (f - f_1)^{-1}([-\frac{2}{3}, -\frac{2}{9}])$. Then A_2 and A_{-2} are disjoint closed subsets of S. By Urysohn's lemma, there exists a continuous function $f_2: S \to [-\frac{2}{9}, \frac{2}{9}]$ such that $f_2(A_2) = \frac{2}{9}$ and $f_2(A_{-2}) = -\frac{2}{9}$. Now, for all $s \in S_1$,

$$|f(s) - f_1(s) - f_2(s)| \le \tfrac{4}{9}.$$

Continuing by induction, we construct functions f_n for all positive integers n, such that

$$f_n: S \to \left[\frac{-2^{n-1}}{3^n}, \frac{2^{n-1}}{3^n} \right]$$

and furthermore, for all $s \in S_1$,

$$\left| f(s) - \sum_{j=1}^{n} f_j(s) \right| \le (\tfrac{2}{3})^n.$$

We may regard the f_n as functions $f_n: S \to [-1, 1]$ such that $|f_n(s)| \le 2^{n-1}/3^n$. By the Weierstrass M-test, the series $\sum_{n=1}^{\infty} f_n$ converges uniformly to a continuous function $g: S \to [-1, 1]$. (We leave to the student the verification that these convergence properties, familiar from analysis on R^1, are also valid in the more general setting considered here.) Since

$$|f(s) - g(s)| = \lim_{n \to \infty} \left| f(s) - \sum_{j=1}^{n} f_j(s) \right| \le \lim_{n \to \infty} (\tfrac{2}{3})^n = 0$$

for all $s \in S_1$, we see that $g(s) = f(s)$ on S_1. $\square$

2.3 More separability

Definition. A topological space S is *locally compact* if, for each $s \in S$, there exists an open set U_s with $s \in U_s$ such that $\overline{U}_s$ is compact.

Remark. A finite product of locally compact spaces is locally compact. For, in fact, if

$$s = (s_1, \ldots, s_k) \in \underbrace{S_1 \times \cdots \times S_k}_{(k\text{-fold product})}$$

then there exist open sets $U_j \subset S_j (j = 1, \ldots, k)$ with $s_j \in U_j$ and $\bar{U}_j$ compact. Thus $s \in U_1 \times \cdots \times U_k$ and $\overline{U_1 \times \cdots \times U_k} = \bar{U}_1 \times \cdots \times \bar{U}_k$ is compact. Note that this argument fails in the case of infinite products since sets of the form $\prod_{w \in W} U_w$ (U_w open in S_w) are in general not open in $\prod_{w \in W} S_w$.

Definition. Let S be a T_1 space. Let ∞ denote a point not in S. The *1-point compactification* of S is the topological space $\tilde{S}$ obtained as follows. As a point set, $\tilde{S} = S \cup \{\infty\}$. Let $\mathscr{U}_S$ denote the topology on S. Let

$$\mathscr{V} = [V \subset \tilde{S}; \ V' \text{ is a compact closed subset of } S]$$
$$= [\tilde{S} - F; \ F \text{ is compact and closed in } S].$$

A basis for the topology on $\tilde{S}$ is then

$$\mathscr{B}_{\tilde{S}} = \mathscr{U}_S \cup \mathscr{V}.$$

Exercise. We leave to the student the verification that $\mathscr{B}_{\tilde{S}}$ is in fact a basis for a topology on $\tilde{S}$. Note that the relative topology on S as a subset of $\tilde{S}$ is the same as the original topology on S.

EXAMPLE 1. Let $S = R^1$. Then $\tilde{S} = R^1 \cup \{\infty\}$, where an open set about ∞ is a complement of a compact subset of R^1. In particular, given an open set U containing ∞, there exists a real number $M_U > 0$ such that $|x| > M_U$ implies $x \in U$. $\tilde{S}$ is homeomorphic to the circle S^1.

Exercise. Why is the last statement true?

EXAMPLE 2. Let $S = R^2$. Then $\tilde{S} = R^2 \cup \{\infty\}$, where the open sets about ∞ are complements of compact sets in R^2. In particular, each open set containing ∞ must contain all points in the exterior of some sufficiently large ball. $\tilde{S}$ is homeomorphic to the 2-sphere S^2. In fact, a homeomorphism $\varphi: S^2 \to \tilde{S}$ is given by stereographic projection (Figure 2.2). We regard the sphere as sitting on the xy plane with its south pole at the origin. Given a point $s \in S^2$, $\varphi(s)$ is the intersection with the xy plane (R^2) of the line through the north pole of S^2 and s. It is clear that φ is a one-to-one correspondence mapping $S^2 - \{\text{north pole}\}$ onto R^2 and mapping the north pole to ∞. That φ is, in fact, a homeomorphism is easily checked.

EXAMPLE 3. Let $S = R^n$. Then $\tilde{S}$ is homeomorphic to the n-sphere S^n. A homeomorphism is given, as in Example 2, by stereographic projection.

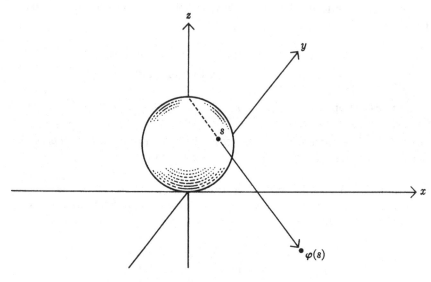

Figure 2.2

Remark. If S is compact, then $\{\infty\}$ is open in $\tilde{S}$. Thus the 1-point compactification is, in this case, uninteresting.

Theorem 1. *Let S be a T_1 space. Then the 1-point compactification $\tilde{S}$ of S is compact and T_1. Furthermore, $\tilde{S}$ is Hausdorff if and only if S is Hausdorff and locally compact.*

PROOF. $\tilde{S}$ is T_1. For if s_1 and $s_2 \in S$, then appropriate separating open sets U_1 and U_2 exist in S (since S is T_1), and U_1 and U_2 are also open in $\tilde{S}$. If $s_1 \in S$ and $s_2 = \infty$, let $U_1 = S$ and $U_2 = \{s_1\}'$. Then U_1 is open in $\tilde{S}$ since it is open in S, and U_2 is open in $\tilde{S}$ since it is the complement of the compact closed set $\{s_1\}$. ($\{s_1\}$ is closed because S is T_1.) Clearly $s_1 \in U_1$, $s_1 \notin U_2$, $s_2 \in U_2$, and $s_2 \notin U_1$. Thus $\tilde{S}$ is T_1.

$\tilde{S}$ is compact. For let $\{U_w\}_{w \in W}$ be an open covering of $\tilde{S}$. Then $\infty \in U_{w_0}$ for some $w_0 \in W$. Claim: U'_{w_0} is compact in S. For $U_{w_0} = \bigcup_{V \in \mathcal{B}_1} V$, where $\mathcal{B}_1$ is some subset of the basis $\mathcal{B}_{\tilde{S}}$ for the topology on $\tilde{S}$. Since $\infty \in U_{w_0}$, at least one element in $\mathcal{B}_1$ is of the form F', where F is a closed compact subset of S. Thus

$$U'_{w_0} = \bigcap_{V \in \mathcal{B}_1} V' \subset F;$$

that is, U'_{w_0} is a closed subset of the compact set F, hence U'_{w_0} is compact in S, as claimed.

Since $\{U_w\}_{w \in W}$ covers $\tilde{S}$, $\{U_w \cap U'_{w_0}\}_{w \in W - \{w_0\}}$ must be a covering of U'_{w_0} by open sets in U'_{w_0}. Since U'_{w_0} is compact, there exists a finite subcollection

$\{U_{w_1} \cap U'_{w_0}, \ldots, U_{w_k} \cap U'_{w_0}\}$ which covers U'_{w_0}. Hence the finite collection $U_{w_0}, U_{w_1}, \ldots, U_{w_k}$ must cover $\tilde{S}$. Thus $\tilde{S}$ is compact.

Now assume $\tilde{S}$ is Hausdorff. We must verify that S is Hausdorff and locally compact. S is certainly Hausdorff, because every subset of a Hausdorff space is Hausdorff in the relative topology. S is locally compact because, for each $s \in S$, there exist disjoint open sets U_1 and U_2 in $\tilde{S}$ such that $s \in U_1$ and $\infty \in U_2$ ($\tilde{S}$ is Hausdorff). Now U'_2 is compact (by the above argument that shows U'_{w_0} compact), and $\bar{U}_1 \subset \bar{U}'_2 = U'_2$, so $\bar{U}_1$ is compact by Theorem 4, Section 1.3.

Conversely, suppose S is Hausdorff and locally compact. Let $s_1, s_2 \in \tilde{S}$. If both s_1 and $s_2 \in S$, then there exist disjoint open sets U_1 and U_2 in S such that $s_1 \in U_1$ and $s_2 \in U_2$. But U_1 and U_2 are also open in $\tilde{S}$, so s_1 and s_2 can be separated in $\tilde{S}$. If $s_1 \in S$ and $s_2 = \infty$, then, since S is locally compact, there exists an open set U in S such that $s_1 \in U$ and $\bar{U}$ is compact (and closed). Then $V = \bar{U}'$ is open in $\tilde{S}$; and $s_1 \in U$, $s_2 \in V$, and $U \cap V = \varnothing$. Thus $\tilde{S}$ is Hausdorff. $\qquad \square$

Definition. A topological space S is *completely regular* if

(1) it is T_1, and
(2) given any $s \in S$ and any closed set C with $s \notin C$, there exists a continuous function $f: S \to [0, 1]$ such that $f(s) = 0$ and $f(C) = 1$.

Remark. Every normal space is completely regular (by Urysohn's lemma) and every completely regular space is regular. The latter is true because if C is closed in a completely regular space S and $s \notin C$, then

$$U_1 = f^{-1}([0, \tfrac{1}{2})) \qquad \text{and} \qquad U_2 = f^{-1}((\tfrac{1}{2}, 1])$$

separate s and C, where f is a function such that $f(s) = 0$ and $f(C) = 1$.

Theorem 2. *Every subset of a completely regular space is completely regular (in the relative topology).*

PROOF. Trivially, every subset of a T_1 space is T_1. Suppose $T \subset S$. Let C be closed in T, and $t \in T$, $t \notin C$. Then $C = T \cap F$ for some closed set F in S. Furthermore, $t \notin F$, for otherwise $t \in T \cap F = C$. Since S is completely regular, there exists a continuous function $f: S \to [0, 1]$ such that $f(t) = 0$ and $f(F) = 1$. The restriction of f to T is continuous and has the required properties. $\qquad \square$

Theorem 3. *Every locally compact Hausdorff space is completely regular.*

PROOF. Let S be locally compact Hausdorff. Then the 1-point compactification $\tilde{S}$ of S is compact and Hausdorff by Theorem 1, hence normal. In particular, $\tilde{S}$ is completely regular; hence so is S by Theorem 2. $\qquad \square$

Remark. We have seen that certain separation properties of a topological space imply certain others. These implications are gathered together in the following diagram:

Compact Hausdorff $\Rightarrow$ Locally compact Hausdorff

Normal $\Rightarrow$ Completely regular $\Rightarrow$ Regular $\Rightarrow$ Hausdorff $\Rightarrow$ T_1 $\Rightarrow$ T_0.

Metric

By Theorem 2, every subset of a completely regular space is completely regular. Since an arbitrary product of compact spaces is compact (Tychonoff theorem) and an arbitrary product of Hausdorff spaces is Hausdorff, we see that an arbitrary product of closed intervals is compact Hausdorff, hence completely regular. Thus every subset of an arbitrary product of closed intervals is completely regular. Theorem 4 below shows that all completely regular spaces may be regarded as such subsets.

Theorem 4. *Every completely regular space is homeomorphic to a subset of a product of closed intervals.*

PROOF. We proceed as in the proof of Theorem 2, Section 1.6. Let S be completely regular. Let W be the set of all continuous functions mapping S into the closed interval $[0, 1]$. For each $w \in W$, let $I_w = [0, 1]$, so that $\{I_w\}_{w \in W}$ is a collection of closed intervals. Let $P = \prod_{w \in W} I_w$, and let $\varphi : S \to P$ be defined by

$$\varphi(s)(w) = w(s) \qquad (w \in W).$$

We shall show that φ is a homeomorphism onto $\varphi(S)$.

It has already been shown in the proof of Theorem 2, Section 1.6, that φ is continuous and injective. (Note that since S is completely regular, W separates points and closed sets. In particular, since points are closed, W separates points.) Thus we need only verify that φ^{-1} is continuous; that is, that φ maps open sets of S onto open sets of $\varphi(S)$.

Let U be open in S. It suffices to show that for each $s \in U$, there exists an open set V in $\varphi(S)$ with $\varphi(s) \in V \subset \varphi(U)$. For then $\varphi(U)$ is the union of these open sets. So suppose $s \in U$. By complete regularity of S, there exists a $w \in W$ such that $w(s) = 0$ and $w(U') = 1$. Then $\pi_w{}^{-1}([0, 1))$ is open in P, where $\pi_w : P \to I_w$ is projection. Let $V = \pi_w{}^{-1}([0, 1)) \cap \varphi(S)$. Then V is open in $\varphi(S)$. $\varphi(s) \in V$ because $\pi_w(\varphi(s)) = \varphi(s)(w) = w(s) = 0$. Moreover, $V \subset \varphi(U)$ because if $f \in V$, then $f = \varphi(t)$ for some $t \in S$, and $\pi_w(f) \in [0, 1)$; that is, $1 > \pi_w(f) = \pi_w(\varphi(t)) = \varphi(t)(w) = w(t)$; that is, $t \in w^{-1}([0, 1)) \subset U$, and $f = \varphi(t) \in \varphi(U)$. $\qquad \square$

Remark. Recall that when we first defined a topological space, we were motivated by properties of more familiar spaces, namely metric spaces. Theorem 4 shows that under the assumption of complete regularity, we are back on familiar ground. We understand closed intervals quite thoroughly, and we have a pretty good feeling for the operation of taking products of spaces. Now we find that all topological spaces satisfying the axiom of complete regularity are subsets of products of closed intervals. However, we are not yet back into the realm of metric spaces, as Theorem 5 will show.

Definition. A topological space S is *first countable* if, for each point $s \in S$, there exists a countable collection $\{U_s(n)\}_{n \in \{\text{positive integers}\}}$ of open sets containing s such that if U is any open set containing s, then there exists an n such that $U_s(n) \subset U$.

Remark. Let (S, ρ) be a metric space. Then S is first countable because the balls $\{B_s(1/n)\}$ have the required property.

Theorem 5. *For each real number* r, *let* $I_r = [0, 1]$. *Let* $P = \prod_{r \in R^1} I_r$. *Then* P *is not first countable. In particular,* P *is not metrizable; that is, there exists no metric* ρ *on* P *such that the metric topology is the product topology on* P.

PROOF. Let $p \in P$ be such that $p(r) = 0$ for all $r \in R^1$. Suppose there exists a countable collection $\{V_n\}$ of open sets containing p such that for each open set U containing p, $V_n \subset U$ for some n. We shall show that this assumption leads to a contradiction, thereby proving that P is not first countable.

Now each V_n is open in P and hence contains a basis element $\tilde{V}_n$ with $p \in \tilde{V}_n$. In particular, for each open set U containing p, there exists an n such that $\tilde{V}_n \subset U$. But, for each n, there exists a finite subset F_n of R^1 such that

$$\tilde{V}_n = \prod_{r \in R^1} T_{n,r} \quad \text{where} \quad T_{n,r} = \begin{cases} I_r & (r \notin F_n) \\ U_{n,r} & (r \in F_n) \end{cases}$$

and $U_{n,r}$ is, for each $r \in F_n$, an open set in I_r. Let $C = \bigcup_{n=1}^{\infty} F_n$. Then C is a countable union of finite sets, and hence is countable. In particular, $C \neq R^1$, so there exists an $r_0 \in R^1 - C$, and $T_{n,r_0} = I_{r_0}$ for each n; that is, the r_0-coordinate is unrestricted in each $\tilde{V}_n$. Thus, if $U_0 \subsetneqq I_{r_0}$, then $U = \prod_{r \in R^1} S_r$, where

$$S_r = \begin{cases} I_r & (r \neq r_0) \\ U_{r_0} & (r = r_0) \end{cases}$$

is open in P, and U contains no $\tilde{V}_n$. $\qquad \square$

2.4 Complete metric spaces

Definitions. Let (S, ρ) be a metric space. Recall that a sequence $\{s_n\}$ of points in S *converges* if there exists an $s \in S$ such that $\lim_{n \to \infty} \rho(s_n, s) = 0$. $\{s_n\}$ is a *Cauchy sequence* if $\lim_{m,n \to \infty} \rho(s_n, s_m) = 0$. S is called a *complete metric space* if every Cauchy sequence converges.

Let (S_1, ρ_1) and (S_2, ρ_2) be two metric spaces. An *isometric embedding* of S_1 into S_2 is a map $f\colon S_1 \to S_2$ such that $\rho_2(f(s), f(t)) = \rho_1(s, t)$ for all $s, t \in S_1$. In particular, if $f\colon S_1 \to S_2$ is an isometric embedding, then f maps S_1 homeomorphically onto $f(S_1)$, where S_1 and S_2 are provided with the metric topology. For any metric space (S, ρ), there exists an isometric embedding f of (S, ρ) into a complete metric space, which is called a *completion* of (S, ρ) if $f(S)$ is dense.

Remark. Unlike most other concepts we have thus far considered, the notion of completeness is not a topological concept; that is, it is not invariant under homeomorphisms. For example, if $S_1 = (-\pi/2, \pi/2)$ and $S_2 = R^1$, then S_1 and S_2 are both metric spaces in the usual way. S_2 is complete, but S_1 is not. However, there is a homeomorphism $f\colon S_1 \to S_2$ given by $f(s) = \tan s$.

We have discussed compactness for topological spaces in general. When we restrict our attention to metric spaces, there are several equivalent ways of saying that a space is compact.

Theorem 1. *Let S be a metric space. The following four conditions on S are equivalent.*

(1) *S is compact. (Heine–Borel property)*

(2) *S is countably compact; that is, every infinite subset of S has a limit point. (Bolzano–Weierstrass property)*

(3) *S is sequentially compact; that is, every sequence in S has a convergent subsequence.*

(4) *S is complete and totally bounded; that is, S is complete, and for every real number $\varepsilon > 0$, there exists a finite number of balls of radius ε, say $B_{s_1}(\varepsilon), \ldots, B_{s_m}(\varepsilon)$, which cover S; that is, $S = \bigcup_{i=1}^{m} B_{s_i}(\varepsilon)$.*

PROOF

(1) $\Rightarrow$ (2) by Theorem 5, Section 1.3.

(2) $\Rightarrow$ (3). Let $\langle s_n \rangle$ be a sequence in S. If an infinite number of the s_n are equal, then they form a convergent subsequence. On the other hand, suppose there exist infinitely many distinct elements in $\langle s_n \rangle$. By (2), this infinite set has a limit point $s_0 \in S$. Consider the ball $B_{s_0}(1)$. Since s_0 is a limit point of $\langle s_n \rangle$, $s_{n_1} \in B_{s_0}(1)$ for some n_1. Next consider $B_{s_0}(\frac{1}{2})$. Then since s_0 is a limit point of $\langle s_n \rangle$, $s_{n_2} \in B_{s_0}(\frac{1}{2})$ for some $n_2 > n_1$. Continuing by induction, we obtain a subsequence $s_{n_1}, s_{n_2}, \ldots$ such that $s_{n_j} \in B_{s_0}(1/2^{j-1})$, $(j = 1, 2, \ldots)$. This subsequence converges to s_0 because

$$\lim_{j \to \infty} \rho(s_{n_j}, s_0) \le \lim_{j \to \infty} \frac{1}{2^{j-1}} = 0.$$

(3) $\Rightarrow$ (4). S is complete because if $\{s_n\}$ is a Cauchy sequence, then by (3), there exists a subsequence $\{s_{n_k}\}$ converging to some point s_0. In fact, s_n converges to s_0 because

$$|s_n - s_0| \le |s_n - s_{n_k}| + |s_{n_k} - s_0| \to 0 \qquad \text{as } n \text{ and } n_k \to \infty.$$

Moreover, S is totally bounded. For otherwise there exists $\varepsilon > 0$ such that no finite collection of balls of radius ε cover S. Let $s_1 \in S$. Then $B_{s_1}(\varepsilon) \ne S$ so there exists $s_2 \notin B_{s_1}(\varepsilon)$; that is, $\rho(s_1, s_2) \ge \varepsilon$. Now $B_{s_1}(\varepsilon) \cup B_{s_2}(\varepsilon) \ne S$ so there exists $s_3 \notin B_{s_1}(\varepsilon) \cup B_{s_2}(\varepsilon)$. Thus $\rho(s_m, s_n) \ge \varepsilon$ for $m \ne n$, $\{m, n\} \subset \{1, 2, 3\}$. Continuing by induction, we obtain a sequence $\{s_n\}$ such that $\rho(s_m, s_n) \ge \varepsilon$ for all m, n $(m \ne n)$. By (3), there exists a convergent subsequence $\{s_{n_j}\}$. Let $s_0 = \lim_{j \to \infty} s_{n_j}$. Then there exists a j_0 such that, for $> j_0$, $s_{n_j} \in B_{s_0}(\varepsilon/2)$. For $j_1, j_2 > j_0$ and $j_1 \ne j_2$, this implies that

$$\rho(s_{n_{j_1}}, s_{n_{j_2}}) \le \rho(s_{n_{j_1}}, s_0) + \rho(s_0, s_{n_{j_2}}) < \frac{\varepsilon}{2} + \frac{\varepsilon}{2} = \varepsilon.$$

But $n_{j_1} \ne n_{j_2}$ implies $\rho(s_{n_{j_1}}, s_{n_{j_2}}) \ge \varepsilon$. This contradiction establishes (4).

(4) $\Rightarrow$ (1). We shall prove this remaining implication in two steps:

(a) (4) $\Rightarrow$ (3),
(b) (4) + (3) $\Rightarrow$ (1).

(a) Let $\{s_n\}$ be a sequence. By (4), there exists a finite number of balls $B_{1,1}, \ldots, B_{1,k_1}$ of radius 1 that cover S. Thus for some j_1, there exists an infinite set J_1 of positive integers such that $n \in J_1$ implies $s_n \in B_{1,j_1}$. Let n_1 be the first integer such that $s_{n_1} \in B_{1,j_1}$.

Now by (4) again, there exists a finite number of balls $B_{2,1}, \ldots, B_{2,k_2}$ of radius $\frac{1}{2}$ which cover S. Hence for some j_2, there exists an infinite subset J_2 of J_1 such that $n \in J_2$ implies $s_n \in B_{2,j_2}$. Let n_2 be the first such integer with $n_2 > n_1$.

Continuing by induction, we obtain a sequence $\{s_{n_k}\}$ such that for all $l \ge k$, $s_{n_l} \in B_{k,j_k}$ (a ball of radius $1/k$). Thus, for $l, m \ge k$,

$$\rho(s_{n_l}, s_{n_m}) < 2/k,$$

and $\{s_{n_k}\}$ is a Cauchy sequence. Since S is complete, $\{s_{n_k}\}$ converges.

(b) Let $\mathcal{V}$ be an open covering of S. To prove compactness, it suffices to show that for some positive integer n, each ball of radius $1/n$ lies in some $V \in \mathcal{V}$. If so, then by total boundedness, there exists a finite number of balls, say $B_1, \ldots, B_k$, of radius $1/n$ which cover S. But for each j there exists a $V_j \in \mathcal{V}$ with $B_j \subset V_j$, so $S = \bigcup_{j=1}^k B_j \subset \bigcup_{j=1}^k V_j$, and $V_1, \ldots, V_k$ is a finite subcovering of S.

Suppose, for each n, there exists a ball $B_{s_n}(1/n)$ of radius $1/n$ which is not contained in any $V \in \mathcal{V}$. We shall show that this assumption leads to a contradiction, thereby establishing the theorem. The sequence $\{s_n\}$ of centers of these balls has a convergent subsequence $\{s_{n_k}\}$ by (3). Let $s_0 = \lim_{k \to \infty} s_{n_k}$. Then $s_0 \in V_0$ for some $V_0 \in \mathscr{V}$. Since V_0 is open, there exists $r > 0$ such that $B_{s_0}(r) \subset V_0$. Since $s_{n_k} \to s_0$, there exists a k_0 such that $s_{n_k} \in B_{s_0}(r/2)$ for all $k \geq k_0$. Let $l \geq k_0$ be such that $1/n_l < r/2$. Then $B_{s_{n_l}}(1/n_l) \subset B_{s_0}(r) \subset V_0$. But this is a contradiction. $\qquad\square$

Remark. Condition (4) of Theorem 1 shows that in a complete metric space, compactness is equivalent to the existence, given any $\varepsilon > 0$, of a finite set of points in S such that each point of S is within ε of one of these.

Theorem 2. *Let (S, ρ) be a complete metric space. Suppose $\{U_n\}$ is a countable collection of open sets each of which is dense in S, that is,*

$$\bar{U}_n = S \quad \text{for} \quad n = 1, 2, \ldots.$$

Then $\bigcap_{n=1}^{\infty} U_n \neq \varnothing$.

PROOF. We shall construct a sequence which will converge to a point in $\bigcap_{n=1}^{\infty} U_n$. Choose $s_1 \in U_1$. Since U_1 is open, there exists a ball $B_{s_1}(r_1) \subset U_1$ for some r_1. Since $s_1 \in S = \bar{U}_2$, $B_{s_1}(r_1) \cap U_2 \neq \varnothing$. Choose $s_2 \in B_{s_1}(r_1) \cap U_2$. Let r_2 be such that $\overline{B_{s_2}(r_2)} \subset B_{s_1}(r_1) \cap U_2$ and $r_2 < \min\{r_1/2, r_1 - \rho(s_1, s_2)\}$. $\overline{B_{s_2}(r_2)} \subset B_{s_1}(r_1)$ because $s \in \overline{B_{s_2}(r_2)}$ implies

$$\rho(s, s_1) \leq \rho(s, s_2) + \rho(s_2, s_1) \leq r_2 + \rho(s_1, s_2) < r_1.$$

We continue by induction to obtain a sequence $\{s_n\}$ of points in S, and about each s_n, a ball $B_{s_n}(r_n)$ such that

$$r_n < \frac{r_1}{2^{n-1}} \quad \text{and} \quad \overline{B_{s_{n+1}}(r_{n+1})} \subset B_{s_n}(r_n) \subset U_n.$$

$\{s_n\}$ is a Cauchy sequence. For, given $\varepsilon > 0$, we can choose n_0 such that $r_{n_0} < \varepsilon/2$. Since $s_n \in B_{s_n}(r_n) \subset B_{s_{n_0}}(r_{n_0})$ for all $n \geq n_0$,

$$\rho(s_m, s_n) \leq \rho(s_m, s_{n_0}) + \rho(s_{n_0}, s_n) < r_{n_0} + r_{n_0} < \varepsilon$$

for all $m, n \geq n_0$. Since S is complete, $\{s_n\}$ converges to some point s_0.

We must verify that $s_0 \in U_n$ for all n. But, for any n, $s_{n+k} \in B_{s_{n+1}}(r_{n+1})$ for all $k \geq 1$. Since $s_0 = \lim_{k \to \infty} s_{n+k}$,

$$s_0 \in \overline{B_{s_{n+1}}(r_{n+1})} \subset B_{s_n}(r_n) \subset U_n,$$

as was to be shown. $\qquad\square$

Theorem 2, stated somewhat differently, has many applications in analysis. We now proceed to recast this theorem into its more standard form.

Definition. Let S be a topological space. A subset T of S is *nowhere dense* if $\overline{T}$ contains no nonempty open set.

Remark. $T \subset S$ is nowhere dense if and only if $(\overline{T})'$ is dense in S.

Definition. A subset T of a topological space S is of the *first category* if it is a countable union of nowhere dense sets. Otherwise, T is said to be of the *second category*.

Corollary to Theorem 2 (Baire Category Theorem). *A complete metric space is of the second category; that is, it is not the union of a countable number of nowhere dense sets.*

PROOF. Let (S, ρ) be a complete metric space. Suppose $S = \bigcup_{n=1}^{\infty} T_n$, where each T_n is nowhere dense. Then $S = \bigcup_{n=1}^{\infty} \overline{T}_n$, so, taking complements, $\varnothing = \bigcap_{n=1}^{\infty} (\overline{T}_n)'$. But each $(\overline{T}_n)'$ is open and dense by the above remark. This contradicts Theorem 2. $\qquad\qquad\square$

Exercise. Theorem 2 and its corollary are also true if the term "complete metric" is replaced by "compact Hausdorff." Can you prove it?

2.5 Applications

Definition. A *normed linear space* is a linear space (vector space) L, over the reals or complexes, together with a real-valued function (denoted by $\| \ \|$) on L satisfying the following conditions for all vectors a and b and all scalars λ:

(1) $\|a\| \geq 0$ and $\|a\| = 0 \Leftrightarrow a = 0$
(2) $\|\lambda a\| = |\lambda| \|a\|$
(3) $\|a + b\| \leq \|a\| + \|b\|$

Remark. Let $(L, \| \ \|)$ be a normed linear space. Define a metric on L by

$$\rho(a, b) = \|a - b\|.$$

Then (L, ρ) is a metric space because

(1) $\rho(a, b) = 0 \Leftrightarrow \|a - b\| = 0 \Leftrightarrow a - b = 0 \Leftrightarrow a = b$;
(2) $\rho(a, b) = \|a - b\| = |-1| \|a - b\| = \|b - a\| = \rho(b, a)$; and
(3) $\rho(a, c) = \|a - c\| = \|a - b + b - c\| \leq \|a - b\| + \|b - c\|$
$\qquad = \rho(a, b) + \rho(b, c)$.

Definition. A normed linear space L is a *Banach space* if (L, ρ) is a *complete* metric space.

Remark. A sequence a_n in L is Cauchy if

$$\lim_{n,m \to \infty} \rho(a_n, a_m) = 0, \quad \text{that is, if} \quad \lim_{n,m \to \infty} \|a_n - a_m\| = 0.$$

EXAMPLE. Let $L = C([0, 1])$ be the space of all continuous real-valued functions on $[0, 1]$. For $f \in L$, define $\|f\| = \max_{x \in [0,1]} |f(x)|$. Then L is a Banach space. That $\| \ \|$ is a norm is easily verified. That (L, ρ) is a complete metric space follows from the theorem that a uniformly convergent sequence of continuous functions converges to a continuous function.

Remark. More generally, if S is any compact Hausdorff space, the set of all continuous real-valued functions on S is a Banach space, where $\|f\| = \max_{s \in S} |f(s)|$.

Theorem 1. *There exists a continuous real-valued function $f \in C([0, 1])$ such that f has a derivative at no point of $[0, 1]$.*

PROOF. For n any positive integer, let

$$C_n = \left[f \in C([0, 1]); \left| \frac{f(t + h) - f(t)}{h} \right| \le n \right.$$
$$\left. \text{for some } t \text{ and all } h \text{ with } t + h \in [0, 1] \right].$$

We shall show that C_n is nowhere dense for each n. Since $C([0, 1])$ is a complete metric space and hence is of the second category, it will then follow that

$$\bigcup_{n=1}^{\infty} C_n \ne C([0, 1]);$$

that is, there exists a function $f \in C([0, 1])$ such that $f \notin C_n$ for any n. This f is the required function because $f \notin C_n$ means that

$$\left| \frac{f(t + h) - f(t)}{h} \right| > n$$

for all $t \in [0, 1]$ and some h (depending on t and n). Note that for each fixed t, $h \to 0$ as $n \to \infty$ because given $\varepsilon > 0$ the difference quotient $|(f(t + h) - f(t))/h|$ is bounded as a function of h for $|h| \ge \varepsilon$. Thus

$$\limsup_{h \to 0} \left| \frac{f(t + h) - f(t)}{h} \right| = \infty,$$

and the derivative of f at t fails to exist for each $t \in [0, 1]$.

To prove that C_n is nowhere dense, we must show that $\bar{C}_n$ contains no nonempty open set. First we show that C_n is closed; that is, that $C_n = \bar{C}_n$. Note that since $C([0, 1])$ is a metric space, $C([0, 1])$ is first countable, and hence any limit point of a set $T \subset C([0, 1])$ is in fact a limit of a countable subset of T. Thus to show C_n is closed, it suffices to show that if $\{f_k\} \subset C_n$ is a sequence which converges in $C([0, 1])$, then the limit $f \in C_n$. But $f_k \in C_n$ implies that there exists $t_k \in [0, 1]$ such that

$$\left| \frac{f_k(t_k + h) - f_k(t_k)}{h} \right| \leq n$$

for all h. Since $[0, 1]$ is compact, the sequence $\{t_k\}$ has a convergent subsequence, which we also denote by $\{t_k\}$. Let $t_0 = \lim_{k \to \infty} t_k$. Then

$$\left| \frac{f(t_0 + h) - f(t_0)}{h} \right| = \left| \frac{f(t_0 + h) - f(t_k + h)}{h} + \frac{f(t_k + h) - f_k(t_k + h)}{h} \right.$$

$$\left. + \frac{f_k(t_k + h) - f_k(t_k)}{h} + \frac{f_k(t_k) - f(t_k)}{h} + \frac{f(t_k) - f(t_0)}{h} \right|$$

$$\leq \left| \frac{f(t_0 + h) - f(t_k + h)}{h} \right| + \left| \frac{f(t_k + h) - f_k(t_k + h)}{h} \right|$$

$$\text{①} \qquad\qquad\qquad\qquad \text{②}$$

$$+ \left| \frac{f_k(t_k + h) - f_k(t_k)}{h} \right| + \left| \frac{f_k(t_k) - f(t_k)}{h} \right|$$

$$\text{③} \qquad\qquad\qquad\qquad \text{④}$$

$$+ \left| \frac{f(t_k) - f(t_0)}{h} \right|.$$

$$\text{⑤}$$

Now fix h. For any $\varepsilon > 0$, if k is large enough, ① and ⑤ are smaller than ε because f is continuous and $t_k \to t_0$; while ② and ④ are smaller than ε because f_k converges uniformly to f. Because of the previous paragraph, ③ $\leq n$. Hence

$$\left| \frac{f(t_0 + h) - f(t_0)}{h} \right| \leq n + 4\varepsilon \qquad \text{(any } \varepsilon > 0\text{)};$$

that is,

$$\left| \frac{f(t_0 + h) - f(t_0)}{h} \right| \leq n \quad \text{and} \quad f \in C_n.$$

Thus each C_n is closed. Now we show that C_n is nowhere dense; that is, given any $g \in C_n = \bar{C}_n$ and any $\varepsilon > 0$, there exists $f \in C[0, 1]$ such that

$\|f - g\| < \varepsilon$ and $f \notin C_n$. Now a typical example of a function in $C[0, 1]$ which is not in C_n is the "sawtooth" function (Figure 2.3). For any n we can find such a function, whose norm is less than or equal to any prescribed

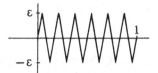

Figure 2.3

$\varepsilon > 0$, and where the slope of each line segment is greater than n in absolute value. To find a function $f \notin C_n$ within ε of g, we need only construct a sawtooth function close to g, as in Figure 2.4.

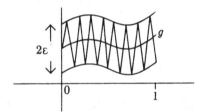

Figure 2.4

We leave this construction to the reader, but include the following hint. Using the uniform continuity of the function g, we can find a continuous function g_1, so that $\|g - g_1\| < \varepsilon/2$ and g_1 is piecewise linear; that is, g_1 looks like the line segments in Figure 2.5. It suffices to find the appropriate sawtooth function for each linear piece of g_1 and then to patch. □

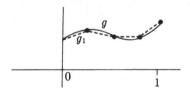

Figure 2.5

Remark. Theorem 1 above is of considerable interest, but more important than the statement is the method of proof. This method is used frequently in analysis and topology to prove the existence of functions with specified properties.

We exhibit a regular topological space which is not normal as another application of the Baire category theorem.

EXAMPLE. Let S be the upper half-plane in R^2 together with the x axis R^1. Let $\mathscr{B}_1$ be the usual open sets in the upper half-plane. Let

$$\mathscr{B}_2 = [B_{(r,\varepsilon)}(\varepsilon) \cup \{r\}; r \in R^1, \varepsilon > 0],$$

so that an element of $\mathscr{B}_2$ is an open disc tangent to R^1 together with r the point of tangency, as in Figure 2.6. Take $\mathscr{B}_1 \cup \mathscr{B}_2$ as basis for the topology on S.

Figure 2.6

Note that in the relative topology, R^1 is discrete. In fact, for each $p \in R^1$, $R^1 - \{p\}$ is closed in S because its complement is clearly a union of open sets. Since $F = \bigcap_{p \in R^1 - F} (R^1 - \{p\})$ for any subset $F \subset R^1$, it follows that every subset of R^1 is closed in S.

S *is regular.* Suppose F is closed in S and $p \notin F$. If $p \notin R^1$, there exists a ball $B_p(\varepsilon)$ contained in the open set F'; and the open sets $B_p(\varepsilon/2)$ and $(\overline{B_p(\varepsilon/2)})'$ separate p and F. If $p \in R^1$, there exists a basic open set V containing p and contained in the open set F'; that is, $V \cap F = \varnothing$. Let V_1 be a basic open set containing p such that $\overline{V}_1 \subset V$ (Figure 2.7). Clearly each point in F (in fact, each point in V') is contained in a basic open set disjoint from V_1. Let U be the union of these. Then V_1 and U separate p and F.

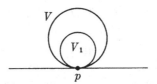

Figure 2.7

S *is not normal.* Let $F_1 \subset R^1$ be the rationals and $F_2 \subset R^1$ be the irrationals. Then F_1 and F_2 are closed sets in S. Moreover, it is not possible to find disjoint open sets U_1 and U_2 in S such that $F_1 \subset U_1$ and $F_2 \subset U_2$. For suppose such sets U_1 and U_2 did exist. Then for each $p \in F_j$ $(j = 1, 2)$, there exists a basic open set $D_p \subset U_j$ with $p \in D_p$ (D_p is the union of $\{p\}$ with a disc tangent at p). $\bigcup_{p \in F_1} D_p \subset U_1$ and $\bigcup_{p \in F_2} D_p \subset U_2$ are also disjoint open sets containing F_1 and F_2 respectively. Let $f: R^1 \to R^1$ be defined by $f(p) =$ radius of D_p. (f is not necessarily continuous.) For each positive integer n, let

$$T_n = \left[p \in F_2; f(p) \geq \frac{1}{n} \right].$$

Then $\bigcup_{n=1}^{\infty} T_n = F_2$ because $p \in F_2$ implies $f(p) \geq 1/n$ for some n. Moreover,

$$R^1 = \left(\bigcup_{n=1}^{\infty} T_n\right) \cup \left(\bigcup_{q \in F_1} \{q\}\right),$$

so we have expressed the set R^1 as a countable union of sets T_n and $\{q\}$. Now consider the usual topology on R^1. Then R^1 is a complete metric space, and hence is of the second category. Since we have expressed R^1 as a union of countably many sets, not all of these sets can be nowhere dense in R^1 (in the usual topology). Each $\{q\}$ is nowhere dense, so $\bar{T}_n$ must contain an open set in R^1 for some n. In particular, $\bar{T}_n$ must contain some interval I. Let $q \in I$ be rational. Since $q \in \bar{T}_n$, there exists a sequence $\{p_k\} \subset T_n$ which converges to q. But $(\bigcup_{k=1}^{\infty} D_{p_k}) \cap D_q \subset U_1 \cap U_2 = \varnothing$. This is impossible since each D_{p_k} has radius $\geq 1/n$; that is, $f(p_k)$ does not tend to zero as $k \to \infty$ and $p_k \to q$. This contradiction completes the proof. $\qquad\square$

Theorem 2 (Uniform boundedness principle). *Let S be a complete metric space. Let $\mathscr{F}$ be a family of continuous real-valued functions on S with the property that for each $s \in S$, there exists a constant M_s such that $|f(s)| \leq M_s$ for all $f \in \mathscr{F}$. Then there exists a nonempty open set $U \subset S$ and a constant M such that $|f(s)| \leq M$ for all $f \in \mathscr{F}$ and all $s \in U$.*

PROOF. For each $f \in \mathscr{F}$ and each positive integer n, let

$$T_{n,f} = [s \in S; |f(s)| \leq n].$$

$T_{n,f}$ is closed because $T_{n,f} = f^{-1}([-n, n])$. Let

$$T_n = \bigcap_{f \in \mathscr{F}} T_{n,f} = [s \in S; |f(s)| \leq n \text{ for all } f \in \mathscr{F}].$$

T_n is closed because it is the intersection of closed sets. $\bigcup_{n=1}^{\infty} T_n = S$ because if $s \in S$, then $s \in T_n$ for all $n > M_s$. By the Baire category theorem, not all T_n are nowhere dense; that is, some $\bar{T}_n = T_n$ must contain a nonempty open set U. Then, for all $s \in U$ and $f \in \mathscr{F}$, $|f(s)| \leq n$. Take $M = n$. $\qquad\square$

Fundamental group and covering spaces 3

3.1 Homotopy

Notation. The letter I will henceforth denote the closed interval $[0, 1]$. A *map* is a function; the two words will be used interchangeably.

Definition. Let X and Y be topological spaces, and let f_0 and f_1 be continuous maps $X \to Y$. f_0 is *homotopic* to f_1 (written $f_0 \simeq f_1$) if there exists a continuous map $F: X \times I \to Y$ such that $F(x, 0) = f_0(x)$ and $F(x, 1) = f_1(x)$ for all $x \in X$. The map F is called a *homotopy* from f_0 to f_1.

Remark. For $(x, t) \in X \times I$, we may regard t as measuring time. Then $f_t(x) = F(x, t)$ is a 1-parameter family of maps $X \to Y$. At time 0 we have the map f_0. At time 1 we have the map f_1. As time increases from 0 to 1, the map f_0 is deformed continuously into the map f_1.

EXAMPLE. Let $i: R^n \to R^n$ be the identity map. Let $c: R^n \to R^n$ be defined by $c(v) = 0$ for all $v \in R^n$. Then $c \simeq i$. For let $F: R^n \times I \to R^n$ be defined by $F(v, t) = tv$. Then F is continuous, and $F(v, 0) = 0 = c(v)$ and $F(v, 1) = v = i(v)$ for all $v \in R^n$. Geometrically, the map F shrinks the image R^n of i to the point $\{0\}$ as t varies from 1 to 0.

Theorem 1. *Homotopy is an equivalence relation; that is, for f, g, h continuous maps $X \to Y$,*

(1) $f \simeq f$,
(2) $f \simeq g$ *implies* $g \simeq f$, *and*
(3) $f \simeq g$ *and* $g \simeq h$ *implies* $f \simeq h$.

PROOF

(1) Let $F: X \times I \to Y$ be defined by $F(x, t) = f(x)$. F is continuous because it is the composition of the continuous maps f and projection onto the first factor.

(2) Given a homotopy $F: X \times I \to Y$ such that $F(x, 0) = f(x)$ and $F(x, 1) = g(x)$, let $G: X \times I \to Y$ be defined by $G(x, t) = F(x, 1 - t)$. G is continuous because $t \to 1 - t$ is continuous. G is a homotopy from g to f.

(3) Given homotopies $F, G: X \times I \to Y$, with $F(x, 0) = f(x)$, $F(x, 1) = g(x) = G(x, 0)$, and $G(x, 1) = h(x)$, let $H: X \times I \to Y$ be defined by

$$H(x, t) = \begin{cases} F(x, 2t) & (0 \le t \le \tfrac{1}{2}) \\ G(x, 2t - 1) & (\tfrac{1}{2} \le t \le 1). \end{cases}$$

Then $H(x, 0) = f(x)$, and $H(x, 1) = h(x)$. H is continuous by the following lemma. $\square$

Glueing (or pasting) lemma. *Let X and Y be topological spaces. Assume $X = A \cup B$, where A and B are closed (open) sets in X. Suppose $f_1: A \to Y$ and $f_2: B \to Y$ are continuous functions such that $f_1(x) = f_2(x)$ for all $x \in A \cap B$. Let $g: X \to Y$ be defined by*

$$g(x) = \begin{cases} f_1(x) & (x \in A) \\ f_2(x) & (x \in B). \end{cases}$$

Then g is continuous.

Remark. The function g of the lemma is thus obtained by "glueing" f_1 and f_2 together along their common domain.

PROOF OF THE GLUEING LEMMA. Assume A and B are both closed. We show that inverse images of closed sets are closed. Let F be a closed subset of Y. Then

$$\begin{aligned} g^{-1}(F) &= g^{-1}(F) \cap (A \cup B) \\ &= (g^{-1}(F) \cap A) \cup (g^{-1}(F) \cap B) \\ &= f_1^{-1}(F) \cup f_2^{-1}(F). \end{aligned}$$

Since f_1 is continuous, $f_1^{-1}(F)$ is closed in A, and hence in X because A is closed in X. Similarly, $f_2^{-1}(F)$ is closed in X. Moreover, the union of two closed sets is closed.

If both A and B are open, the same argument shows that inverse images of open sets are open. $\square$

Theorem 2. *Let X, Y, Z be topological spaces. Suppose that f_0 and f_1 are homotopic maps $X \to Y$ and that g_0 and g_1 are homotopic maps $Y \to Z$. Then $g_0 \circ f_0$ and $g_1 \circ f_1$ are homotopic maps $X \to Z$.*

PROOF. We break the proof into two steps:

(a) $g_0 \circ f_0 \simeq g_0 \circ f_1$
(b) $g_0 \circ f_1 \simeq g_1 \circ f_1$.

Then $g_0 \circ f_0 \simeq g_1 \circ f_1$ by (3) of Theorem 1.

(a) Let $F: X \times I \to Y$ be a homotopy from f_0 to f_1. Let $G = g_0 \circ F$: $X \times I \to Z$. Then G is a homotopy from $g_0 \circ f_0$ to $g_0 \circ f_1$.

(b) Let $H: Y \times I \to Z$ be a homotopy from g_0 to g_1. Let $\tilde{f}_1: X \times I \to Y \times I$ be defined by $\tilde{f}_1(x, t) = (f_1(x), t)$. $\tilde{f}_1$ is easily seen to be continuous. Let

$$K = H \circ \tilde{f}_1: X \times I \to Z.$$

Then K is a homotopy from $g_0 \circ f_1$ to $g_1 \circ f_1$. □

Definition. Two spaces X and Y are of the same *homotopy type* if there exist continuous maps $f: X \to Y$ and $g: Y \to X$ such that $g \circ f \simeq i_X$ and $f \circ g \simeq i_Y$, where i_X and i_Y are the identity maps on X and Y respectively.

Remark. It is easy to verify that "same homotopy type" is an equivalence relation. Thus, the collection of all topological spaces is partitioned into equivalence classes. Two spaces are in the same class if and only if they are of the same homotopy type. Clearly homeomorphic spaces are of the same homotopy type. Much of algebraic topology is concerned with the study of those properties of topological spaces which are invariants of homotopy type, that is, those properties which, when possessed by *one* topological space X, are possessed by *every* topological space of the same homotopy type as X.

Definition. A topological space X is *contractible* if the identity map $i_X: X \to X$ is homotopic to a constant map; that is, if $i_X \simeq c$, where $c: X \to \{x_0\}$ for some $x_0 \in X$.

Theorem 3. *A space X is contractible if and only if X is of the same homotopy type as a single point.*

PROOF. Suppose X is contractible. Then $i_X \simeq c$ for some $c: X \to \{x_0\}$ ($x_0 \in X$). Let $f = c: X \to \{x_0\}$ and let $g: \{x_0\} \to X$ be defined by $g(x_0) = x_0$. Then clearly $g \circ f = c \simeq i_X$, and $f \circ g = i_{\{x_0\}} \simeq i_{\{x_0\}}$, so X and $\{x_0\}$ have the same homotopy type.

Conversely, suppose $Y = \{y\}$ consists of a single point, and X has the same homotopy type as Y. Then there exist continuous maps $f: X \to Y$ and $g: Y \to X$ such that $g \circ f \simeq i_X$ and $f \circ g \simeq i_Y$. Let $x_0 = g(y)$, and let $c: X \to \{x_0\}$. Then $c = g \circ f \simeq i_X$, so X is contractible. □

Remark. Both Theorem 3 and the example at the beginning of this section show that R^n has same homotopy type as a single point. Thus from the viewpoint of homotopy theory, R^n is a trivial space.

3.2 Fundamental group

Definition. Let X be a topological space. A *path* in X from x_0 to x_1 (with origin x_0 and end x_1) is a continuous map $\alpha\colon I \to X$ such that $\alpha(0) = x_0$ and $\alpha(1) = x_1$.

Remark. Note that a path is a function and *not* a set of points. A path is a "parameterized curve."

Definition. A space X is *arcwise connected* if, given any two points x_0 and x_1 in X, there exists a path with origin x_0 and end x_1.

Theorem 1. *If a topological space is arcwise connected, then it is connected.*

PROOF. Let X be arcwise connected. Suppose U_0 and U_1 are nonempty disjoint open sets with $U_0 \cup U_1 = X$. Let $x_0 \in U_0$ and $x_1 \in U_1$. Since X is arcwise connected, there exists a path α from x_0 to x_1. Clearly $\alpha^{-1}(U_0)$ and $\alpha^{-1}(U_1)$ are disjoint open sets in I with $\alpha^{-1}(U_0) \cup \alpha^{-1}(U_1) = I$. Moreover, $0 \in \alpha^{-1}(U_0)$ and $1 \in \alpha^{-1}(U_1)$, so these sets are nonempty. But I is connected, so this is impossible. $\square$

Remark. The converse of Theorem 1 is false, as is shown by the following example.

EXAMPLE. Let $X = A \cup B \subset R^2$ with the relative topology, where

$$A = [(0, y) \in R^2; |y| \le 1],$$
$$B = [(x, \sin 1/x) \in R^2; 0 < x \le 1].$$

X is connected. For let U_1 and U_2 be disjoint open sets in X with $U_1 \cup U_2 = X$. The point $(0, 1)$ is in one of these two sets, say $(0, 1) \in U_1$. We shall show that $U_2 = \varnothing$, and hence X is connected.

First, $(0, 1) \in U_1 \cap A$, so $U_1 \cap A \ne \varnothing$. Since A is connected, and

$$A = (U_1 \cap A) \cup (U_2 \cap A),$$

it follows that $U_2 \cap A = \varnothing$. Next, consider $U_1 \cap B$. Since any ball in R^2 about $(0, 1)$ must contain points of the form $(x, \sin 1/x)$, $U_1 \cap B \ne \varnothing$. But B is connected—it is the continuous image of the connected set $(0, 1]$—and $B = (U_1 \cap B) \cup (U_2 \cap B)$. Thus $U_2 \cap B = \varnothing$. Hence $U_2 = (U_2 \cap A) \cup (U_2 \cap B) = \varnothing$.

X is *not* arcwise connected. For let α be a path in X with origin $(0, 1)$. We shall show that $\alpha(I) \subset A$, and hence no point in B can be joined to $(0, 1)$ by a path in X. Consider $\alpha^{-1}(A)$. Since A is closed in X, $\alpha^{-1}(A)$ is closed in I. $\alpha^{-1}(A) \ne \varnothing$ because $0 \in \alpha^{-1}(A)$. Hence it suffices to show that $\alpha^{-1}(A)$ is open in I, for then $\alpha^{-1}(A) = I$ because I is connected. Suppose $t_0 \in \alpha^{-1}(A)$. Then $\alpha(t_0) \in A$. Let $U = X \cap B_{\alpha(t_0)}(\frac{1}{2})$. Then U is open in X. Since α is continuous, there exists an $\varepsilon > 0$ such that $\alpha(t) \in U$ whenever $|t - t_0| < \varepsilon$. Claim: $\alpha((t_0 - \varepsilon, t_0 + \varepsilon)) \subset A$. For suppose $|t_1 - t_0| < \varepsilon$ and $\alpha(t_1) \in B$. Now $U \cap B$ is a union of disjoint arcs (homeomorphic images of open intervals), and the arc containing $\alpha(t_1)$ is both open and closed in U. This contradicts the connectedness of $\alpha((t_0 - \varepsilon, t_0 + \varepsilon))$. Thus $(t_0 - \varepsilon, t_0 + \varepsilon) \subset \alpha^{-1}(A)$ and $\alpha^{-1}(A)$ is open in I (see Figure 3.1).

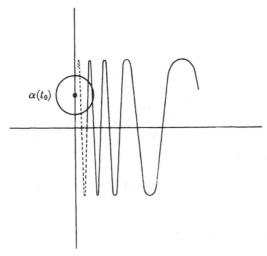

Figure 3.1

Definition. Let α be a path from x_0 to x_1 and let β be a path from x_1 to x_2. The *product* of α and β is the path $\alpha\beta$ from x_0 to x_2 defined by

$$\alpha\beta(t) = \begin{cases} \alpha(2t) & (0 \leq t \leq \frac{1}{2}) \\ \beta(2t - 1) & (\frac{1}{2} \leq t \leq 1). \end{cases}$$

The *inverse* of α is the path α^{-1} from x_1 to x_0 defined by $\alpha^{-1}(t) = \alpha(1 - t)$.

Remark. That the product of two paths is continuous is a consequence of the "Glueing lemma" (Section 3.1).

Definition. Two paths α and β from x_0 to x_1 are *homotopic* (written $\alpha \simeq \beta$) if there exists a continuous map $F: I \times I \to X$ such that

$$F(0, t_2) = x_0 \quad \text{and} \quad F(1, t_2) = x_1 \qquad \text{for all } t_2 \in I;$$
$$F(t_1, 0) = \alpha(t_1) \quad \text{and} \quad F(t_1, 1) = \beta(t_1) \qquad \text{for all } t_1 \in I.$$

(See Figure 3.2).

Remark. Thus a homotopy of paths is a homotopy in the usual sense together with the additional requirement that the end points remain fixed

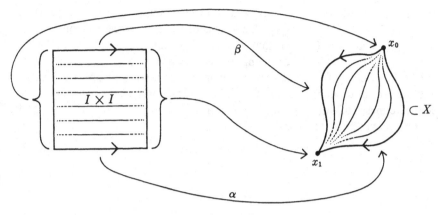

Figure 3.2

throughout the homotopy. Note that without this additional requirement every path is homotopic to a constant path.

Remark. The relation $\dot{\simeq}$ is an equivalence relation. The proof is similar to the corresponding proof for $\simeq$ (see Theorem 1).

Theorem 2. *Suppose $\alpha_0 \dot{\simeq} \alpha_1$ and $\beta_0 \dot{\simeq} \beta_1$ are paths such that $\alpha_0\beta_0$ is defined. Then $\alpha_1\beta_1$ is defined and $\alpha_1\beta_1 \dot{\simeq} \alpha_0\beta_0$.*

PROOF. $\alpha_1\beta_1$ is defined because

$$\text{end of } \alpha_1 = \text{end of } \alpha_0 = \text{origin of } \beta_0 = \text{origin of } \beta_1.$$

Let F and G be homotopies from α_0 to α_1 and from β_0 to β_1 respectively. Then a homotopy H from $\alpha_0\beta_0$ to $\alpha_1\beta_1$ is given by

$$H(t_1, t_2) = \begin{cases} F(2t_1, t_2) & (0 \leq t_1 \leq \tfrac{1}{2}) \\ G(2t_1 - 1, t_2) & (\tfrac{1}{2} \leq t_1 \leq 1). \end{cases}$$

H is continuous by the "glueing lemma." $\qquad\square$

Theorem 3. *Suppose α_0 and α_1 are homotopic paths. Then $\alpha_0^{-1} \dot{\simeq} \alpha_1^{-1}$.*

PROOF. Let F be a homotopy from α_0 to α_1. A homotopy H from α_0^{-1} to α_1^{-1} is given by

$$H(t_1, t_2) = F(1 - t_1, t_2). \qquad\square$$

Notation. Let $\langle\alpha\rangle$ denote the $\dot{\simeq}$ equivalence class of α; that is, $\langle\alpha\rangle$ is the set of all paths homotopic to α. Since homotopic paths have the same end points, the origin and end of $\langle\alpha\rangle$ are defined.

Definition. The *product* and *inverse* of $\simeq$ equivalence classes are defined by

$$\langle\alpha\rangle\langle\beta\rangle = \langle\alpha\beta\rangle \quad \text{(if } \alpha\beta \text{ is defined)},$$
$$\langle\alpha\rangle^{-1} = \langle\alpha^{-1}\rangle.$$

These are well defined by Theorems 2 and 3 above.

Theorem 4. *For each $x \in X$, let $e_x: I \to X$ be the path defined by $e_x(t) = x$ for all $t \in I$. Then:*

(1) *If $\langle\alpha\rangle$ has origin x_0, then $\langle e_{x_0}\rangle\langle\alpha\rangle = \langle\alpha\rangle$.*
(2) *If $\langle\alpha\rangle$ has end x_1, then $\langle\alpha\rangle\langle e_{x_1}\rangle = \langle\alpha\rangle$.*
(3) *If $\langle\alpha\rangle$ has origin x_0 and end x_1, then*

$$\langle\alpha\rangle\langle\alpha^{-1}\rangle = \langle e_{x_0}\rangle \quad \text{and} \quad \langle\alpha^{-1}\rangle\langle\alpha\rangle = \langle e_{x_1}\rangle.$$

(4) $(\langle\alpha\rangle\langle\beta\rangle)\langle\gamma\rangle = \langle\alpha\rangle(\langle\beta\rangle\langle\gamma\rangle)$ *(if $(\alpha\beta)\gamma$ is defined).*

Corollary. *Let X be a topological space and let $x_0 \in X$. The set of $\simeq$ equivalence classes of paths with origin $=$ end $= x_0$ forms a group under the operations of multiplication and inverse as defined above. This group is denoted by $\pi_1(X, x_0)$ and is called the* fundamental group, *or* first homotopy group, *of the pair (X, x_0).*

PROOF OF THEOREM 4

(1) We must show that $e_{x_0}\alpha \simeq \alpha$. Thus we want to construct a continuous mapping $F: I \times I \to X$ such that

$$F|_{I \times \{0\}} = e_{x_0}\alpha,$$
$$F|_{I \times \{1\}} = \alpha,$$
$$F|_{\{0\} \times I} = x_0,$$
$$F|_{\{1\} \times I} = x_1 = \alpha(1).$$

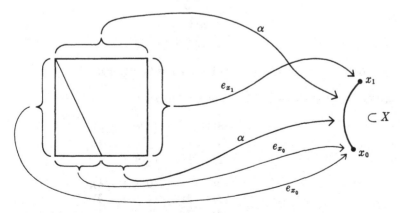

Figure 3.3

55

We do this by defining F on the triangle in Figure 3.3 to be the constant map into x_0, and by requiring that F, restricted to each horizontal line segment in the trapezoid of the figure, be equal to α (after suitably parameterizing the line segment). Explicitly,

$$F(t_1, t_2) = \begin{cases} x_0 & (2t_1 \le 1 - t_2), \\ \alpha\left(\dfrac{2t_1 - 1 + t_2}{1 + t_2}\right) & (1 - t_2 \le 2t_1). \end{cases}$$

(2) We must show that $\alpha e_{x_1} \simeq \alpha$. The proof is similar to the proof of (1) (see Figure 3.4).

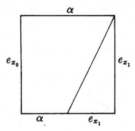

Figure 3.4

(3) It suffices to prove that $\langle\alpha\rangle\langle\alpha^{-1}\rangle = \langle e_{x_0}\rangle$, for then we may interchange the roles of α and α^{-1}. Thus we must show that $\alpha\alpha^{-1} \simeq e_{x_0}$. We do this by "pulling the end point x_1 in to x_0 along the path α" (see Figure 3.5).

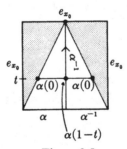

Figure 3.5

Exercise. Find the analytic expression for this homotopy.

(4) We must show that $(\alpha\beta)\gamma \simeq \alpha(\beta\gamma)$. Now

$$((\alpha\beta)(\gamma))(t) = \begin{cases} \alpha(4t) & (0 \le t \le \tfrac{1}{4}), \\ \beta(4t - 1) & (\tfrac{1}{4} \le t \le \tfrac{1}{2}), \\ \gamma(2t - 1) & (\tfrac{1}{2} \le t \le 1); \end{cases}$$

$$(\alpha(\beta\gamma))(t) = \begin{cases} \alpha(2t) & (0 \le t \le \tfrac{1}{2}), \\ \beta(4t - 2) & (\tfrac{1}{2} \le t \le \tfrac{3}{4}), \\ \gamma(4t - 3) & (\tfrac{3}{4} \le t \le 1). \end{cases}$$

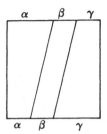

Figure 3.6

The homotopy is (see Figure 3.6)

$$F(t_1, t_2) = \begin{cases} \alpha\left(\dfrac{4t_1}{t_2 + 1}\right) & (4t_1 - 1 \le t_2), \\ \beta(4t_1 - t_2 - 1) & (4t_1 - 2 \le t_2 \le 4t_1 - 1), \\ \gamma\left(\dfrac{4t_1 - t_2 - 2}{2 - t_2}\right) & (t_2 \le 4t_1 - 2). \end{cases} \qquad \square$$

Remark. Given two base points x_0 and x_1 in X, we cannot in general expect any relationship between $\pi_1(X, x_0)$ and $\pi_1(X, x_1)$. For example, if x_0 and x_1 do not lie in a common connected subset of X, there can be no relationship. Consider the disjoint union X of a circle and a point x_1 (Figure 3.7). For x_0 in the circle, $\pi_1(X, x_0)$ has nontrivial elements, whereas $\pi_1(X, x_1)$ is clearly trivial.

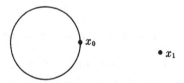

Figure 3.7

Similarly, if X is not arcwise connected, we can expect no relationship between these groups. Consider the space $X = X_1 \cup X_2 \cup X_3 \subset R^2$, where X_1 is the graph of $\sin 1/x$ $(0 < x < 1)$, $X_2 = \{0\} \times I$, and X_3 is a circle tangent to the y axis at $(0, 1)$ with center to the left of the y axis (Figure 3.8). For $x_0 \in X_1$, $\pi_1(X, x_0)$ is trivial. But for $x_1 = (0, 1)$, $\pi_1(X, x_1)$ has nontrivial elements.

For arcwise connected spaces, the situation is better.

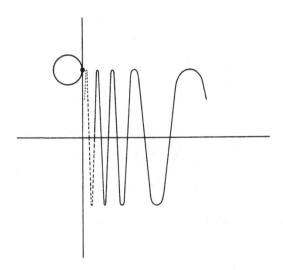

Figure 3.8

Theorem 5. *Let X be an arcwise connected topological space. Let $x_0, x_1 \in X$. Then there exists a group isomorphism of $\pi_1(X, x_0)$ onto $\pi_1(X, x_1)$.*

PROOF. Let γ be a path from x_0 to x_1. Let $\gamma_\#: \pi_1(X, x_0) \to \pi_1(X, x_1)$ be defined by

$$\gamma_\#(\langle \alpha \rangle) = \langle \gamma \rangle^{-1} \langle \alpha \rangle \langle \gamma \rangle = \langle \gamma^{-1} \alpha \gamma \rangle \qquad \text{for } \langle \alpha \rangle \in \pi_1(X, x_0).$$

$\gamma_\#$ is a homomorphism because

$$\begin{aligned}
\gamma_\#(\langle \alpha \rangle) \gamma_\#(\langle \beta \rangle) &= \langle \gamma \rangle^{-1} \langle \alpha \rangle \langle \gamma \rangle \langle \gamma \rangle^{-1} \langle \beta \rangle \langle \gamma \rangle \\
&= \langle \gamma \rangle^{-1} \langle \alpha \rangle \langle e_{x_0} \rangle \langle \beta \rangle \langle \gamma \rangle \\
&= \langle \gamma \rangle^{-1} \langle \alpha \rangle \langle \beta \rangle \langle \gamma \rangle \\
&= \gamma_\#(\langle \alpha \rangle \langle \beta \rangle).
\end{aligned}$$

$\gamma_\#$ is an isomorphism because it has an inverse, namely, $\gamma_\#^{-1} = (\gamma^{-1})_\#$. $\square$

Remark. Recall that for a group G and $a \in G$, the *inner automorphism* of G *due to* a is the isomorphism i_a of G onto itself given by $i_a(b) = aba^{-1}$ for $b \in G$. Now, given two paths in X from x_0 to x_1, we obtain two isomorphisms $\pi_1(X, x_0) \to \pi_1(X, x_1)$. The above proof actually shows that these isomorphisms differ by an inner automorphism of $\pi_1(X, x_0)$.

Corollary. *Let γ_1 and γ_2 be two paths in X from x_0 to x_1. Then*

$$(\gamma_2)_\# = (\gamma_1)_\# \circ i_a,$$

where i_a is the inner automorphism of $\pi_1(X, x_0)$ due to $a = \langle \gamma_1 \gamma_2^{-1} \rangle \in \pi_1(X, x_0)$.

PROOF. For $\langle \alpha \rangle \in \pi_1(X, x_0)$,

$$(\gamma_1)_\#^{-1} \circ (\gamma_2)_\#(\langle \alpha \rangle) = \langle \gamma_1 \rangle \langle \gamma_2 \rangle^{-1} \langle \alpha \rangle \langle \gamma_2 \rangle \langle \gamma_1 \rangle^{-1}$$
$$= \langle \gamma_1 \gamma_2^{-1} \rangle \langle \alpha \rangle \langle \gamma_1 \gamma_2^{-1} \rangle^{-1}$$
$$= i_a \langle \alpha \rangle;$$

that is, $(\gamma_1)_\#^{-1} \circ (\gamma_2)_\# = i_a$, or $(\gamma_2)_\# = (\gamma_1)_\# \circ i_a$. $\qquad\square$

Remark. Theorem 5 shows that all fundamental groups of an arcwise connected space X are isomorphic; that is, associated with the space X is a certain abstract group, the fundamental group of X. However, the above corollary shows that unless this group is commutative, no natural way exists of identifying the groups arising from different base points. (The isomorphism depends on the homotopy class of the path joining the base points.) Hence we shall regard the fundamental group as being a concrete group, computed with respect to a given base point. The importance of the base point will become clearer as we study the behavior of the fundamental group relative to continuous maps.

Definition. Let X and Y be arcwise connected spaces. Let $f: X \to Y$ be continuous. For $x_0 \in X$, let $f_*: \pi_1(X, x_0) \to \pi_1(Y, f(x_0))$ be defined by

$$f_*(\langle \alpha \rangle) = \langle f \circ \alpha \rangle$$

for $\langle \alpha \rangle \in \pi_1(X, x_0)$. Note that this definition makes sense because if α and β are two paths in X—both beginning and ending at x_0 with $\alpha \simeq \beta$—then $f \circ \alpha \simeq f \circ \beta$. (A homotopy from $f \circ \alpha$ to $f \circ \beta$ is given by $f \circ F$ where F is a homotopy from α to β.)

Remark. $f_*: \pi_1(X, x_0) \to \pi_1(Y, f(x_0))$ is a homomorphism, because for $\langle \alpha \rangle$ and $\langle \beta \rangle \in \pi_1(X, x_0)$,

$$f_*(\langle \alpha \rangle \langle \beta \rangle) = f_*(\langle \alpha \beta \rangle) = \langle f \circ (\alpha \beta) \rangle$$
$$= \langle (f \circ \alpha)(f \circ \beta) \rangle = \langle f \circ \alpha \rangle \langle f \circ \beta \rangle$$
$$= f_*(\langle \alpha \rangle) f_*(\langle \beta \rangle).$$

f_* is called the *homomorphism induced by f*.

Theorem 6. *Let X, Y, Z be arcwise connected, and let $x_0 \in X$. Then:*
(1) *If $f: X \to Y$ and $g: Y \to Z$ are continuous, then $(g \circ f)_* = g_* \circ f_*$.*
(2) *If f_0 and $f_1: X \to Y$ are homotopic maps and $F: X \times I \to Y$ is a homotopy from f_0 to f_1, then $(f_1)_* = \sigma_\# \circ (f_0)_*$, where σ is the path in Y from $f_0(x_0)$ to $f_1(x_0)$ given by*

$$\sigma(t) = F(x_0, t).$$

Remark. Part (2) of Theorem 6 says that homotopic maps induce the same homomorphism on fundamental groups, up to an inner automorphism that compensates for the fact that the two maps may send the base point in X into different points in Y.

PROOF OF THEOREM 6

(1) Let $\langle \alpha \rangle \in \pi_1(X, x_0)$. Then

$$(g \circ f)_*(\langle \alpha \rangle) = \langle g \circ f \circ \alpha \rangle = g_*(\langle f \circ \alpha \rangle) = g_* \circ f_*(\langle \alpha \rangle).$$

(2) Let $\langle \alpha \rangle \in \pi_1(X, x_0)$. We want to show that

$(f_1)_*\langle \alpha \rangle = \sigma_\#((f_0)_*\langle \alpha \rangle));$ that is, $\langle f_1 \circ \alpha \rangle = \sigma_\#(\langle f_0 \circ \alpha \rangle) = \langle \sigma^{-1}(f_0 \circ \alpha)\sigma \rangle.$

Thus we must show that

$$f_1 \circ \alpha \simeq \sigma^{-1}(f_0 \circ \alpha)\sigma.$$

For this, consider the map $G: I \times I \to Y$ defined by

$$G(t_1, t_2) = F(\alpha(t_1), t_2) \qquad (t_1, t_2 \in I).$$

G is continuous because F and α are. Moreover,

$$\begin{aligned}
G(t_1, 0) &= F(\alpha(t_1), 0) = f_0(\alpha(t_1)) = f_0 \circ \alpha(t_1), \\
G(t_1, 1) &= F(\alpha(t_1), 1) = f_1(\alpha(t_1)) = f_1 \circ \alpha(t_1), \\
G(0, t_2) &= F(\alpha(0), t_2) = F(x_0, t_2) = \sigma(t_2), \\
G(1, t_2) &= F(\alpha(1), t_2) = F(x_0, t_2) = \sigma(t_2).
\end{aligned}$$

Thus the boundary of $I \times I$ is mapped by G as indicated in Figure 3.9.

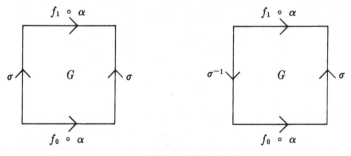

Figure 3.9

The required homotopy H from $f_1 \circ \alpha$ to $\sigma^{-1}(f_0 \circ \alpha)\sigma$ is then obtained by deforming G as indicated in Figure 3.10.

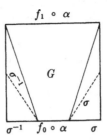

Figure 3.10

Analytically, H is given by

$$H(t_1, t_2) = \begin{cases} \sigma^{-1}(2t_1) & \left(t_1 \le \dfrac{1 - t_2}{2}\right), \\[2ex] G\left(\dfrac{4t_1 + 2t_2 - 2}{3t_2 + 1}, t_2\right) & \left(\dfrac{1 - t_2}{2} \le t_1 \le \dfrac{t_2 + 3}{4}\right), \\[2ex] \sigma(4t_1 - 3) & \left(t_1 \ge \dfrac{t_2 + 3}{4}\right). \end{cases} \qquad \square$$

Corollary 1. *If X and Y are arcwise connected spaces of the same homotopy type, then their fundamental groups are isomorphic.*

PROOF. Since X and Y are of the same homotopy type, there exist maps $f: X \to Y$ and $g: Y \to X$ such that $g \circ f \simeq i_X$ and $f \circ g \simeq i_Y$. Choose a base point $x_0 \in X$ in the image of g, say $x_0 = g(y_0)$. We shall show that

$$f_*: \pi_1(X, x_0) \to \pi_1(Y, f(x_0))$$

is an isomorphism. Now,

$$g_*: \pi_1(Y, f(x_0)) \to \pi_1(X, g \circ f(x_0))$$

and, by Parts (1) and (2) of Theorem 6,

$$g_* \circ f_* = (g \circ f)_* = \sigma_\# \circ (i_X)_* = \sigma_\#$$

where $\sigma_\#: \pi_1(X, x_0) \to \pi_1(X, g \circ f(x_0))$ is an isomorphism. Thus f_* is injective.

On the other hand, we may consider the homomorphism, also denoted by g_*, mapping $\pi_1(Y, y_0) \to \pi_1(X, x_0)$. Then

$$f_* \circ g_* = (f \circ g)_* = (\sigma_1)_\# \circ (i_Y)_* = (\sigma_1)_\#$$

where $(\sigma_1)_\#: \pi_1(Y, y_0) \to \pi_1(Y, f \circ g(y_0))$ is an isomorphism. Thus f_* is surjective also, hence an isomorphism. $\qquad \square$

Corollary 2. *If X is contractible, then $\pi_1(X, x_0) = (e)$; that is, the fundamental group of X consists of the identity element only.*

PROOF. By definition, X contractible means that X is of the same homotopy type as a one-point space. Thus Corollary 1 implies the result. $\square$

Corollary 3. $\pi_1(R^n, 0) = (e)$.

3.3 Covering spaces

All spaces throughout this section are Hausdorff topological spaces.

Definition. A space X is *locally connected* if for each point $x \in X$ and each open set V containing x, there exists a connected open set U such that $x \in U \subset V$. A space X is *locally arcwise connected* if for each point $x \in X$ and each open set V containing x, there exists an open set U, with $x \in U \subset V$, such that whenever $x_1, x_2 \in U$, there exists a path α from x_1 to x_2 with $\alpha(I) \subset V$.

Remark. Not every arcwise connected space is locally arcwise connected. For let X be the union of the graph of $\sin 1/x$, $x \in (0, 1]$, with an arc connecting $(1, 0)$ and $(0, 1)$ (Figure 3.11). Then X is arcwise connected, but if B is any ball in R^2 about $(0, 1)$ of radius < 1, then $V = B \cap X$ is open in X, yet V contains no open set U with the property required for local arcwise connectedness.

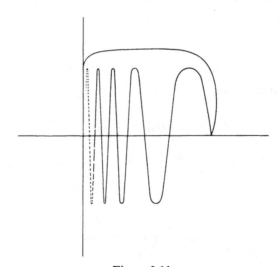

Figure 3.11

Definition. Let X and $\tilde{X}$ be arcwise connected, locally arcwise connected spaces, and let $p: \tilde{X} \to X$ be continuous. The pair $(\tilde{X}, p)$ is called a *covering space* of X if

(1) p is surjective, and
(2) for each $x \in X$, there exists an open set U in X containing x such that $p^{-1}(U)$ is a disjoint union of open sets, each of which is mapped homeomorphically onto U by p. Such an open set U will be called *admissible*.

EXAMPLE 1. Let

$$X = S^1 = [z; z \text{ a complex number with } |z| = 1].$$

Let $\tilde{X} = R^1$, and let $p: \tilde{X} \to X$ be given by $p(r) = e^{2\pi i r}$. Then $(\tilde{X}, p)$ is a covering space of X. For $x \in X$, let U be a small open arc containing x. Then $p^{-1}(U)$ is a disjoint union of open intervals, each a translate of every other by some integer (Figure 3.12). $\tilde{X}$ may be looked at as an infinite spiral over S^1, with p being ordinary projection (Figure 3.13).

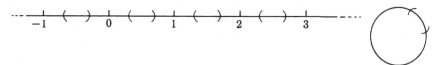

Figure 3.12

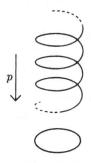

Figure 3.13

EXAMPLE 2. Let

$$X = P^2 = \text{the set of all lines through the origin in } R^3.$$

The topology on P^2 is generated by "open cones" of lines through the origin (Figure 3.14). P^2 is called the real *projective plane*. Let $\tilde{X} = S^2 \subset R^3$, and let $p: \tilde{X} \to X$ be defined by

$$p(\tilde{x}) = \text{the line through the origin which passes through } \tilde{x} \quad (\tilde{x} \in \tilde{X}).$$

Then $(\tilde{X}, p)$ is covering space of X, in fact, a "double covering" of X; that is $p^{-1}(x)$ consists of two points for each $x \in X$.

63

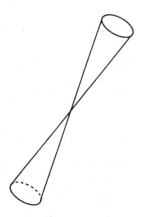

Figure 3.14

EXAMPLE 3. Let $X = \tilde{X} = S^1 = [z; |z| = 1]$. Let $p: \tilde{X} \to X$ be given by $p(z) = z^2$. Then $(\tilde{X}, p)$ is a double covering of X.

EXAMPLE 4. Let $X = S^1 \times S^1 =$ the torus. Let $\tilde{X} = R^2$ and $p: \tilde{X} \to X$ be defined by

$$p(r_1, r_2) = (e^{2\pi i r_1}, e^{2\pi i r_2}).$$

Then $(\tilde{X}, p)$ is a covering space of X. For a "patch" U on X, $p^{-1}(U)$ is a union of "patches" in R^2, each a translate of every other by a vector with integer coordinates (Figure 3.15). (Note that unit square gets mapped *onto* $S^1 \times S^1$ by p, and, in fact, $S^1 \times S^1$ may be regarded as a closed unit square with one pair of opposite edges identified to form a cylinder and the other pair of opposite edges (now circles) identified to form a tire.)

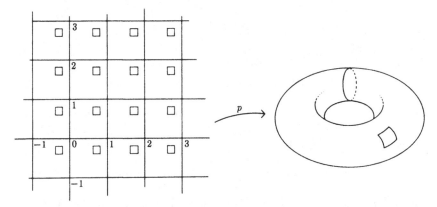

Figure 3.15

EXAMPLE 5 (A "nonexample"). Let $X = S^1$, and let $\tilde{X}$ be a finite open spiral over S^1 with projection $p\colon \tilde{X} \to X$ (Figure 3.16). $(\tilde{X}, p)$ is *not* a covering space of X because if x lies under an "end" of $\tilde{X}$, then there exists no open set U about x satisfying (2) of the definition.

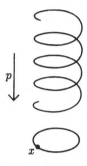

Figure 3.16

EXAMPLE 6 (Another "nonexample"). Let $X = S^2$ and $\tilde{X} = $ an infinite cylinder circumscribed about S^2; and let p be radial projection (Figure 3.17). Then $(\tilde{X}, p)$ is *not* a covering space because p is not surjective.

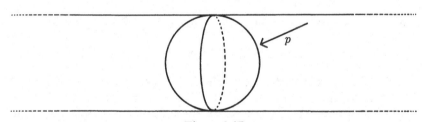

Figure 3.17

Theorem 1. *Let $(\tilde{X}, p)$ be a covering space of X. Then $p\colon \tilde{X} \to X$ is an open mapping; that is, $p(\tilde{U})$ is open in X for each open set $\tilde{U} \subset \tilde{X}$.*

PROOF. Let $\tilde{U}$ be open in $\tilde{X}$. For $x \in p(\tilde{U})$, let V be an admissible open set containing x, so that $p^{-1}(V)$ is a disjoint union of open sets, each mapped homeomorphically onto V by p. Let $\tilde{x} \in \tilde{U}$ be such that $p(\tilde{x}) = x$. Then $\tilde{x}$ lies in one of these open sets mapped homeomorphically onto V. Call it W. Then $W \cap \tilde{U}$ is open in W, so $p(W \cap \tilde{U})$ is open in V since $p|_W$ is a homeomorphism. But V is open in X, so $p(W \cap \tilde{U})$ is an open set in X with $x \in p(W \cap \tilde{U}) \subset p(\tilde{U})$. Since $x \in p(\tilde{U})$ was arbitrary, $p(\tilde{U})$ is a union of open sets, and hence is open. □

Remark. Given two spaces Y and X, a continuous map $f\colon Y \to X$, and a covering space $(\tilde{X}, p)$ of X, we find it often of interest to know whether we can "lift" the map f to a map $\tilde{f}\colon Y \to \tilde{X}$; that is, whether we can construct a map $\tilde{f}\colon Y \to \tilde{X}$ such that $p \circ \tilde{f} = f$. The map $\tilde{f}$ is also said to *cover f*. The next theorem (Theorem 2) shows that under certain mild conditions, any two such maps (if they exist) either agree everywhere or agree nowhere. The following theorem (Theorem 3) and its corollaries tell us that certain maps can, in fact, be "lifted."

Theorem 2. *Let* $(\tilde{X}, p)$ *be a covering space of a space* X, *and let* Y *be a connected and locally connected space. Suppose* $\alpha, \beta\colon Y \to \tilde{X}$ *are continuous maps such that*

(1) $p \circ \alpha = p \circ \beta$, *and*
(2) $\alpha(y_0) = \beta(y_0)$ *for some* $y_0 \in Y$.

Then $\alpha = \beta$.

PROOF. Let

$$Z = [y \in Y; \alpha(y) = \beta(y)].$$

We must show that $Z = Y$. Since Y is connected, and $Z \neq \varnothing$ $(y_0 \in Z)$, it suffices to show that Z is both open and closed.

Z is closed: For consider the map $\alpha \times \beta\colon Y \to \tilde{X} \times \tilde{X}$ defined by

$$(\alpha \times \beta)(y) = (\alpha(y), \beta(y)).$$

Then $\alpha \times \beta$ is continuous. Let D be the diagonal in $\tilde{X} \times \tilde{X}$; that is,

$$D = [(\tilde{x}, \tilde{x}); \tilde{x} \in \tilde{X}].$$

D is closed because $\tilde{X}$ is Hausdorff. Hence $Z = (\alpha \times \beta)^{-1}(D)$ is closed in Y. (Note that this argument shows that for *any* two continuous maps from *any* topological space S into *any* Hausdorff space T, the set of points where the two maps agree will be closed in S.)

Z is open: Suppose $z \in Z$. Let $x = p \circ \alpha(z) = p \circ \beta(z)$. Let U be an admissible open set about x. Then $p^{-1}(U)$ is a union of disjoint open sets one of which, call it W, contains $\alpha(z) = \beta(z)$. Now W is both open and closed in $p^{-1}(U)$. Hence any connected subset of $p^{-1}(U)$ which has points in W must lie entirely in W. Now $z \in (p \circ \alpha)^{-1}(U)$, an open set in Y. Since Y is locally connected, there exists a connected open set V_1 in Y with $z \in V_1 \subset (p \circ \alpha)^{-1}(U)$. Since α is continuous, $\alpha(V_1)$ is connected in $p^{-1}(U)$ with $\alpha(z) \in W$; so $\alpha(V_1) \subset W$. Similarly, there exists a connected open set V_2 in Y containing z, such that $\beta(V_2) \subset W$. Since $p|_W$ is injective, and since $p \circ \alpha = p \circ \beta$, it follows that $\alpha = \beta$ on $V_1 \cap V_2$. Thus $V_1 \cap V_2$ is contained in Z and is an open set containing z. Thus Z is open. $\qquad\square$

Theorem 3 (Covering homotopy theorem). *Let* $(\tilde{X}, p)$ *be a covering space of a space* X, *and let* Y *be a compact, connected, and locally connected space. Let*

$f: Y \to \tilde{X}$, and let $F: Y \times I \to X$ be a homotopy with $F(y, 0) = p \circ f(y)$
for $y \in Y$. Then there exists a homotopy $G: Y \times I \to \tilde{X}$ such that

(1) $G(y, 0) = f(y)$, and
(2) $p \circ G = F$.

Moreover, G may be chosen to be stationary; *that is, whenever $y \in Y$ is such
that $F(y, t)$ is constant for t in some interval, then $G(y, t)$ is also constant for
t in that interval.*

Remark. Before proving Theorem 3, we derive several consequences.

Corollary 1. *Let $(\tilde{X}, p)$ be a covering space of X, and let $\tilde{x} \in \tilde{X}$. Then
$p_*: \pi_1(\tilde{X}, \tilde{x}) \to \pi_1(X, p(\tilde{x}))$ is injective.*

PROOF. Suppose $\langle \tilde{\alpha} \rangle \in \pi_1(\tilde{X}, \tilde{x})$ is in the kernel of p_*. Then

$$\langle p \circ \tilde{\alpha} \rangle = p_*(\langle \tilde{\alpha} \rangle) = e = \langle e_x \rangle; \quad \text{that is,} \quad p \circ \tilde{\alpha} \simeq e_x,$$

where $x = p(\tilde{x})$. We want to show that $\langle \tilde{\alpha} \rangle$ is the identity element in $\pi_1(\tilde{X}, \tilde{x})$,
that is, that $\tilde{\alpha} \simeq e_x$.

Let $F: I \times I \to X$ be a homotopy from $p \circ \tilde{\alpha}$ to e_x. By the covering homotopy theorem, with $Y = I$, there exists a homotopy $G: I \times I \to \tilde{X}$ with
$p \circ G = F$ such that $G(t_1, 0) = \tilde{\alpha}(t_1)$ for all $t_1 \in I$. Furthermore, since
$F(0, t_2) = x$ and $F(1, t_2) = x$ are constant for $t_2 \in I$, G may be chosen so that

$$G(0, t_2) = \tilde{\alpha}(0) = \tilde{x} \quad \text{and} \quad G(1, t_2) = \tilde{\alpha}(1) = \tilde{x}$$

for all $t_2 \in I$. Since $p \circ G = F$ and since $F(t_1, 1) = e_x(t_1)$, the paths $t_1 \to G(t_1, 1)$
and $t_1 \to e_{\tilde{x}}(t_1)$ are both mapped by p into e_x. Moreover, $G(0, 1) = \tilde{x} = e_{\tilde{x}}(0)$,
so these paths agree at a point. Thus, by Theorem 2, $G(t_1, 1) = e_{\tilde{x}}(t_1)$ for all
$t_1 \in I$, and G is a homotopy from $\tilde{\alpha}$ to $e_{\tilde{x}}$. □

Corollary 2. *Let $(\tilde{X}, p)$ be a covering space of X. Let α be a path in X. Let
$x_0 = \alpha(0)$, and choose $\tilde{x}_0 \in \tilde{X}$ with $p(\tilde{x}_0) = x_0$. Then there exists a unique
path $\tilde{\alpha}$ in $\tilde{X}$, with origin $\tilde{x}_0$, covering α, that is, with $p \circ \tilde{\alpha} = \alpha$.*

PROOF. The uniqueness is a consequence of Theorem 2 above. To prove
existence, we apply the covering homotopy theorem to the case where
$Y = \{y_0\}$ is a 1-point space and $f: Y \to \tilde{X}$ is defined by $f(y_0) = \tilde{x}_0$. Since
projection on I is a homeomorphism of $Y \times I$ onto I, we may identify these
two spaces through this homeomorphism and regard α as mapping $Y \times I \to X$.
Then $\alpha(y_0, 0) = x_0 = p \circ f(y_0)$, so by the covering homotopy theorem, there
exists a homotopy $G: Y \times I \to \tilde{X}$ such that $p \circ G = \alpha$ and

$$G(y_0, 0) = f(y_0) = \tilde{x}_0.$$

Let $\tilde{\alpha}: I \to \tilde{X}$ be defined by $\tilde{\alpha}(t) = G(y_0, t)$. Then $\tilde{\alpha}$ has the required properties. □

Corollary 3. *Let $(\tilde{X}, p)$ be a covering space of X. Let $x \in X$, and $\tilde{x} \in \tilde{X}$ with $p(\tilde{x}) = x$. Then there exists a "natural" one-to-one correspondence between $p^{-1}(\{x\})$ and the coset space $\pi_1(X, x)/p_*\pi_1(\tilde{X}, \tilde{x})$.*

PROOF. We define $c: \pi_1(X, x) \to p^{-1}(\{x\})$ as follows. Let $g \in \pi_1(X, x)$. Suppose α and β are two representations of g; that is, $g = \langle\alpha\rangle = \langle\beta\rangle$. By Corollary 2, α and β have unique lifts $\tilde{\alpha}$ and $\tilde{\beta}$ to $\tilde{X}$, with origin $\tilde{x}$. Now $\alpha \simeq \beta$, so by the covering homotopy theorem, $\tilde{\alpha} \simeq \tilde{\beta}$. In particular, $\tilde{\alpha}$ and $\tilde{\beta}$ have the same end point; that is, $\tilde{\alpha}(1) = \tilde{\beta}(1)$. Thus we can define

$$c(g) = \tilde{\alpha}(1),$$

where α is any representative of g.

Since $\tilde{X}$ is arcwise connected, $\tilde{x}$ can be joined to each element of $\pi^{-1}(\{x\})$ by a path in $\tilde{X}$, which is in turn the lift of a closed path (its projection) in X. Thus $c: \pi_1(X, x) \to p^{-1}(\{x\})$ is surjective.

We now show that c is constant on cosets of $p_*\pi_1(\tilde{X}, \tilde{x})$ in $\pi_1(X, x)$. Suppose $\langle\alpha\rangle$ and $\langle\beta\rangle$ lie in the same coset. Then $\langle\beta\rangle = \langle\gamma\rangle\langle\alpha\rangle = \langle\gamma\alpha\rangle$ for some $\langle\gamma\rangle \in p_*\pi_1(\tilde{X}, \tilde{x})$. Thus

$$c(\langle\beta\rangle) = c(\langle\gamma\alpha\rangle) = \widetilde{\gamma\alpha}(1).$$

But $\widetilde{\gamma\alpha} = \tilde{\gamma}\tilde{\alpha}$ by Corollary 2, so that

$$c(\langle\beta\rangle) = \tilde{\gamma}\tilde{\alpha}(1) = \tilde{\alpha}(1) = c(\langle\alpha\rangle),$$

and the restriction of c to each coset is constant, as claimed.

Thus c defines a map $\bar{c}: \pi_1(X, x)/p_*\pi_1(\tilde{X}, \tilde{x}) \to p^{-1}(\{x\})$ by

$$\bar{c}(H\langle\alpha\rangle) = c(\langle\alpha\rangle), \qquad \langle\alpha\rangle \in \pi_1(X, x),$$

where $H\langle\alpha\rangle$ is the coset of $H = p_*\pi_1(\tilde{X}, \tilde{x})$ containing $\langle\alpha\rangle$. $\bar{c}$ is surjective because c is. $\bar{c}$ is injective because if $\bar{c}(H\langle\alpha\rangle) = \bar{c}(H\langle\beta\rangle)$, then $c(\langle\alpha\rangle) = c(\langle\beta\rangle)$; that is, $\tilde{\alpha}(1) = \tilde{\beta}(1)$, so that $\langle\tilde{\alpha}\tilde{\beta}^{-1}\rangle \in \pi_1(\tilde{X}, \tilde{x})$. Letting $h = p_*(\langle\tilde{\alpha}\tilde{\beta}^{-1}\rangle)$, we have

$$h\langle\beta\rangle = \langle p \circ (\tilde{\alpha}\tilde{\beta}^{-1})\rangle\langle\beta\rangle = \langle\alpha\beta^{-1}\rangle\langle\beta\rangle = \langle\alpha\rangle,$$

so that $\langle\alpha\rangle$ and $\langle\beta\rangle$ are in the same coset of H; that is, $H\langle\alpha\rangle = H\langle\beta\rangle$. $\square$

EXAMPLE 1. Consider the line R^1 as a covering space for the circle S^1, with $p(r) = e^{2\pi i r}$. Take as base point in S^1 the point $z = 1$. Then $p^{-1}(\{1\})$ is the set of integers. Since $\pi_1(R^1, 0)$ is trivial, $p_*\pi_1(R^1, 0) = (e)$, so $\pi_1(S^1, 1)$ is in one-to-one correspondence with the integers. (We shall see later than this correspondence is, in fact, a group isomorphism.)

EXAMPLE 2. Consider the plane R^2 as a covering space for the torus $S^1 \times S^1$, with $p(r_1, r_2) = (e^{2\pi i r_1}, e^{2\pi i r_2})$. Then $\pi_1(R^2, 0)$ is trivial so $\pi_1(S^1 \times S^1, (1, 1))$ is in one-to-one correspondence with $p^{-1}\{(1, 1)\}$, the Cartesian product of the set of integers with itself. (This correspondence also turns out to be a group isomorphism, of $\pi_1(S^1 \times S^1, (1, 1))$ with the direct product of the integers with itself.)

EXAMPLE 3. The sphere S^2 is a covering space of the projective plane P^2 (see Example 2 at the beginning of this section). Let $n \in P^2$ be the z axis (n is a line through the origin in R^2 and hence is a point in P^2). Then

$$p^{-1}(\{n\}) = \{\text{north pole, south pole}\} \subset S^2.$$

Later we shall prove that $\pi_1(S^2, \text{north pole}) = \{e\}$. Assuming this fact for the moment, we get from Corollary 3 that $\pi_1(P^2, n)$ has two elements; that is, $\pi_1(P^2, n) = Z_2$, the cyclic group of order 2. In particular, there is a path α in P^2 which is not homotopic to a constant, but its square α^2 *is* homotopic to a constant. The curve in P^2 defined by α thus has the property that the path obtained by traveling around once cannot be shrunk to a point, and yet the path obtained by traveling around twice *can* be shrunk to a point. Geometrically, given a great circle on S^2 through the north and south poles, let $x(t) \in S^2$ move along this great circle from the north to the south pole as t varies from 0 to 1. Let $\alpha(t)$ be the line through the origin which passes through $x(t)$. Then α is a path in P^2 with the above property. Note that the lift $t \to x(t)$ of α to S^2 is *not* a closed path in S^2, whereas the lift of α^2 to S^2 *is* closed. This amounts to traveling all the way around the great circle.

PROOF OF THEOREM 3. Since Y and I are compact, $Y \times I$ is compact, hence so is $F(Y \times I)$. Thus $F(Y \times I)$ can be covered by finitely many admissible open sets $U_1, \ldots, U_r$ (cover $F(Y \times I)$ by admissible open sets and take a finite subcovering). Since $\{F^{-1}(U_i)\}_{i=1}^r$ covers $Y \times I$, and since a basis of open sets in $Y \times I$ is given by (open sets of Y) $\times$ (open intervals in I), we can find a finite covering $\{V_\alpha\}$ of Y of connected, open sets and a decomposition of the unit interval $0 = t_0 < t_1 < \cdots < t_k = 1$ such that each $F(V_\alpha \times [t_i, t_{i+1}]) \subset$ some U_l.

We construct G by constructing $G_i: Y \times [t_{i-1}, t_i] \to \tilde{X}$ $(i = 1, \ldots, k)$ with the properties that (1) $p \circ G_i = F|_{Y \times [t_{i-1}, t_i]}$; (2) G_i is continuous; and (3) $G_i = G_{i+1}$ on the closed set

$$Y \times [t_{i-1}, t_i] \cap Y \times [t_i, t_{i+1}] = [(y, t_i); y \in Y].$$

By the glueing lemma for closed sets, G will be continuous, and, of course, $p \circ G = F$.

By induction, assume the G_i have been defined for $i \leq j$ so that they satisfy (1), (2) (for $i = 1, \ldots, j$), and (3) (for $i = 1, \ldots, j - 1$). We construct G_{j+1}.

In order for (3) to be satisfied, $G_{j+1}(y, t_j)$ must equal $G_j(y, t_j)$. Let $G_{j+1}^\alpha: V_\alpha \times [t_j, t_{j+1}] \to \tilde{X}$ be defined as follows. $F(V_\alpha \times [t_j, t_{j+1}]) \subset U_l$, an admissible open set in X. Since V_α is connected, the set $G_j(V_\alpha \times \{t_j\})$ is connected. But

$$p \circ G_j(V_\alpha \times \{t_j\}) = F(V_\alpha \times \{t_j\}) \subset U_l.$$

Hence, $G_j(V_\alpha \times \{t_j\})$ lies in one of the open sets $W \subset p^{-1}(U_l)$ on which $p: W \to U_l$ is a homeomorphism. Define $G_{j+1}^\alpha = (p|_W)^{-1} \circ F|_{V_\alpha \times [t_j, t_{j+1}]}$. Note that $G_{j+1}^\alpha = G_j$ on $V_\alpha \times \{t_j\}$. To construct the continuous map G_{j+1} on the topological space

$$Y \times [t_j, t_{j+1}],$$

we paste the maps G_{j+1}^α defined on the open sets $V_\alpha \times [t_j, t_{j+1}]$ of the space $Y \times [t_j, t_{j+1}]$. By the glueing lemma for open sets, we need only verify that G_{j+1}^α and G_{j+1}^β agree on $(V_\alpha \cap V_\beta) \times [t_j, t_{j+1}]$, which we can assume is not empty.

Suppose $G_j(V_\gamma \times \{t_j\})$ lies in the open set W_γ $(\gamma = \alpha, \beta)$, with $p: W_\alpha \to U_l$ and $p: W_\beta \to U_m$ homeomorphisms. Since $G_{j+1}^\gamma = G_j$ on $V_\gamma \times \{t_j\}$ $(\gamma = \alpha, \beta)$, then $G_{j+1}^\alpha = G_{j+1}^\beta$ on $(V_\alpha \cap V_\beta) \times \{t_j\}$. Since any point of $(V_\alpha \cap V_\beta) \times [t_j, t_{j+1}]$ can be connected to $(V_\alpha \cap V_\beta) \times \{t_j\}$ by an arc in $(V_\alpha \cap V_\beta) \times [t_j, t_{j+1}]$, we must have

$$G_{j+1}^\gamma((V_\alpha \cap V_\beta) \times [t_j, t_{j+1}]) \subset W_\alpha \cap W_\beta \qquad (\gamma = \alpha, \beta).$$

But $p \circ G_{j+1}^\gamma = F|_{V_\gamma \times [t_j, t_{j+1}]}$ for $\gamma = \alpha, \beta$, and $p|_{W_\alpha} = p|_{W_\beta}$ on $W_\alpha \cap W_\beta$. Hence $G_{j+1}^\alpha = G_{j+1}^\beta$ on $(V_\alpha \cap V_\beta) \times [t_j, t_{j+1}]$.

Note that this induction argument also works to start the induction, that is, for the construction of G_1, because we are given G_1 on $Y \times \{t_0\}$ equal to f.

We leave to the reader the verification that the above construction automatically makes G stationary. $\qquad\qquad \square$

Definitions. Let X be an arcwise connected, locally arcwise connected space. X is *simply connected* if its fundamental group is trivial, or equivalently, if every closed path in X is homotopic to a constant.

A covering space $(\tilde{X}, p)$ of a space X is called a *universal covering space* if $\tilde{X}$ is simply connected.

Remark. The line, the plane, and the sphere are universal covering spaces, respectively, of the circle, the torus, and the projective plane. That the sphere S^2 is simply connected, however, remains to be shown.

Definition. A space X is *locally simply connected* if for each $x \in X$, there exists an open set V containing x such that any path α in V, with $\alpha(0) = \alpha(1) = x$, is homotopic in X to the constant path e_x.

Theorem 4. *Let X be arcwise connected, locally arcwise connected, and locally simply connected. Let H be a subgroup of $\pi_1(X, x)$. Then there exists a covering space $(\tilde{X}, p)$ such that $p_*\pi_1(\tilde{X}, \tilde{x}) = H$, where $\tilde{x} \in \tilde{X}$ with $p(\tilde{x}) = x$. In particular, if $H = (e)$, then X is simply connected, so each such X has a universal covering space.*

PROOF. From the covering homotopy theorem and its corollaries, we know that each $\tilde{\alpha}$ in $\tilde{X}$ is the *unique* lift starting at $\tilde{\alpha}(0)$ of the path $p \circ \tilde{\alpha}$ in X; moreover, $\tilde{\alpha}$ is a closed path if and only if $\langle p \circ \tilde{\alpha} \rangle \in p_*\pi_1(X, x)$ (Corollary 3). Hence the point $\tilde{\alpha}(1)$ in $\tilde{X}$ is determined by $\langle p \circ \tilde{\alpha} \rangle$. So it makes sense to try and construct the points of $\tilde{X}$ from paths in X.

Let Ω denote the set of all paths in X beginning at x. We define an equivalence relation $\overset{H}{\equiv}$ on Ω by $\alpha \overset{H}{\equiv} \beta$ if and only if $\alpha(1) = \beta(1)$ and $\langle \alpha\beta^{-1} \rangle \in H$. This is an equivalence relation:

(1) $\alpha \overset{H}{\equiv} \alpha$ because $\langle \alpha\alpha^{-1} \rangle = \langle e_x \rangle \in H$.
(2) If $\alpha \overset{H}{\equiv} \beta$, then $\langle \beta\alpha^{-1} \rangle = \langle \alpha\beta^{-1} \rangle^{-1} \in H$, so $\beta \overset{H}{\equiv} \alpha$.
(3) If $\alpha \overset{H}{\equiv} \beta$ and $\beta \overset{H}{\equiv} \gamma$, then $\alpha(1) = \beta(1) = \gamma(1)$, and

$$\langle \alpha\gamma^{-1} \rangle = \langle \alpha\beta^{-1}\beta\gamma^{-1} \rangle = \langle \alpha\beta^{-1} \rangle\langle \beta\gamma^{-1} \rangle \in H,$$

so $\alpha \overset{H}{\equiv} \gamma$.

Let $\tilde{X}$ be the set of all $\overset{H}{\equiv}$ equivalence classes. For $\alpha \in \Omega$, let $[\alpha]$ denote the $\overset{H}{\equiv}$ equivalence class of α. Define $p: \tilde{X} \to X$ by $p([\alpha]) = \alpha(1)$. p is surjective because X is arcwise connected.

We define a topology on $\tilde{X}$ as follows. For $[\alpha] \in \tilde{X}$, let U be an open set in X containing $\alpha(1)$. Let (see Figure 3.18)

$$([\alpha], U) = [[\alpha\beta]; \beta \text{ a path in } U \text{ beginning at } \alpha(1)].$$

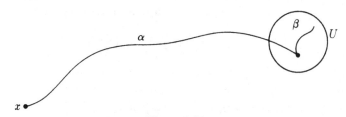

Figure 3.18

The collection of all such $([\alpha], U)$, together with $\varnothing$, forms a basis for the topology on $\tilde{X}$. It is a basis because

(1) $\tilde{X} = ([e_x], X)$ and
(2) for $([\alpha_1], U_1)$ and $([\alpha_2], U_2)$ two such sets, if $[\gamma] \in ([\alpha_1], U_1) \cap ([\alpha_2], U_2)$, then $([\gamma], U_1 \cap U_2) \subset ([\alpha_1], U_1) \cap ([\alpha_2], U_2)$.

71

The topology on $\tilde{X}$ is Hausdorff. For suppose $[\alpha_1]$ and $[\alpha_2] \in \tilde{X}$ with $[\alpha_1] \neq [\alpha_2]$. If $\alpha_1(1) \neq \alpha_2(1)$, then there exist disjoint open sets U_1 and U_2 in X with $\alpha_1(1) \in U_1$ and $\alpha_2(1) \in U_2$ (X is Hausdorff), and clearly $([\alpha_1], U_1) \cap [\alpha_2], U_2)$ $= \varnothing$. So suppose $\alpha_1(1) = \alpha_2(1)$. Since X is locally simply connected, there exists an open set U containing $\alpha_1(1) = \alpha_2(1)$ such that any closed path in U is homotopic in X to the constant path. Then $([\alpha_1], U) \cap ([\alpha_2], U) = \varnothing$, for otherwise there would exist paths β and γ in U starting at $\alpha_1(1) = \alpha_2(1)$ with $[\alpha_1\beta] = [\alpha_2\gamma]$; that is, with $\alpha_1\beta \overset{H}{\equiv} \alpha_2\gamma$. Thus $\langle \alpha_1\beta\gamma^{-1}\alpha_2{}^{-1}\rangle \in H$. But $\beta\gamma^{-1}$ is a closed path in U, so $\beta\gamma^{-1} \simeq e_{\alpha_1(1)}$ and

$$\langle \alpha_1\alpha_2{}^{-1}\rangle = \langle \alpha_1 e_{\alpha_1(1)}\alpha_2{}^{-1}\rangle = \langle \alpha_1\beta\gamma^{-1}\alpha_2{}^{-1}\rangle \in H;$$

that is, $\alpha_1 \overset{H}{\equiv} \alpha_2$ and $[\alpha_1] = [\alpha_2]$, which is a contradiction.

$p \colon \tilde{X} \to X$ is continuous. For if U is open in X, then for each $[\alpha] \in p^{-1}(U)$, $([\alpha], U)$ is open in $\tilde{X}$ and is contained in $p^{-1}(U)$.

$\tilde{X}$ is arcwise connected. For let $[\alpha]$ and $[\beta] \in \tilde{X}$. To construct a path in $\tilde{X}$ from $[\alpha]$ to $[\beta]$, we shrink α in X by pulling its end in to its origin, and then snake out along β toward its end, and then take $\overset{H}{\equiv}$ equivalence classes. Explicitly, an $f \colon I \to \tilde{X}$, with $f(0) = [\alpha]$ and $f(1) = [\beta]$, is defined as follows:

$$f(t_2) = \begin{cases} [\alpha_{t_2}] & (0 \leq t_2 \leq \tfrac{1}{2}) \\ [\beta_{t_2}] & (\tfrac{1}{2} \leq t_2 \leq 1), \end{cases}$$

where

$$\alpha_{t_2}(t_1) = \alpha((1 - 2t_2)t_1) \qquad (t_2 \leq \tfrac{1}{2})$$

and

$$\beta_{t_2}(t_1) = \beta((2t_2 - 1)t_1) \qquad (t_2 \geq \tfrac{1}{2}).$$

f is continuous, because if $([\gamma], U)$ is a basic open set in $\tilde{X}$, then

$$f^{-1}([\gamma], U) = [t_2 \in I; f(t_2) \in ([\gamma], U)] = A \cup B$$

where

$$A = [t_2 \in [0, \tfrac{1}{2}]; [\alpha_{t_2}] \in ([\gamma], U)]$$

and

$$B = [t_2 \in [\tfrac{1}{2}, 1]; [\beta_{t_2}] \in ([\gamma], U)].$$

We show first that A is open. Suppose $\bar{t}_2 \in A$. Then $\alpha_{\bar{t}_2}(1) \in U$. Since $t_2 \to \alpha_{t_2}(1)$ is continuous, there exists an open interval J about $\bar{t}_2$ such that $\alpha_{t_2}(1) \in U$ for

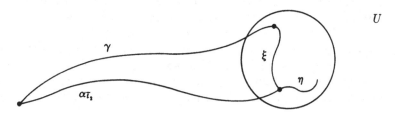

U

Figure 3.19

all $t_2 \in J$. Moreover, for $t_2 \in J$, $\langle \alpha_{t_2} \rangle = \langle \alpha_{\bar{t}_2} \eta \rangle$, where η is the path along α from $\alpha_{\bar{t}_2}(1)$ to $\alpha_{t_2}(1)$. Furthermore, since $\bar{t}_2 \in A$, $[\alpha_{\bar{t}_2}] = [\gamma \xi]$ for some path ξ in U from $\gamma(1)$ to $\alpha_{\bar{t}_2}(1)$; that is, $\langle \alpha_{\bar{t}_2} \xi^{-1} \gamma^{-1} \rangle \in H$. (Figure 3.19). But

$$\langle \alpha_{t_2}(\gamma \xi \eta)^{-1} \rangle = \langle \alpha_{t_2} \eta^{-1} \xi^{-1} \gamma^{-1} \rangle = \langle \alpha_{\bar{t}_2} \xi^{-1} \gamma^{-1} \rangle \in H.$$

Thus

$$\alpha_{t_2} \stackrel{\text{H}}{\equiv} \gamma(\xi \eta),$$

where $\xi \eta$ is a path in U from $\gamma(1)$ to $\alpha_{t_2}(1)$; that is,

$$[\alpha_{t_2}] = [\gamma(\xi \eta)] \in ([\gamma], U)$$

for all $t_2 \in J$. Hence J is an open set about $\bar{t}_2$, contained in A; and A is open. Similarly B is open, and hence so is $f^{-1}([\gamma], U)$. This proves that f is continuous, thus completing the proof that $\tilde{X}$ is arcwise connected.

($\tilde{X}, p$) is a covering space. For given $x_1 \in X$, let U be an arcwise connected open set containing x_1, with the property that each path in U is homotopic in X to a constant. As in the proof above that $\tilde{X}$ is Hausdorff, we see that if $[\alpha_1] \neq [\alpha_2]$ is such that $p([\alpha_1]) = p([\alpha_2])$, that is, $\alpha_1(1) = \alpha_2(1)$, then $([\alpha_1], U) \cap ([\alpha_2], U) = \varnothing$. Also, since U is arcwise connected, $p|_{([\alpha], U)}$: $([\alpha], U) \to U$ is surjective for any $[\alpha]$ with $p([\alpha]) = x_1$. Moreover, $p|_{([\alpha], U)}$ is injective. For suppose $p([\alpha \beta]) = p([\alpha \gamma])$ for some paths β and γ in U starting at x_1. Then $\beta(1) = \gamma(1)$, so $\beta \gamma^{-1}$ is a closed path in U and hence is homotopic to the constant path e_{x_1}; that is,

$$\langle (\alpha \beta)(\alpha \gamma)^{-1} \rangle = \langle \alpha(\beta \gamma^{-1}) \alpha^{-1} \rangle = \langle \alpha e_{x_1} \alpha^{-1} \rangle = \langle e_{x_1} \rangle \in H$$

and $[\alpha \beta] = [\alpha \gamma]$.

Now $p^{-1}(U) = \bigcup_\alpha ([\alpha], U)$ where the union is over all α with $p([\alpha]) = x_1$. For if γ is a path, with $p([\gamma]) = \gamma(1) \in U$, let β be a path in U from $\gamma(1)$ to x_1. Then $p([\gamma \beta]) = x_1$, and $[\gamma] \in ([\gamma \beta], U)$.

To complete the proof that ($\tilde{X}, p$) is a covering space, it remains only to show that p is an open map; that is, that it maps open sets onto open sets.

For this, suppose $\tilde{V}$ is an open set in $\tilde{X}$. Since $\tilde{V}$ is a union of basic open sets $\tilde{U} = ([\alpha], U)$, it suffices to show that $p(\tilde{U})$ is open for such $\tilde{U}$. So let $x_1 \in p(\tilde{U})$; say $x_1 = p([\beta]) = \beta(1)$ for $[\beta] \in \tilde{U}$. Let U_1 be the set of all points in U which can be joined to x_1 by paths in U. Since X is locally arcwise connected, it is easy to verify that U_1 is an open set. Clearly $x_1 \in U_1$. We claim that $U_1 \subset p(\tilde{U})$, implying that $p(\tilde{U})$ is open. In fact, $U_1 = p([\beta], U_1)$. Moreover, $[\beta] = [\alpha\gamma]$, where γ is a path contained in U. Thus any element $[\eta] \in ([\beta], U_1)$ is of the form

$$[\eta] = [\beta\delta] = [\alpha\gamma\delta] \in ([\alpha], U) = \tilde{U},$$

where δ is a path in U_1 from $\beta(1)$ to $\eta(1)$, and $([\beta], U_1) \subset \tilde{U}$.

Finally, to complete the proof of the theorem, we must show that $p_*(\pi_1(\tilde{X}, \tilde{x})) = H$, where $\tilde{x} = [e_x]$. For this, let α be any path in X starting at x. Let $\tilde{\alpha}$ be the path in $\tilde{X}$ starting at $\tilde{x}$ defined by $\tilde{\alpha}(t_2) = [\alpha_{t_2}]$, where $\alpha_{t_2}(t_1) = \alpha(t_2 t_1)$. $\tilde{\alpha}$ is continuous: see the argument used above for f in the proof that $\tilde{X}$ is arcwise connected. $\tilde{\alpha}$ covers α because $p(\tilde{\alpha}(t_2)) = p([\alpha_{t_2}]) = \alpha_{t_2}(1) = \alpha(t_2)$. Now suppose α is closed; that is, suppose $\langle\alpha\rangle \in \pi_1(X, x)$. Then

$$\langle\alpha\rangle \in p_*(\pi_1(\tilde{X}, \tilde{x})) \Leftrightarrow \tilde{\alpha} \text{ is a closed path in } \tilde{X}$$
$$\Leftrightarrow \tilde{\alpha}(1) = [e_x]$$
$$\Leftrightarrow [\alpha] = [e_x]$$
$$\Leftrightarrow \alpha \overset{H}{\equiv} e_x$$
$$\Leftrightarrow \langle\alpha\rangle \in H.$$

Thus $p_*(\pi_1(\tilde{X}, \tilde{x})) = H$, as required. $\square$

Remark. Recall that, for a group G, two subgroups H_1 and H_2 are said to be *conjugate* if

$$H_2 = gH_1g^{-1} = [gh_1g^{-1}; h_1 \in H_1] \text{ for some } g \in G.$$

Theorem 5. *Let $(\tilde{X}, p)$ be a covering space of a space X, let $x \in X$, and let $\tilde{x}_1, \tilde{x}_2 \in p^{-1}(\{x\})$. Then $p_*\pi_1(\tilde{X}, \tilde{x}_1)$ and $p_*\pi_1(\tilde{X}, \tilde{x}_2)$ are conjugate in $\pi_1(X, x)$.*

PROOF. Let $\tilde{\gamma}$ be a path $\tilde{X}$ from $\tilde{x}_1$ to $\tilde{x}_2$. Then, by Theorem 5, Section 3.2,

$$\pi_1(\tilde{X}, \tilde{x}_1) = [\langle\tilde{\gamma}\rangle\langle\tilde{\alpha}\rangle\langle\tilde{\gamma}^{-1}\rangle; \langle\tilde{\alpha}\rangle \in \pi_1(\tilde{X}, \tilde{x}_2)].$$

Projecting,

$$p_*\pi_1(\tilde{X}, \tilde{x}_1) = [\langle p \circ \tilde{\gamma}\rangle h\langle p \circ \tilde{\gamma}^{-1}\rangle; h \in p_*\pi_1(\tilde{X}, \tilde{x}_2)],$$

and $\langle p \circ \tilde{\gamma}\rangle \in \pi_1(X, x)$. $\square$

Theorem 6. *Let X be locally simply connected and let $(\tilde{X}, p)$ be a covering space of X. Suppose $(\tilde{\tilde{X}}, \tilde{p})$ is a covering space of $\tilde{X}$. Then $(\tilde{\tilde{X}}, p \circ \tilde{p})$ is a covering space of X.*

PROOF. We leave the proof as an exercise.

Theorem 7. *Let X be locally simply connected and let $(\tilde{X}_1, p_1)$ and $(\tilde{X}_2, p_2)$ be covering spaces of X. Suppose $\tilde{x}_j \in \tilde{X}_j$ $(j = 1, 2)$, with $p_1(\tilde{x}_1) = p_2(\tilde{x}_2)$, and*

$$(p_1)_* \pi_1(\tilde{X}_1, \tilde{x}_1) \subset (p_2)_* \pi_1(\tilde{X}_2, \tilde{x}_2).$$

Then there exists a unique map $\tilde{p}: \tilde{X}_1 \to \tilde{X}_2$ with $\tilde{p}(\tilde{x}_1) = \tilde{x}_2$ such that $(\tilde{X}_1, \tilde{p})$ is a covering space of $\tilde{X}_2$. Furthermore, $p_2 \circ \tilde{p} = p_1$.

PROOF. For $y \in \tilde{X}_1$, we define $\tilde{p}(y)$ as follows. Let γ_1 be a path in $\tilde{X}_1$ from $\tilde{x}_1$ to y. Then $p_1 \circ \gamma_1$ is a path in X with origin $p_1(\tilde{x}_1) = p_2(\tilde{x}_2)$. Let γ_2 be the lift of $p_1 \circ \gamma_1$ to a path in $\tilde{X}_2$ starting at $\tilde{x}_2$. Set $\tilde{p}(y) = \gamma_2(1)$. Then $\tilde{p}$ is well-defined because if β_1 is another path in $\tilde{X}_1$ from $\tilde{x}_1$ to y, then $\langle \gamma_1 \beta_1^{-1} \rangle \in \pi_1(\tilde{X}_1, \tilde{x}_1)$ so that

$$(p_1)_*(\langle \gamma_1 \beta_1^{-1} \rangle) \in (p_1)_* \pi_1(\tilde{X}_1, \tilde{x}_1) \subset (p_2)_* \pi_1(\tilde{X}_2, \tilde{x}_2),$$

and hence $p_1 \circ \gamma_1 \beta_1^{-1}$ lifts to a closed path in $\tilde{X}_2$; that is, $\gamma_2(1) = \beta_2(1)$.

The proofs that $\tilde{p}$ is continuous and unique and $(\tilde{X}_1, \tilde{p})$ is in fact a covering space of $\tilde{X}_2$ involve no new techniques and are left as exercises for the student. □

Definition. Two covering spaces $(\tilde{X}_1, p_1)$ and $(\tilde{X}_2, p_2)$ of a space X are *isomorphic* if there exists a homeomorphism $h: \tilde{X}_1 \to \tilde{X}_2$ such that $p_2 \circ h = p_1$.

Remark. According to Theorem 4, given a subgroup H of the fundamental group of a space, there exists a covering space whose fundamental group is mapped isomorphically onto H by the projection map. The following result asserts that this covering space is unique up to isomorphism.

Theorem 8. *Let X be locally simply connected and let $(\tilde{X}_1, p_1)$ and $(\tilde{X}_2, p_2)$ be covering spaces of X. Suppose $\tilde{x}_j \in \tilde{X}_j$ $(j = 1, 2)$, with $p_1(\tilde{x}_1) = p_2(\tilde{x}_2)$ and*

$$(p_1)_* \pi_1(\tilde{X}_1, \tilde{x}_1) = (p_2)_* \pi_1(\tilde{X}_2, \tilde{x}_2).$$

Then $(\tilde{X}_1, \tilde{p}_1)$ and $(\tilde{X}_2, \tilde{p}_2)$ are isomorphic.

PROOF. By Theorem 7, there exist maps $\tilde{p}: \tilde{X}_1 \to \tilde{X}_2$ and $\tilde{q}: \tilde{X}_2 \to \tilde{X}_1$ such that $(\tilde{X}_1, \tilde{p})$ is a covering space of $\tilde{X}_2$, $(\tilde{X}_2, \tilde{q})$ is a covering space of $\tilde{X}_1$, $p_1 = p_2 \circ \tilde{p}$, $p_2 = p_1 \circ \tilde{q}$, $\tilde{p}(\tilde{x}_1) = \tilde{x}_2$, and $\tilde{q}(\tilde{x}_2) = \tilde{x}_1$. Thus

$$p_1 = p_2 \circ \tilde{p} = p_1 \circ (\tilde{q} \circ \tilde{p})$$

and $(\tilde{X}_1, \tilde{q} \circ \tilde{p})$ is a covering space of $\tilde{X}_1$ by Theorem 6. From the uniqueness part of Theorem 7, it follows that $\tilde{q} \circ \tilde{p} = i_{\tilde{X}_1}$. Similarly, $\tilde{p} \circ \tilde{q} = i_{\tilde{X}_2}$. Therefore $\tilde{q} = \tilde{p}^{-1}$ and $\tilde{p}$ is a homeomorphism with $p_2 \circ \tilde{p} = p_1$; that is, $(\tilde{X}_1, p_1)$ and $(\tilde{X}_2, p_2)$ are isomorphic. □

Definition. Let $(\tilde{X}, p)$ be a covering space of a space X. A *covering transformation*, or *deck transformation*, of $(\tilde{X}, p)$ is a homeomorphism $h\colon \tilde{X} \to \tilde{X}$ such that $p \circ h = p$.

Remark. The set of covering transformations is a group under composition. It is called the group of deck transformations and is denoted by $\mathcal{G}(\tilde{X}, p)$. Note that a deck transformation h permutes the "decks" of $\tilde{X}$; that is, if U is an admissible open set in X, then h permutes the copies of U in $p^{-1}(U)$.

Definition. A covering space $(\tilde{X}, p)$ is called a *regular covering space* of X if $p_*\pi_1(\tilde{X}, \tilde{x})$ is a normal subgroup of $\pi_1(X, p(\tilde{x}))$ for some $\tilde{x} \in \tilde{X}$. Note that since a normal subgroup equals all its conjugate subgroups, the condition of regularity of a covering space is independent of the base point $\tilde{x}$.

Theorem 9. *Let $(\tilde{X}, p)$ be a regular covering space of a locally simply connected space X. Then the group $\mathcal{G}(\tilde{X}, p)$ of deck transformations is isomorphic to the quotient group $\pi_1(X, p(\tilde{x}))/p_*\pi_1(\tilde{X}, \tilde{x})$.*

Corollary 1. *If $(\tilde{X}, p)$ is a universal covering space, then $\pi_1(X, x) \cong \mathcal{G}(\tilde{X}, p)$; that is, $\pi_1(X, x)$ is isomorphic to $\mathcal{G}(\tilde{X}, p)$.*

Corollary 2. $\pi_1(S^1, 1) \cong \mathcal{I}$, *where $\mathcal{I}$ is the group of integers.*

PROOF OF COROLLARY 2. The universal covering space of S^1 is (R^1, p) where $p(r) = e^{2\pi i r}$. A deck transformation of (R^1, p) is a homeomorphism $h\colon R^1 \to R^1$ such that $p \circ h = p$, that is, such that $e^{2\pi i h(r)} = e^{2\pi i r}$. Thus $h(r) - r$ is an integer for all $r \in R^1$. But $r \to h(r) - r$ is a continuous map $R^1 \to \mathcal{I}$. Since R^1 is connected, so is its image under this map. Hence $h(r) - r = k$ for some fixed k; that is, $h(r) = r + k$ for some k, and h is translation by the integer k. $\qquad\square$

Corollary 3. $\pi_1(S^1 \times S^1, (1, 1)) \cong \mathcal{I} \oplus \mathcal{I}$.

PROOF OF THEOREM 9. Let $H = p_*\pi_1(\tilde{X}, \tilde{x})$. By Theorem 8, we may replace $(\tilde{X}, p)$ by any covering space whose fundamental group projects onto H. In particular, we may assume that $(\tilde{X}, p)$ is the covering space constructed in the proof of Theorem 4 and that $\tilde{x} = [e_x]$, $x \in X$. We shall construct a homomorphism $\varphi\colon \pi_1(X, x) \to \mathcal{G}(\tilde{X}, p)$ which is surjective and has kernel H. The theorem then follows from group theory.

For $\langle\alpha\rangle \in \pi_1(X, x)$, let $\varphi(\langle\alpha\rangle)$ be the deck transformation defined by

$$\varphi(\langle\alpha\rangle)([\beta]) = [\alpha\beta] \quad \text{for} \quad [\beta] \in \tilde{X}.$$

This map φ is well-defined because if $\alpha \overset{\simeq}{} \alpha_1$ and $\beta \overset{\text{H}}{\equiv} \beta_1$ then $\langle\alpha\rangle = \langle\alpha_1\rangle$ and $\langle\beta\beta_1^{-1}\rangle \in H$ so that

$$\langle\alpha\beta(\alpha_1\beta_1)^{-1}\rangle = \langle\alpha\rangle\langle\beta\beta_1^{-1}\rangle\langle\alpha_1\rangle^{-1} = \langle\alpha\rangle\langle\beta\beta_1^{-1}\rangle\langle\alpha\rangle^{-1} \in H$$

since H is normal; thus $\alpha\beta \overset{\text{H}}{\equiv} \alpha_1\beta_1$ and $[\alpha\beta] = [\alpha_1\beta_1]$.

Note that $\varphi(\langle\alpha\rangle)$ is injective and surjective because its inverse is $\varphi(\langle\alpha^{-1}\rangle)$. $\varphi(\langle\alpha\rangle)$ is continuous because clearly $\varphi(\langle\alpha\rangle)^{-1}(([\beta], U)) = \varphi\langle\alpha^{-1}\rangle(([\beta], U)) = ([\alpha^{-1}\beta], U)$ for any $([\beta], U)$. Similarly $\varphi(\langle\alpha\rangle)^{-1}$ is continuous so $\varphi(\langle\alpha\rangle)$ is a homeomorphism. Since

$$p((\varphi\langle\alpha\rangle)([\beta])) = \alpha\beta(1) = \beta(1) = p([\beta]),$$

$\varphi(\langle\alpha\rangle)$ is a deck transformation for each $\langle\alpha\rangle$.

φ is a homomorphism because

$$\begin{aligned} \varphi(\langle\alpha_1\rangle\langle\alpha_2\rangle)([\beta]) &= \varphi(\langle\alpha_1\alpha_2\rangle)([\beta]) \\ &= [\alpha_1\alpha_2\beta] \\ &= \varphi(\langle\alpha_1\rangle)([\alpha_2\beta]) \\ &= \varphi(\langle\alpha_1\rangle) \circ \varphi(\langle\alpha_2\rangle)([\beta]) \end{aligned}$$

for all $[\beta] \in \tilde{X}$; that is, $\varphi(\langle\alpha_1\rangle\langle\alpha_2\rangle) = \varphi(\langle\alpha_1\rangle) \circ \varphi(\langle\alpha_2\rangle)$ for all $\langle\alpha_1\rangle, \langle\alpha_2\rangle \in \pi_1(X, x)$.

The kernel of φ is H because

$$\begin{aligned} \varphi\langle\alpha\rangle([\beta]) = [\beta] &\Leftrightarrow [\alpha\beta] = [\beta] \\ &\Leftrightarrow \alpha\beta \overset{H}{\equiv} \beta \\ &\Leftrightarrow \langle\alpha\rangle = \langle\alpha\beta\beta^{-1}\rangle \in H. \end{aligned}$$

φ is surjective. For suppose $h: \tilde{X} \to \tilde{X}$ is a deck transformation. Let $[\alpha] = h([e_x])$. Then $\alpha(1) = p([\alpha]) = x$, so α is a closed path in X; that is, $\langle\alpha\rangle \in \pi_1(X, x)$. Now $\varphi(\langle\alpha\rangle) = h$. For, in fact, $\varphi(\langle\alpha\rangle)$ and h are both maps $\tilde{X} \to \tilde{X}$, with

$$p \circ (\varphi\langle\alpha\rangle) = p \circ h = p$$

and with $\varphi(\langle\alpha\rangle)([e_x]) = [\alpha e_x] = [\alpha] = h([e_x])$, so, by Theorem 2, $\varphi(\langle\alpha\rangle) = h$. $\qquad\square$

(Note that $\tilde{X}$ is locally connected because it is locally arcwise connected: for $y \in \tilde{X}$ and V open in $\tilde{X}$ containing y, the set of all points in V which can be joined to x by a path is a connected open set in V.)

4 Simplicial complexes

The goal of this chapter is to develop some machinery which will enable us to compute the fundamental group of a large class of spaces. These spaces are the ones which can be obtained by piecing together in a nice way basic topological building blocks called simplices. A 0-dimensional simplex is a point, a 1-dimensional simplex a line segment, a 2-dimensional simplex a triangle, a 3-dimensional simplex a tetrahedron, and so on. All the spaces which will occupy our attention in the coming chapters will be homeomorphic to spaces built up from simplices.

Given a decomposition (triangulation) of two spaces into small enough simplices, we shall show that any continuous map from one space to the other can be approximated by a map which is linear on each simplex. Moreover, this approximating map will be in the same homotopy class as the original one. Thus we will have reduced difficult topological problems of mappings to more accessible algebraic problems of "piecewise linear" maps and spaces.

Figure 4.1 illustrates these ideas. It shows two triangulations of the unit

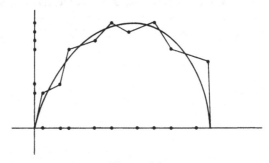

Figure 4.1

interval I and a piecewise linear approximation to the function $f: I \to I$ defined by

$$f(x) = 4x - 4x^2.$$

4.1 Geometry of simplicial complexes

Definition. Let V be a vector space over R^1 and let C be a subset of V. C is *convex* if

$$c_1, c_2 \in C \Rightarrow tc_1 + (1 - t)c_2 \in C$$

for all $t \in I$.

Definition. A set $\{v_0, v_1, \ldots, v_k\}$ of vectors in a vector space V is *convex-independent*, or *c-independent*, if the set $\{v_1 - v_0, v_2 - v_0, \ldots, v_k - v_0\}$ is linearly independent. Note that this definition does not depend on which vector is called v_0.

EXAMPLE. In R^2, $\{v_0, v_1, v_2\}$ is c-independent if and only if v_0, v_1, and v_2 are not collinear.

Theorem 1. *Suppose $\{v_0, v_1, \ldots, v_k\}$ is a c-independent set. Let C be the convex set generated by $\{v_0, v_1, \ldots, v_k\}$; that is, C is the smallest convex set containing $\{v_0, v_1, \ldots, v_k\}$. Then C consists of all vectors of the form $\sum_{i=0}^{k} a_i v_i$, where $a_i \geq 0$ for all i and $\sum_{i=0}^{k} a_i = 1$. Furthermore, each $v \in C$ is uniquely expressible in this form.*

PROOF. First note that the intersection of convex sets is a convex set, so C exists; C is the intersection of all convex sets containing $\{v_0, v_1, \ldots, v_k\}$.

Now let

$$C_1 = \left[v; v = \sum_{i=0}^{k} a_i v_i, a_i \geq 0, \sum_{i=0}^{k} a_i = 1 \right].$$

C_1 is convex because if $v = \sum_{i=0}^{k} a_i v_i$ and $w = \sum_{i=0}^{k} b_i v_i$, then

$$tv + (1 - t)w = \sum_{i=0}^{k} [ta_i + (1 - t)b_i] v_i$$

and

$$\sum_{i=0}^{k} [ta_i + (1 - t)b_i] = t \sum_{i=0}^{k} a_i + (1 - t) \sum_{i=0}^{k} b_i = t + (1 - t) = 1.$$

Thus C_1 is a convex set containing $\{v_0, v_1, \ldots, v_k\}$; hence $C_1 \supset C$.

Conversely, $C_1 \subset C$. For certainly $\sum_{i=0}^{k} a_i v_i \in C$ whenever all but one a_i are zero.

We proceed by induction. Assume $\sum_{i=0}^{k} a_i v_i \in C$ whenever, at most, n of the a_i are nonzero ($n < k + 1$). Let $\sum_{i=0}^{k} a_i v_i$ have $n + 1$ nonzero a_i, which we may assume (by relabelling if necessary) are $a_0, a_1, \ldots, a_n$. $a_n \neq 1$ for otherwise all other $a_i = 0$. Thus

$$\sum_{i=0}^{k} a_i v_i = (1 - a_n) \sum_{i=0}^{n-1} \frac{a_i}{1 - a_n} v_i + a_n v_n.$$

Since

$$\sum_{i=0}^{n-1} \frac{a_i}{1 - a_n} = \frac{1}{1 - a_n} \sum_{i=0}^{n-1} a_i = \frac{1}{1 - a_n} (1 - a_n) = 1,$$

$\sum_{i=0}^{n-1} [a_i/(1 - a_n)] v_i \in C$ by the induction assumption. Hence

$$t \sum_{i=0}^{n-1} \frac{a_i}{1 - a_n} v_i + (1 - t) v_n \in C$$

for $t \in I$ since C is convex. Let $t = 1 - a_n$. It follows that $\sum_{i=0}^{k} a_i v_i \in C$; that is, $C_1 \subset C$, and $C_1 = C$.

We proceed to show uniqueness. Suppose

$$v = \sum_{i=0}^{k} a_i v_i = \sum_{i=0}^{k} b_i v_i,$$

where $\sum_{i=0}^{k} a_i = \sum_{i=0}^{k} b_i = 1$. Then

$$0 = \sum_{i=0}^{k} (a_i - b_i) v_i$$

$$= \sum_{i=0}^{k} (a_i - b_i) v_i - \left(\sum_{i=0}^{k} a_i - \sum_{i=0}^{k} b_i \right) v_0$$

$$= \sum_{i=0}^{k} (a_i - b_i)(v_i - v_0).$$

Since $\{v_1 - v_0, v_2 - v_0, \ldots, v_k - v_0\}$ is linearly independent, $a_i - b_i = 0$ for all $i > 0$. Then clearly $a_0 = b_0$ also. $\qquad\square$

Definition. Let V be a vector space over R^1. A convex set generated by c-independent vectors $\{v_0, v_1, \ldots, v_k\}$ is called a (closed) k-*simplex* and is denoted by $[v_0, v_1, \ldots, v_k]$. k is called the *dimension* of the simplex. If $v \in [v_0, v_1, \ldots, v_k]$, then the coefficients a_i, with $a_i \geq 0$ and $\sum_{i=0}^{k} a_i = 1$ such that $v = \sum_{i=0}^{k} a_i v_i$, are called the *barycentric coordinates* of v.

EXAMPLES. For $\{v_0, v_1\}$ vectors in R^1, the simplex $[v_0, v_1]$ is the closed interval $[v_0, v_1]$. For $\{v_0, v_1, v_2\} \subset R^2$, $[v_0, v_1, v_2]$ is the triangle with vertices v_0, v_1, and v_2. The centroid of this triangle is the point with barycentric coordinates $(\frac{1}{3}, \frac{1}{3}, \frac{1}{3})$. For $V = R^n$, the simplex $[v_0, v_1, \ldots, v_k]$ is a compact metric space (it is closed and bounded) in the relative topology. In fact, using barycentric coordinates, it is not difficult to see that $[v_0, v_1, \ldots, v_k]$ is homeomorphic to a product of k unit intervals. However, this homeomorphism is *not* an isometry.

Definitions. Let $\{v_0, v_1, \ldots, v_k\}$ be a c-independent set. The set

$$[v \in [v_0, v_1, \ldots, v_k]; a_i(v) > 0, i = 0, 1, \ldots, k]$$

is called an *open simplex* and is denoted by $(v_0, v_1, \ldots, v_k)$. We shall also denote an open simplex by (s) and the corresponding closed simplex by $[s]$.

Let $[s] = [v_0, v_1, \ldots, v_k]$ be a closed simplex. The *vertices* of $[s]$ are the points $v_0, v_1, \ldots, v_k$. The *closed faces* of $[s]$ are the closed simplices $[v_{j_0}, v_{j_1}, \ldots, v_{j_h}]$ where $\{j_0, j_1, \ldots, j_h\}$ is a nonempty subset of $\{0, 1, \ldots, k\}$. The *open faces* of the simplex $[s]$ are the open simplices $(v_{j_0}, v_{j_1}, \ldots, v_{j_h})$.

Remarks

(1) A vertex is a 0-dimensional closed face. It is also an open face.
(2) An open simplex (s) is an open set in the closed simplex $[s]$. Its closure is $[s]$.
(3) The closed simplex $[s]$ is the union of its open faces.
(4) Distinct open faces of a simplex are disjoint.
(5) The open simplex (s) is the *interior* of the closed simplex $[s]$; that is, it is the closed simplex minus its proper open faces (faces $\neq (s)$).

Definition. A *simplicial complex K* (Euclidean) is a finite set of open simplices in some R^n such that

(1) if $(s) \in K$, then all open faces of $[s] \in K$;
(2) if $(s_1), (s_2) \in K$ and $(s_1) \cap (s_2) \neq \varnothing$, then $(s_1) = (s_2)$.

The *dimension* of K is the maximum dimension of the simplices of K.

Remarks. If K is a simplicial complex, let $[K]$ denote the point set union of the open simplices of K. Then $[K]$ is compact, and $[K] = \bigcup_{(s) \in K} (s) = \bigcup_{(s) \in K} [s]$.

If $[s]$ is a closed simplex, the collection of its open faces is a simplicial complex which we denote by s.

EXAMPLES. Figure 4.2 shows examples of simplicial complexes. Those shown

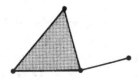
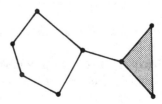

Figure 4.2

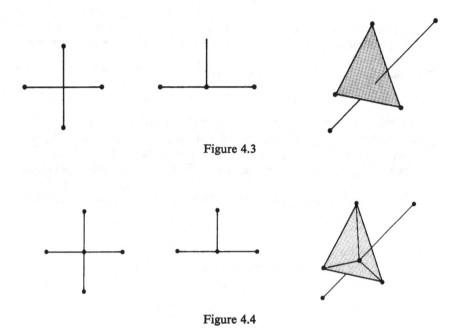

Figure 4.3

Figure 4.4

in Figure 4.3 are *not* simplicial complexes. By adding simplices, however, the point sets in Figure 4.3 can be made into complexes (Figure 4.4). Note that a complex is more than just a point set. It is a set with additional structure. It is possible to have two different complexes with the same point set, as in Figure 4.5.

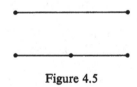

Figure 4.5

Definition. A *subcomplex* of a simplicial complex K is a simplicial complex L such that $(s) \in L$ implies $(s) \in K$.

Remark. For each $(s) \in K$, the simplicial complex s is a subcomplex of K.

Definition. Let K be a complex. Let r be an integer less than or equal to $\dim K$. The *r-skeleton* K^r of K is the collection $K^r = [(s) \in K; \dim s \leq r]$.

Remark. The *r-skeleton* K^r is a subcomplex of K.

4.2 Barycentric subdivisions

Definition. Let $v \in R^n$ and let $A \subset R^n$. The pair (v, A) is in *general position* if $v \notin A$ and, for each $a_1, a_2 \in A$ with $a_1 \neq a_2$, $[v, a_1] \cap [v, a_2] = \{v\}$.

EXAMPLES The points and sets in the plane shown in Figure 4.6 are in general position. Those shown in Figure 4.7 are *not* in general position. Note that if A

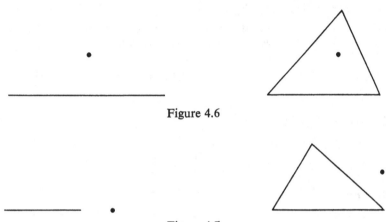

Figure 4.6

Figure 4.7

is a triangle with its interior, then there is no $v \in R^2$ such that (v, A) is in general position.

Definition. Let (v, A) be in general position. The *cone* with vertex v and base A (or the *join* of v with A), denoted by $v * A$ (Figure 4.8), is the set

$$v * A = \bigcup_{a \in A} [v, a].$$

Theorem 1. *Let* $[s] = [v_0, v_1, \ldots, v_k]$ *be a k-simplex. Let* $v \in (s)$. *Then* $(v, [s^{k-1}])$ *is in general position, and* $v * [s^{k-1}] = [s]$.

Figure 4.8

PROOF. Let $a_1, a_2 \in [s^{k-1}]$. Suppose there exists $w \in [v, a_1] \cap [v, a_2]$ with $w \neq v$. We must then show that $a_1 = a_2$. Consider the expressions for $a_1, a_2,$ and v in terms of barycentric coordinates in $[s]$:

$$a_1 = \sum_{i=0}^{k} \alpha_{1i} v_i, \qquad a_2 = \sum_{i=0}^{k} \alpha_{2i} v_i, \qquad v = \sum_{i=0}^{k} \beta_i v_i.$$

Since $a_1 \notin (s)$ and $a_2 \notin (s)$, $\alpha_{1 i_1} = 0$ and $\alpha_{2 i_2} = 0$ for some i_1 and $i_2 \leq k$. Moreover, since $v \in (s)$, $\beta_i \neq 0$ for all i. Since $w \in [v, a_1]$, $w = t_1 v + (1 - t_1) a_1$ for some $t_1 \in I$. Thus $w = \sum_{i=0}^{k} [t_1 \beta_i + (1 - t_1) \alpha_{1i}] v_i$.

Similarly, since $w \in [v, a_2]$, we have $w = \sum_{i=0}^{k} [t_2 \beta_i + (1 - t_2) \alpha_{2i}] v_i$ for some $t_2 \in I$. By the uniqueness of barycentric coordinates,

$$t_1 \beta_i + (1 - t_1) \alpha_{1i} = t_2 \beta_i + (1 - t_2) \alpha_{2i} \qquad (i = 0, 1, \ldots, k).$$

Hence $t_1 - t_2 = (1/\beta_i)[(1 - t_2) \alpha_{2i} - (1 - t_1) \alpha_{1i}]$.
Taking $i = i_1$ yields

$$t_1 - t_2 = \frac{1}{\beta_{i_1}} (1 - t_2) \alpha_{2 i_1} \geq 0.$$

Taking $i = i_2$ yields

$$t_1 - t_2 = -\frac{1}{\beta_{i_2}} (1 - t_1) \alpha_{1 i_2} \leq 0.$$

Hence $t_1 - t_2 = 0$, $t_1 = t_2$, and

$$(1 - t_1) \alpha_{1i} = (1 - t_1) \alpha_{2i}$$

for all i. Now $t_1 \neq 1$ since $w \neq v$, so this implies that $\alpha_{1i} = \alpha_{2i}$ for all i; hence $a_1 = a_2$, completing the proof that $(v, [s^{k-1}])$ is in general position.

Now $v * [s^{k-1}] \subset [s]$ because $[s]$ is convex. Conversely, $[s] \subset v * [s^{k-1}]$. For if $w \in [s^{k-1}]$, then certainly $w \in v * [s^{k-1}]$. So suppose $w \in (s)$. We may assume $w \neq v$. In barycentric coordinates,

$$w = \sum_{i=0}^{k} \alpha_i v_i, \qquad v = \sum_{i=0}^{k} \beta_i v_i \qquad (\text{all } \alpha_i, \beta_i > 0).$$

Since

$$\sum_{i=0}^{k} (\alpha_i - \beta_i) = \sum_{i=0}^{k} \alpha_i - \sum_{i=0}^{k} \beta_i = 1 - 1 = 0,$$

and since $\alpha_i - \beta_i \neq 0$ for some i, $\alpha_j - \beta_j < 0$ for some j. For each such j, let $f_j(t) = \beta_j + t(\alpha_j - \beta_j)$. Since $f_j(1) > 0$ and $f_j(t) < 0$ for large t, there

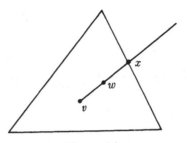

Figure 4.9

exists a $t_j > 1$ such that $\beta_j + t_j(\alpha_j - \beta_j) = 0$. Choose i_0 from among the numbers j so that $t_{i_0} \le t_j$, for all such j. Then $\beta_{i_0} + t_{i_0}(\alpha_{i_0} - \beta_{i_0}) = 0$, and $\beta_i + t_{i_0}(\alpha_i - \beta_i) \ge 0$ for all i. Hence $v + t_{i_0}(w - v) = x \in [s^{k-1}]$ (see Figure 4.9). Also,

$$w = \frac{1}{t_{i_0}} x + \frac{t_{i_0} - 1}{t_{i_0}} v = t^1 x + (1 - t^1)v$$

with $t^1 = 1/t_{i_0} < 1$. Hence $w \in v * [s^{k-1}]$. $\qquad\square$

Exercise. Let $[s] = [v_0, v_1, \ldots, v_k]$. Prove that $(v, [s])$ is in general position if and only if $\{v_0, v_1, \ldots, v_k, v\}$ is c-independent, in which case

$$v * [s] = [v_0, v_1, \ldots, v_k, v].$$

Definitions. Let s be a k-simplex. The *barycenter* of s, denoted $b(s)$, is the point in (s) with barycentric coordinates $(1/(k + 1), \ldots, 1/(k + 1))$; that is, if

$$(s) = (v_0, v_1, \ldots, v_k),$$

then

$$b(s) = \frac{1}{k + 1} \sum_{j=0}^{k} v_i.$$

Let K be a simplicial complex. A *subdivision* of K is a simplicial complex $K^\dagger$ such that

(1) $[K^\dagger] = [K]$
(2) if $s \in K^\dagger$, then $(s) \subset$ some open simplex of K.

EXAMPLES. Each of the complexes in Figure 4.10, second column, is a subdivision of the corresponding complex in the first column. Note that although the second and third complexes in the second column have the same point set, neither is a subdivision of the other.

85

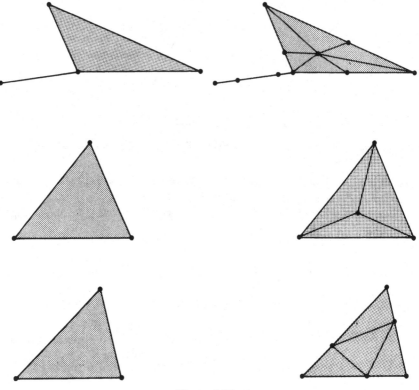

Figure 4.10

Theorem 2. *Let s be a k-simplex. Let $K^\dagger$ be a subdivision of s^{k-1}. Let $v \in (s)$. Then $(v, [K^\dagger])$ is in general position. Furthermore, $v * [K^\dagger]$ is the point set of a complex $\tilde{K}$ defined by $\tilde{K} = K^\dagger \cup (\bigcup_{s^\dagger \in K^\dagger} (s^\dagger, v)) \cup (v)$ (see Figure 4.11). Here, for*

$$(s^\dagger) = (v_0, v_1, \ldots, v_r) \in K^\dagger,$$

$(s^\dagger, v) = (v_0, v_1, \ldots, v_r, v)$. The complex $\tilde{K}$ is a subdivision of s.

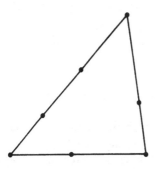

Figure 4.11

PROOF. By Theorem 1, $(v, [s^{k-1}])$ is in general position and $v * [s^{k-1}] = [s]$. But $[K^\dagger] = [s^{k-1}]$, so $(v, [K^\dagger])$ is in general position, and $v * [K^\dagger] = [s]$.

We must show that $\tilde{K}$ is a simplicial complex. It is a set of open simplices. Moreover, each simplex $\neq (v)$ in $\tilde{K}$ is either in $K^\dagger$ or is of the form $(s^\dagger, v)$. If it is in $K^\dagger$, then all its open faces are in $K^\dagger$, hence in $\tilde{K}$. If it is of the form $(s^\dagger, v)$, then its open faces are $s^\dagger$, (v), and $[(s_1{}^\dagger, v); s_1{}^\dagger$ an open face of $s^\dagger]$. Thus, in each case, all open faces are in $\tilde{K}$, and Condition (1) for a complex is satisfied. To verify Condition (2) for a complex, we must show that distinct open simplices have void intersection. Since $K^\dagger$ is a complex, this is certainly satisfied by pairs of simplices each of which is in $K^\dagger$. Moreover, if $s^\dagger \in K^\dagger$ then $(s^\dagger) \cap (s_1{}^\dagger, v) = \varnothing$ for all $s_1{}^\dagger \in K^\dagger$ because, in fact, $(s_1{}^\dagger, v) \subset (s)$. Clearly, (v) meets no other open simplices of $\tilde{K}$. So now suppose $s_1{}^\dagger, s_2{}^\dagger \in K^\dagger$ have $(s_1{}^\dagger, v) \cap (s_2{}^\dagger, v) \neq \varnothing$. Let $w \in (s_1{}^\dagger, v) \cap (s_2{}^\dagger, v)$. Since these are open simplices, $w \neq v$. Now there exists a unique $x \in [s^{k-1}] = [K^\dagger]$ such that $w \in [v, x]$ (see Theorem 1). Since $[s_1{}^\dagger, v] = v * [s_1{}^\dagger]$, it follows that $x \in (s_1{}^\dagger)$. Similarly $x \in (s_2{}^\dagger)$, so $(s_1{}^\dagger) \cap (s_2{}^\dagger) \neq \varnothing$. Therefore $s_1{}^\dagger = s_2{}^\dagger$ since $K^\dagger$ is a complex, and $(s_1{}^\dagger, v) = (s_2{}^\dagger, v)$. Thus $\tilde{K}$ is a complex.

The point set of $\tilde{K}$ is

$$[\tilde{K}] = \bigcup_{\tilde{s} \in \tilde{K}} [\tilde{s}] = \bigcup_{s^\dagger \in K^\dagger} v * [s^\dagger] = v * [K^\dagger] = [s].$$

Since each open simplex of $\tilde{K}$ is contained in an open simplex of s—those in $K^\dagger$ are, because $K^\dagger$ is a subdivision of s^{k-1}; the rest are contained in (s)— $\tilde{K}$ is a subdivision of s. $\qquad\square$

Definition. Let K be a simplicial complex. A partial ordering is defined on K by $s_1 \leq s_2$ if and only if s_1 is a face of s_2. The notation $s_1 < s_2$ shall mean $s_1 \leq s_2$ and $s_1 \neq s_2$.

Theorem 3. *Let K be a simplicial complex. Let*

$$K^{(1)} = [(b(s_0), b(s_1), \ldots, b(s_k)); s_0 < s_1 < \cdots < s_k; s_0, s_1, \ldots, s_k \in K].$$

Then $K^{(1)}$ is a subdivision of K. Furthermore, for each $s_0, s_1, \ldots, s_r \in K$ with $s_0 < s_1 < \cdots < s_r$, $(b(s_0), \ldots, b(s_r)) \subset (s_r)$.

Remark. The subdivision $K^{(1)}$ is called the *first barycentric subdivision* of K. Iterating, $K^{(n)} = \underbrace{(((K^{(1)})^{(1)}) \cdots)^{(1)}}_{n \text{ times}}$ is the *nth barycentric subdivision* of K.

PROOF OF THEOREM 3. (By induction on dim K.) For dim $K = 0$, $K^{(1)} = K$ so there is nothing to prove. Assume the theorem is true for all simplicial complexes of dimension $\leq n - 1$. Let dim $K = n$. Then the $(n - 1)$-skeleton K^{n-1} is a complex of dimension $\leq n - 1$; hence the theorem is true for K^{n-1}. In particular, if

$$s_0, s_1, \ldots, s_r \in K,$$

$s_0 < s_1 < \cdots < s_r$, and $\dim s_r \leq n - 1$, then $\{b(s_0), b(s_1), \ldots, b(s_r)\}$ is c-independent and gives an open simplex $(b(s_0), b(s_1), \ldots, b(s_r))$ in $(K^{n-1})^{(1)}$. Furthermore,

$$(b(s_0), b(s_1), \ldots, b(s_r)) \subset (s_r).$$

Now suppose $s_0, s_1, \ldots, s_r \in K$ $(s_0 < \cdots < s_r)$, and $\dim s_r = n$. Since $s_{r-1} < s_r$, $\dim s_{r-1} \leq n - 1$, so that $(b(s_0), b(s_1), \ldots, b(s_{r-1}))$ is a simplex $\subset (s_{r-1})$, which is a face of s_r. Since $b(s_r) \in (s_r)$, Theorem 1 implies that $(b(s_r), (b(s_0), \ldots, b(s_{r-1})))$ is in general position. Hence $(b(s_0), b(s_1), \ldots, b(s_r))$ is an open simplex, the interior of the closed simplex

$$[b(s_0), b(s_1), \ldots, b(s_r)] = b(s_r) * [b(s_0), \ldots, b(s_{r-1})] \subset [s_r].$$

Thus $(b(s_0), \ldots, b(s_r)) \subset (s_r)$.

Thus far we know that $K^{(1)}$ is a collection of open simplices. It is in fact a simplicial complex. Condition (1) is clearly satisfied: any face of $(b(s_0), b(s_1), \ldots, b(s_r))$ is of the form $(b(s_{j_0}), b(s_{j_1}), \ldots, b(s_{j_h}))$ and is hence in $K^{(1)}$. We now verify Condition (2). Suppose $s_0 < \cdots < s_r$, $\bar{s}_0 < \cdots < \bar{s}_q$, and

$$(b(s_0), \ldots, b(s_r)) \cap (b(\bar{s}_0), \ldots, b(\bar{s}_q)) \neq \varnothing.$$

Let w belong to the intersection. Then $w \in (s_r) \cap (\bar{s}_q)$. Since K is a complex, $s_r = \bar{s}_q$ and $b(s_r) = b(\bar{s}_q)$. Moreover,

$$(b(s_0), \ldots, b(s_{r-1})) \subset (s_{r-1}) \quad \text{and} \quad (b(\bar{s}_0), \ldots, b(\bar{s}_{q-1})) \subset (\bar{s}_{q-1})$$

where (s_{r-1}) and $(\bar{s}_{q-1})$ are both faces of (s_r); hence $(b(s_0), \ldots, b(s_{r-1}))$ and $(b(\bar{s}_0), \ldots, b(\bar{s}_{q-1})) \in s_r^{(1)}$. Since

$$w \in (b(s_0), \ldots, b(s_{r-1}), b(s_r)) \cap (b(\bar{s}_0), \ldots, b(\bar{s}_{q-1}), b(s_r))$$
$$\subset b(s_r) * (b(s_0), \ldots, b(s_{r-1})) \cap b(s_r) * (b(\bar{s}_0), \ldots, b(\bar{s}_{q-1})),$$

we conclude by Theorem 2 and the induction assumption that

$$(b(s_0), \ldots, b(s_{r-1})) = (b(\bar{s}_0), \ldots, b(\bar{s}_{q-1})).$$

This shows that $K^{(1)}$ is a simplicial complex. To complete the proof of the induction step and of the theorem, we must show that $[K^{(1)}] = [K]$. Clearly, $[K^{(1)}] \subset [K]$. Also, $[K^{(1)}] \supset [(K^{n-1})^{(1)}] = [K^{n-1}]$, using the induction assumption. Hence we must show that

$$[K^{(1)}] \supset [K] - [K^{n-1}].$$

So suppose $w \in [K] - [K^{n-1}]$. Then w must lie in some open simplex (s) of dimension n. Thus

$$w \in (s) \subset [s] = b(s) * [s^{n-1}].$$

Now $[s^{n-1}] \subset [K^{n-1}] = [(K^{n-1})^{(1)}]$ so $w \in b(s) * (s_1)$ for some

$$(s_1) = (b(s_0), \ldots, b(s_k)) \in (K^{n-1})^{(1)}.$$

If $w = b(s)$, then w is a vertex in $K^{(1)}$. If $w \neq b(s)$, then

$$w \in (b(s_0), \ldots, b(s_k), b(s)) \subset [K^{(1)}].$$ □

Definitions. Let (S, ρ) be a metric space, and let T be a compact subset of S. The *diameter* of T is

$$\text{diam } T = \sup_{t_1, t_2 \in T} \rho(t_1, t_2).$$

Since T is compact and ρ is continuous, the maximum is assumed, and we may write:

$$\text{diam } T = \max_{t_1, t_2 \in T} \rho(t_1, t_2).$$

Let K be a simplicial complex in R^n, where R^n is provided with the usual metric. The *mesh* of K is the maximum diameter of simplices of K:

$$\text{mesh } K = \max_{s \in K} \text{diam } [s].$$

Lemma. *If s is a simplex in R^n, then* $\text{diam } [s] = \rho(v_1, v_2)$ *for some pair v_1, v_2 of vertices of s. If K is a simplicial complex, then* $\text{mesh } K = \rho(v_1, v_2)$, *where v_1 and v_2 are vertices of some simplex of K.*

PROOF. Let s be a simplex and let v_1, $v_2 \in [s]$ be such that $\text{diam } [s] = \rho(v_1, v_2)$. Suppose, say, that v_2 is not a vertex. Then v_2 is in some open simplex of dimension ≥ 1. In particular, there exist w_1, $w_2 \in [s]$ with $w_1 \neq w_2$, such that

$$v_2 = tw_1 + (1 - t)w_2$$

for some t with $0 < t < 1$. But the convex function f defined on I by

$$f(t) = \rho(v_1, tw_1 + (1 - t)w_2),$$

has no maximum for $0 < t < 1$, contradicting the maximality of ρ at (v_1, v_2).

The second statement of the lemma follows immediately from the definition of mesh K. □

Theorem 4. *Let K be a simplicial complex of dimension m. Then* $\text{mesh } K^{(1)} \leq (m/(m + 1)) \text{ mesh } K$. *In particular,* $\lim_{n \to \infty} \text{mesh } K^{(n)} = 0$.

PROOF. By the lemma, there exists a simplex $(b(s_0), b(s_1), \ldots, b(s_r)) \in K^{(1)}$ such that $\text{mesh } K^{(1)} = \rho(b(s_k), b(s_h))$ with $s_k < s_h$. By renumbering vertices if necessary, $s_k = (v_0, \ldots, v_p)$, $s_h = (v_0, \ldots, v_p, v_{p+1}, \ldots, v_q)$, and

$$\text{mesh } K^{(1)} = \|b(s_k) - b(s_h)\|$$

$$= \left\| \frac{1}{p+1} \sum_{i=0}^{p} v_i - \frac{1}{q+1} \sum_{j=0}^{q} v_j \right\|$$

$$= \frac{1}{q+1} \left\| \frac{q+1}{p+1} \sum_{i=0}^{p} v_i - \sum_{j=0}^{q} v_j \right\|$$

$$= \frac{1}{q+1} \left\| \sum_{j=0}^{q} \left(\frac{1}{p+1} \sum_{i=0}^{p} v_i - v_j \right) \right\|$$

$$= \frac{1}{p+1} \frac{1}{q+1} \left\| \sum_{j=0}^{q} \sum_{i=0}^{p} (v_i - v_j) \right\|$$

$$\leq \frac{1}{p+1} \frac{1}{q+1} \sum_{j=0}^{q} \sum_{i=0}^{p} \| v_i - v_j \|.$$

But $\| v_i - v_j \| \leq \text{diam } [s_h] \leq \text{mesh } K$. Moreover, whenever $i = j$, the (i, j)th term in this summation is zero. There are $p + 1$ such terms. The number of nonzero terms is therefore

$$(p + 1)(q + 1) - (p + 1) = (p + 1)q.$$

Since each term in the summation is $\leq \text{mesh } K$, and since $q \leq m$,

$$\text{mesh } K^{(1)} \leq \frac{q}{q+1} \text{ mesh } K \leq \frac{m}{m+1} \text{ mesh } K. \qquad \square$$

4.3 Simplicial approximation theorem

Definition. Let K and L be simplicial complexes. A map $\varphi \colon [K] \to [L]$ is a *simplicial map* if

(1) for each vertex v of K, $\varphi(v)$ is a vertex of L,
(2) for each simplex $(v_0, v_1, \ldots, v_k) \in K$, the vertices $\varphi(v_0), \varphi(v_1), \ldots, \varphi(v_k)$ all lie in some closed simplex of L, and
(3) for each $(s) = (v_0, v_1, \ldots, v_k) \in K$ and $p = \sum_{i=0}^{k} a_i v_i \in (s)$, the image of p is given by

$$\varphi(p) = \sum_{i=0}^{k} a_i \varphi(v_i).$$

Remark. Condition (1) says that φ must map the 0-skeleton of K into the 0-skeleton of L. Condition (2) says that for each simplex $(v_0, v_1, \ldots, v_k) \in K$, the set $\varphi(v_0), \varphi(v_1), \ldots, \varphi(v_k)$, with redundancies removed, is the set of vertices of some simplex in L. Condition (3) says that the mapping φ is linear on each simplex.

Since a simplicial map depends on K and L, not just $[K]$ and $[L]$, we will denote it by $\varphi \colon K \to L$.

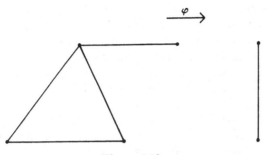

Figure 4.12

EXAMPLES. In Figure 4.12, projection is a simplicial map. However, in Figure 4.13, projection is *not* a simplicial map, even though conditions (1) and (3) are satisfied.

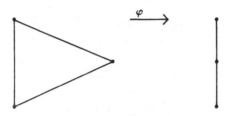

Figure 4.13

Remarks. With the glueing lemma, it is easy to check that a simplicial map is continuous.

A simplicial map is, by Condition (3), determined by its effect on vertices. Conversely, a vertex map $K^0 \to L^0$ from the vertices of K into the vertices of L can be extended to a simplicial map $K \to L$ if and only if Condition (2) above is satisfied.

Definition. Let K be a simplicial complex and let v be a vertex of K. The *star* of v is the point set

$$\mathrm{St}(v) = \bigcup_{\substack{v \in [s] \\ (s) \in K}} (s)$$

Theorem 1. *Let K be a simplicial complex. For v a vertex of K, $\mathrm{St}(v)$ is an open set in $[K]$ containing v, and v is the only vertex of K which lies in $\mathrm{St}(v)$. The collection $\{\mathrm{St}(v)\}_{v \in K^0}$ is an open covering of $[K]$.*

PROOF. We shall show that the complement of $\mathrm{St}(v)$ in $[K]$ is closed.

$$\mathrm{St}(v)' = \bigcup_{v \notin [s]} (s).$$

91

Since $v \notin [s]$ implies $v \notin$ face of s, we have $(s) \subseteq$ St $(v)'$ implies $[s] \subseteq$ St $(v)'$. Since $[s]$ is compact, $[s]$ is closed. Hence St $(v)' = \bigcup_{(s) \subset St(v)'} [s]$ is closed.

Next, v is the only vertex in St (v) because the only open simplex containing a vertex is the 0-simplex consisting of that vertex alone.

Finally, $\bigcup_{v \in K^0}$ St $(v) = [K]$ because if $p \in [K]$, then $p \in (s)$ for some $(s) \in K$, and hence $p \in$ St (v) for any vertex v of (s). $\qquad \square$

Definition. Let K and L be simplicial complexes. Let $f: [K] \to [L]$ be continuous. A simplicial map $\varphi: K \to L$ is a *simplicial approximation* to f if $f(\text{St } (v)) \subset \text{St } (\varphi(v))$ for each vertex v of K.

Theorem 2. *Suppose* $\varphi: K \to L$ *is a simplicial approximation to* $f: [K] \to [L]$. *Then, for any* $p \in [K]$, $f(p)$ *and* $\varphi(p)$ *lie in a common closed simplex of* $[L]$.

PROOF. Let $p \in [K]$. Then $p \in (s)$ for some simplex $(s) = (v_0, v_1, \ldots, v_r) \in K$, and

$$f(p) \in f((s)) \subset f(\text{St } (v_j)) \subset \text{St } (\varphi(v_j))$$

for all $j \in \{0, 1, \ldots, r\}$. Now $f(p) \in (t)$ for some simplex $(t) \in L$, so for (t), $(t) \cap \text{St } (\varphi(v_j)) \neq \varnothing$ for all j. But, since L is a complex and St $(\varphi(v_j))$ is a union of open simplices, $(t) \subset$ St $(\varphi(v_j))$ for all j; that is, $\varphi(v_j)$ is a vertex of (t) for all j. In terms of barycentric coordinates in s, $p = \sum_{j=0}^{r} a_j v_j$, and

$$\varphi(p) = \sum_{j=0}^{r} a_j \varphi(v_j) \in [t].$$

This completes the proof. $\qquad \square$

Corollary. *Suppose* $\varphi: K \to L$ *is a simplicial approximation to* $f: [K] \to [L]$. *Then*

$$d(f, \varphi) \leq \text{mesh } L,$$

where $d(f, \varphi) = \sup_{p \in [K]} \rho(f(p), \varphi(p))$.

Theorem 3. *Let* φ *be a simplicial approximation to* $f: [K] \to [L]$. *Let* K_1 *be a subcomplex of* K, *and suppose that the restriction of* f *to* $[K_1]$ *is a simplicial map. Then there exists a homotopy between* f *and* φ *which is stationary on* $[K_1]$.

PROOF. Define $F: [K] \times I \to [L]$ by

$$F(p, t) = t\varphi(p) + (1 - t)f(p).$$

F *does* map into $[L]$ because, by Theorem 2, $f(p)$ and $\varphi(p)$ lie in a common simplex that, being convex, also contains the line joining them. It is easily verified that F is continuous, and clearly $F(p, 0) = f(p)$ and $F(p, 1) = \varphi(p)$ for all $p \in [K]$.

F is stationary on $[K_1]$ because $\varphi|_{[K_1]}$ is a simplicial approximation to $f|_{[K_1]}$ (since $f(\mathrm{St}_{K_1}(v)) \subset f(\mathrm{St}_K(v)) \subset \mathrm{St}(\varphi(v))$ for each vertex $v \in K_1$), and hence $f = \varphi$ on $[K_1]$ by the following lemma. □

Lemma. *Suppose $f: K \to L$ is a simplicial map and that φ is a simplicial approximation to f. Then $\varphi = f$.*

PROOF. For each vertex $v \in K$,

$$f(v) \in f(\mathrm{St}(v)) \subset \mathrm{St}(\varphi(v)).$$

But, since f is a simplicial map, $f(v)$ is a vertex and, by Theorem 1, $f(v) = \varphi(v)$. Thus f and φ agree on vertices; hence they agree everywhere since both are simplicial maps. □

Theorem 4. *Let $f: [K] \to [L]$ be continuous and $\varphi: K^0 \to L^0$ be a vertex map. φ can be extended to a simplicial approximation to f if and only if $f(\mathrm{St}(v)) \subset \mathrm{St}(\varphi(v))$ for all $v \in K^0$.*

PROOF. The implication in one direction is obvious. For the other direction we need only verify that φ can be extended to a simplicial map $K \to L$; that is, we need only show that if $(s) = (v_0, v_1, \ldots, v_r)$ is a simplex in K, then $\varphi(v_0), \varphi(v_1), \ldots, \varphi(v_r)$ are vertices of a common simplex of L. But, in fact, $f((s)) \subset f(\mathrm{St}(v_j)) \subset \mathrm{St}(\varphi(v_j))$ for all $j \in \{0, 1, \ldots, r\}$ so $\bigcap_{j=0}^{r} \mathrm{St}(\varphi(v_j)) \neq \varnothing$. This implies there exists an open simplex $(t) \subset \mathrm{St}(\varphi(v_j))$ for all j. $\varphi(v_j)$ must then be a vertex of (t) for all j. □

Theorem 5. *Let $f: [K] \to [L]$ be continuous. Let $\{K_n\}$ be a sequence of subdivisions of K such that $\lim_{n \to \infty} \mathrm{mesh}\, K_n = 0$. Then, for n sufficiently large, there exists a simplicial map $\varphi: K_n \to L$ such that φ is a simplicial approximation to f.*

PROOF. By Theorem 1, $\{\mathrm{St}(w)\}_{w \in L^0}$ is an open covering of $[L]$. Since f is continuous, $\{f^{-1}(\mathrm{St}(w))\}_{w \in L^0}$ is an open covering of $[K]$. Since $[K]$ is a compact metric space, there exists a $\delta > 0$ such that any ball of radius δ lies in some open set of this covering. Choose n large enough so that $\mathrm{mesh}\, K_n < \delta/2$. Then $\mathrm{diam}[s] \leq \delta/2$ for each $s \in K_n$. Hence, for each vertex v in K_n, $\mathrm{St}(v) \subset B_v(\delta)$. But $B_v(\delta) \subset f^{-1}(\mathrm{St}(w))$ for some $w \in L^0$, so for each $v \in (K_n)^0$, $\mathrm{St}(v) \subset f^{-1}(\mathrm{St}(w))$ for some $w \in L^0$. For each $v \in (K_n)^0$, define $\varphi(v)$ to be any such vertex $w \in L^0$. (There are only finitely many such w; pick any one.) Then $\varphi: (K_n)^0 \to L^0$ has the property that $\mathrm{St}(v) \subset f^{-1}(\mathrm{St}(\varphi(v)))$; that is, $f(\mathrm{St}(v)) \subset \mathrm{St}(\varphi(v))$ for each $v \in (K_n)^0$. By Theorem 4, φ can be extended to a simplicial approximation to f. □

Corollary. *Let $f: [K] \to [L]$ be continuous. Then, for any $\varepsilon > 0$, there exist subdivisions K_n of K and L_m of L, and a simplicial approximation $\varphi: K_n \to L_m$ to f such that $d(f, \varphi) < \varepsilon$.*

PROOF. By Theorem 4 of Section 4.2, there exist subdivisions with arbitrarily small mesh. Given $\varepsilon > 0$, Let L_m be a subdivision of L such that mesh $L_m < \varepsilon$. Then $f: [K] \to [L_m]$. By Theorem 5, there exists a subdivision K_n of K and a simplicial approximation $\varphi: K_n \to L_m$ to f. By the corollary to Theorem 2,

$$d(f, \varphi) \leq \text{mesh } L_m < \varepsilon. \qquad \square$$

4.4 Fundamental group of a simplicial complex

Definition. Let K and L be simplicial complexes. Let φ_1 and $\varphi_2: K \to L$ be simplicial maps. φ_1 and φ_2 are *contiguous* if, for each simplex $(v_0, v_1, \ldots, v_k) \in K$, there exists a simplex $t \in L$ such that

$$\varphi_1(v_0), \varphi_1(v_1), \ldots, \varphi_1(v_k) \quad \text{and} \quad \varphi_2(v_0), \varphi_2(v_1), \ldots, \varphi_2(v_k)$$

are vertices of t.

EXAMPLE. Let K be the complex of a 3-dimensional simplex, with vertices $\{v_0, v_1, v_2, v_3\}$. Let L be a one dimensional complex with three vertices $\{w_0, w_1, w_2\}$ and two 1-simplices (w_0, w_1) and (w_1, w_2), as in Figure 4.14.

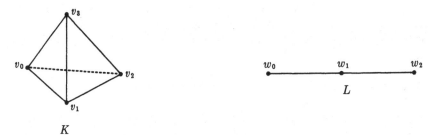

Figure 4.14

Define simplicial maps $\varphi_1, \varphi_2, \varphi_3: K \to L$ by prescribing them on vertices as follows:

$$\varphi_1(v_0) = \varphi_1(v_1) = w_0, \qquad \varphi_1(v_2) = \varphi_1(v_3) = w_1;$$
$$\varphi_2(v_0) = \varphi_2(v_1) = \varphi_2(v_2) = \varphi_2(v_3) = w_1;$$
$$\varphi_3(v_0) = w_2, \qquad \varphi_3(v_1) = \varphi_3(v_2) = \varphi_3(v_3) = w_1.$$

It is easily checked that φ_1 and φ_2 are contiguous and that φ_2 and φ_3 are contiguous. Note, however, that φ_1 and φ_3 are *not* contiguous. Hence the property of being contiguous is *not* an equivalence relation.

Definition. Two simplicial maps $\varphi, \psi: K \to L$ are *contiguous equivalent*, denoted $\overset{c}{\simeq}$, if there exists a finite sequence $\varphi_0, \varphi_1, \ldots, \varphi_k$ of simplicial maps $K \to L$ such that $\varphi_0 = \varphi$, $\varphi_k = \psi$, and φ_i is contiguous with φ_{i-1} for each $i \in \{1, 2, \ldots, k\}$.

Remark. It is easily checked that $\overset{c}{\simeq}$ is an equivalence relation.

Theorem 1. *Let K and L be simplicial complexes and let $f: [K] \to [L]$. Suppose $\varphi_1, \varphi_2 : K \to L$ are both simplicial approximations to f. Then φ_1 and φ_2 are contiguous.*

PROOF. Let $(s) = (v_0, \ldots, v_k)$ be a simplex of K. Then

$$f((s)) \subset f(\mathrm{St}\,(v_j)) \subset \mathrm{St}\,(\varphi_i(v_j))$$

for all $j \in \{0, \ldots, k\}$ and $i \in \{1, 2\}$. Thus

$$f((s)) \subset \bigcap_{j=0}^{k} \mathrm{St}\,(\varphi_1(v_j)) \cap \bigcap_{j=0}^{k} \mathrm{St}\,(\varphi_2(v_j)).$$

Let (t) be an open simplex in L such that $f((s)) \cap (t) \neq \varnothing$. Then

$$(t) \subset \bigcap_{j=0}^{k} \mathrm{St}\,(\varphi_1(v_j)) \cap \bigcap_{j=0}^{k} \mathrm{St}\,(\varphi_2(v_j)),$$

and hence $\varphi_1(v_0), \ldots, \varphi_1(v_k)$ and $\varphi_2(v_0), \ldots, \varphi_2(v_k)$ are vertices of (t). $\qquad\square$

Theorem 2. *Suppose φ_1 and $\varphi_2 : K \to L$ are contiguous simplicial maps. Then φ_1 and φ_2 are homotopic.*

PROOF. Note first that for each $p \in [K]$, $\varphi_1(p)$ and $\varphi_2(p)$ lie in a common simplex of L. For $p \in (s)$ with $(s) = (v_0, \ldots, v_k) \in K$. In barycentric coordinates, $p = \sum_{i=0}^{k} a_i v_i$. Since φ_1 and φ_2 are contiguous, $\varphi_1(v_0), \ldots, \varphi_1(v_k)$ and $\varphi_2(v_0), \ldots, \varphi_2(v_k)$ are vertices of some simplex $(t) \in L$. Then $\varphi_j(p) = \sum_{i=0}^{k} a_i \varphi_j(v_i) \in (t)$ for $j \in \{1, 2\}$.

Now define $F: [K] \times I \to [L]$ by

$$F(p, t) = (1 - t)\varphi_1(p) + t\varphi_2(p), \qquad (p \in [K]; t \in I).$$

F makes sense: since $\varphi_1(p)$ and $\varphi_2(p)$ lie in a common simplex for each $p \in K$, so does the line segment joining them. F is a homotopy from φ_1 to φ_2. $\qquad\square$

Corollary. *Contiguous equivalent simplicial maps are homotopic.*

Theorem 3. *Let K be a simplicial complex. Let α_0 and $\alpha_1 : I \to [K]$ be paths in $[K]$. Suppose $\alpha_0 \simeq \alpha_1$. Then there exists a subdivision I' of I and simplicial maps φ_0 and $\varphi_1 : I' \to K$ such that*

(1) *φ_j is a simplicial approximation to α_j ($j \in \{0, 1\}$) and*
(2) *$\varphi_0 \overset{c}{\simeq} \varphi_1$.*

Moreover, the subdivision I' can be chosen to be finer than any given simplicial subdivision of I.

Remark. More generally, if K and L are two simplicial complexes and

$$f_0, f_1 : [K] \to [L]$$

95

are homotopic maps, then there exists a subdivision K' of K and simplicial maps $\varphi_0, \varphi_1 \colon K' \to L$ satisfying Conditions (1) and (2) above. The proof of this is a generalization of the following proof.

PROOF OF THEOREM 3. Let $F \colon I \times I \to [K]$ be a homotopy from α_0 to α_1. Since $\{\mathrm{St}\,(w)\}_{w \in K^0}$ is an open covering of $[K]$, $\{F^{-1}(\mathrm{St}\,(w))\}_{w \in K^0}$ is an open covering of $I \times I$. Since $I \times I$ is a compact metric space, there exists a $\delta > 0$ such that each ball of radius δ is contained in $F^{-1}(\mathrm{St}\,(w))$ for some $w \in K^0$.

Choose a subdivision I' of I, with vertices $v_0 = 0, v_1, \ldots, v_s = 1$, and another subdivision I'' of I with vertices $l/2^k$, $(l = 0, \ldots, 2^k)$. Then $I' \times I''$ can be made into a simplicial complex M with vertices $v_r^l = (v_r, l/2^k)$ and 2-simplices of the form $(v_r^l, v_{r+1}^l, v_{r+1}^{l+1})$ or $(v_r^l, v_r^{l+1}, v_{r+1}^{l+1})$ (see Figure 4.15).

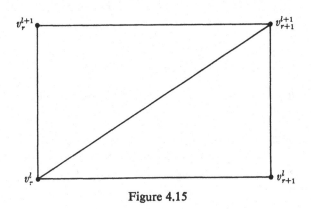

Figure 4.15

The subdivisions can be chosen fine enough so that

$$\mathrm{St}\,(v_r) \times [(l-1)/2^k, (l+1)/2^k]$$

is contained in a ball of radius δ and therefore contained in $F^{-1}(\mathrm{St}\,(w))$ for some $w \in K^0$. Since $\mathrm{St}\,(v_r') \subset \mathrm{St}\,(v_r) \times [(l-1)/2^k, (l+1)/2^k] \subset F^{-1}(\mathrm{St}\,(w))$, there exists, by Theorem 4 of Section 4.3, a simplicial map $\Phi \colon M \to K$, which is a simplicial approximation to F and for which

$$\mathrm{St}\,(v_r) \times [(l-1)/2^k, (l+1)/2^k] \subset F^{-1}(\mathrm{St}\,\Phi(v_r')).$$

Let $\varphi_i = \Phi|_{I \times \{i\}}$, $(i = 0, 1)$, so that φ_i is a simplicial approximation to $F|_{I \times \{i\}} = \alpha_i$.

We now show that $\varphi_0 \overset{c}{\simeq} \varphi_1$. Let $\psi_l = \Phi|_{I \times l/2^k}$ so that $\psi_0 = \varphi_0$ and $\psi_{2^k} = \varphi_1$. It suffices to show that ψ_l and ψ_{l+1} are contiguous for $l = 0, \ldots, 2^k - 1$; that is, for each simplex $(v_r, v_{r+1}) \in I'$, the vertices

$$\psi_l(v_r) = \Phi(v_r^l), \qquad \psi_l(v_{r+1}) = \Phi(v_{r+1}^l),$$

$$\psi_{l+1}(v_r) = \Phi(v_r^{l+1}), \quad \text{and} \quad \psi_{l+1}(v_{r+1}) = \Phi(v_{r+1}^{l+1})$$

lie on a simplex of K. But

$$\bigcap_{i,j=0}^{1} \text{St}\left(\Phi(v_{r+j}^{l+i})\right) \supset F\left(\text{St}(v_r) \times \left[\frac{l}{2^k}, \frac{l+1}{2^k}\right]\right) \cap F\left(\text{St}(v_{r+1}) \times \left[\frac{l}{2^k}, \frac{l+1}{2^k}\right]\right)$$

$$\supset F\left((v_r, v_{r+1}) \times \left[\frac{l}{2^k}, \frac{l+1}{2^k}\right]\right),$$

which is not empty and hence contains a simplex (t) of K. Hence the four vertices in question are vertices of (t). □

Definitions. Let K be a simplicial complex. An *edge* in K is an ordered pair $e = |v_1 v_2|$ of vertices of K, such that v_1 and v_2 lie in some simplex of K. v_1 is the *origin* of e, and v_2 is the *end* of e. If $e = |v_1 v_2|$, the edge $|v_2 v_1|$ is denoted by e^{-1}. A *route* in K is a finite sequence $\omega = e_1 e_2 \cdots e_k$ of edges in K such that, for each $i \in \{1, \ldots, k-1\}$, the end of e_i equals the origin of e_{i+1}. The *origin* of ω is the origin of e_1, and the *end* of ω is the end of e_k. Given two routes $\omega = e_1 \cdots e_k$ and $\tau = e_1' \cdots e_m'$ with end of ω equal to origin of τ, their *product* $\omega\tau$ is defined by

$$\omega\tau = e_1 \cdots e_k e_1' \cdots e_m'.$$

The *inverse* of a route $\omega = e_1 \cdots e_k$ is the route

$$\omega^{-1} = e_k^{-1} \cdots e_1^{-1}.$$

An equivalence relation on the set of all routes in K is defined as follows. If $e = |v_1 v_2|$ and $f = |v_2 v_3|$ are such that v_1, v_2, v_3 are vertices of a simplex, then the product ef is *edge equivalent* to the edge $|v_1 v_3|$. Two routes ω and τ are *edge equivalent*, denoted $\omega \stackrel{E}{\simeq} \tau$, if τ can be obtained from ω by a sequence of such elementary edge equivalences.

EXAMPLE. Suppose K is the complex in Figure 4.16. Then $|v_0 v_1||v_1 v_2| \stackrel{E}{\simeq} |v_0 v_3||v_3 v_2|$ because

$$|v_0 v_1||v_1 v_2| \stackrel{E}{\simeq} |v_0 v_2| \stackrel{E}{\simeq} |v_0 v_3||v_3 v_2|.$$

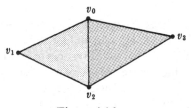

Figure 4.16

Remark. Edge equivalence is an equivalence relation. Moreover, if ω is a route with origin v, then $\omega\omega^{-1} \stackrel{E}{\simeq} |vv|$. Also, if $v_1, v_2, \ldots, v_k$ are vertices of a simplex, then $|v_1 v_2||v_2 v_3| \cdots |v_{k-1} v_k| \stackrel{E}{\simeq} |v_1 v_k|$.

Theorem 4. *Let K be a simplicial complex, and let v_0 be a vertex of K. Let $E(K, v_0)$ be the set of edge equivalence classes of routes in K with origin v_0 and end v_0. Then $E(K, v_0)$ is a group, with identity $|v_0 v_0|$, under the operations of multiplication and inverse defined above for routes. $E(K, v_0)$ is called the* edge path group *of (K, v_0).*

PROOF. Routine.   □

Remark. The edge path group of a complex K is a purely "combinatorial" object; that is, it depends on only the vertices of K and those subsets which are vertices of a simplex. Its definition does not use the topological properties of the space $[K]$.

Indeed, we can define an "abstract" simplicial complex as follows. Let V be a finite set, elements of which we shall call vertices. Let A be a collection of subsets of V, called (abstract) simplices, such that

(1) if $v \in V$, then $\{v\} \in A$;
(2) if $S \in A$, then each nonempty subset of S is also in A.

Such a collection A is called an *abstract simplicial complex.*

Note that every simplicial complex determines an abstract simplicial complex. Conversely, it can be shown that every abstract simplicial complex A has a "realization" as a simplicial complex; that is, there exists a simplicial complex whose abstract complex is A. (Note, however, that each abstract complex corresponds to many (nonisometric) simplicial complexes.)

The edge path group can then be defined for abstract complex A. It is the same as the edge path group of any realization of A. It is in this sense that we mean $E(K, v_0)$ is a purely combinatorial object. In contrast, the mesh of a complex is *not* a combinatorial concept.

Theorem 5. *Let K be a simplicial complex, and let v_0 be a vertex of K. Then $E(K, v_0)$ is isomorphic with $\pi_1([K], v_0)$.*

PROOF. We construct an isomorphism $h: E(K, v_0) \to \pi_1([K], v_0)$ as follows. Let ω be a route in K beginning and ending at v_0. Then $\omega = |v_0 v_1||v_1 v_2| \cdots |v_{k-1} v_k|$ for some set $\{v_1, v_2, \ldots, v_k\}$ of vertices in K, with $v_k = v_0$. Now regard the interval I as the space of a simplicial complex with vertices $\{0, 1/k, 2/k, \ldots, (k-1)/k, 1\}$. Consider the vertex map $\varphi_\omega: I^0 \to K^0$ defined by $\varphi_\omega(j/k) = v_j$, $(j \in \{0, 1, \ldots, k\})$. Since $|v_0 v_1| \cdots |v_{k-1} v_k|$ is a route, φ_ω extends to a simplicial map $\varphi_\omega: I \to K$. Set $h(\omega) = \langle \varphi_\omega \rangle$.

Note that if $\omega \overset{E}{\simeq} \tau$, then $\varphi_\omega \simeq \varphi_\tau$, so $h(\omega) = h(\tau)$. Thus h is well defined. h is a homomorphism because if $\omega = e_1 \cdots e_k$ and $\tau = e_1' \cdots e_m'$ are routes with origin and end v_0, then a homotopy between $\varphi_{\omega\tau}$ and $\varphi_\omega \varphi_\tau$ is obtained from Figure 4.17.

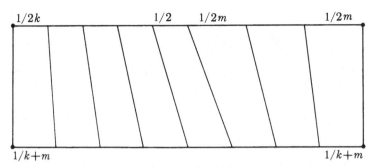

Figure 4.17

h is surjective. For suppose $\langle\alpha\rangle \in \pi_1([K], v_0)$. Then, by the simplicial approximation theorem, there exists a subdivision I' of I and a simplicial approximation $\varphi: I' \to K$ to α. Moreover $\varphi \overset{.}{\simeq} \alpha$, so $\langle\varphi\rangle = \langle\alpha\rangle$. Let $t_0 < t_1 < \cdots < t_k$ be the vertices of I'. Let ω be the route

$$|\varphi(t_0)\varphi(t_1)||\varphi(t_1)\varphi(t_2)|\cdots|\varphi(t_{k-1})\varphi(t_k)|$$

in K. Then $h(\omega) = \langle\varphi\rangle = \langle\alpha\rangle$.

h is injective. For this we must show that if ω is a route such that $\varphi_\omega \overset{.}{\simeq} e_{v_0}$, then $\omega \overset{E}{\simeq} |v_0 v_0|$. But, by Theorem 3, there exists a subdivision I' of I and simplicial maps $\varphi_0, \varphi_1: I' \to K$ such that φ_0 and φ_1 are simplicial approximations to φ_ω and e_{v_0} respectively, and such that $\varphi_0 \overset{C}{\simeq} \varphi_1$. (This subdivision I' can be chosen finer than the subdivision of I used to define φ_ω.) Now, since e_{v_0} is a simplicial map and φ_1 is a simplicial approximation to it, $\varphi_1 = e_{v_0}$. Then to show that

$$\omega \overset{E}{\simeq} |v_0 v_0|,$$

it suffices to prove the following.

(1) If φ and ψ are contiguous equivalent simplicial maps $I' \to K$, then $\omega_\varphi \overset{E}{\simeq} \omega_\psi$, where ω_φ and ω_ψ are the routes associated to φ and ψ.
(2) If $\psi: I \to K$ is a simplicial map and $\varphi: I' \to K$ is a simplicial approximation to ψ on a finer subdivision of I, then $\omega_\psi \overset{E}{\simeq} \omega_\varphi$.

For, by (2), $\omega = \omega_{\varphi_\omega} \overset{E}{\simeq} \omega_{\varphi_0}$, and, by (1),

$$\omega_{\varphi_0} \overset{E}{\simeq} \omega_{\varphi_1} = \omega_{e_{v_0}} = |v_0 v_0|.$$

PROOF OF (1). Since $\overset{E}{\simeq}$ is an equivalence relation, it suffices to prove that contiguous simplicial maps have this property. So suppose $\varphi, \psi: I' \to K$ are contiguous. Now

$$\omega_\varphi = |\varphi(t_0)\varphi(t_1)||\varphi(t_1)\varphi(t_2)|\cdots|\varphi(t_{k-1})\varphi(t_k)|,$$

and

$$\omega_\psi = |\psi(t_0)\psi(t_1)||\psi(t_1)\psi(t_2)|\cdots|\psi(t_{k-1})\psi(t_k)|.$$

Hence

$$\omega_\varphi \omega_\psi{}^{-1} = |\varphi(t_0)\varphi(t_1)| \cdots |\varphi(t_{k-1})\varphi(t_k)||\psi(t_k)\psi(t_{k-1})| \cdots |\psi(t_1)\psi(t_0)|.$$

But since φ and ψ are contiguous, $\varphi(t_{k-1})$, $\varphi(t_k)$, $\psi(t_{k-1})$, and $\psi(t_k)$ are vertices of a common simplex. Moreover $\varphi(t_k) = \psi(t_k) = v_0$. Hence

$$|\varphi(t_{k-1})\varphi(t_k)||\psi(t_k)\psi(t_{k-1})| \overset{\mathrm{E}}{\simeq} |\varphi(t_{k-1})\psi(t_{k-1})|$$

and

$$\omega_\varphi \omega_\psi{}^{-1} \overset{\mathrm{E}}{\simeq} |\varphi(t_0)\varphi(t_1)| \cdots |\varphi(t_{k-2})\varphi(t_{k-1})||\varphi(t_{k-1})\psi(t_{k-1})||\psi(t_{k-1})\psi(t_{k-2})|$$
$$\cdots |\psi(t_1)\psi(t_0)|.$$

Similarly, since φ and ψ are contiguous,

$$|\varphi(t_{k-2})\varphi(t_{k-1})||\varphi(t_{k-1})\psi(t_{k-1})| \overset{\mathrm{E}}{\simeq} |\varphi(t_{k-2})\psi(t_{k-1})|$$

and

$$|\varphi(t_{k-2})\psi(t_{k-1})||\psi(t_{k-1})\psi(t_{k-2})| \overset{\mathrm{E}}{\simeq} |\varphi(t_{k-2})\psi(t_{k-2})|,$$

so that

$$\omega_\varphi \omega_\psi{}^{-1} \overset{\mathrm{E}}{\simeq} |\varphi(t_0)\varphi(t_1)| \cdots |\varphi(t_{k-3})\varphi(t_{k-2})||\varphi(t_{k-2})\psi(t_{k-2})||\psi(t_{k-2})\psi(t_{k-3})|$$
$$\cdots |\psi(t_1)\psi(t_0)|.$$

Continuing in this way we find, by induction, that

$$\omega_\varphi \omega_\psi{}^{-1} \overset{\mathrm{E}}{\simeq} |\varphi(t_0)\psi(t_0)| = |v_0 v_0|.$$

PROOF OF (2). Since the restriction of ψ to the subcomplex of I' consisting of the vertices $\{t_0, t_1, \ldots, t_k\}$ is a simplicial map, $\varphi(t_i) = \psi(t_i)$ for $i \in \{0, 1, \ldots, k\}$. Moreover, since $\psi: I \to K$ is a simplicial map, $(\psi(t_i, t_{i+1}))$ is a simplex (s) in K of dimension 0 or 1.

Claim: For each vertex u of I' with $t_i < u < t_{i+1}$, $\varphi(u)$ is a vertex of (s). For, in fact, since φ is a simplicial approximation to ψ,

$$\psi(u) \in \psi(\mathrm{St}_{I'}(u)) \subset \mathrm{St}_K(\varphi(u)).$$

Since $\psi(u) \in (s)$, $(s) \cap \mathrm{St}\,\varphi(u) \neq \varnothing$, so $(s) \subset \mathrm{St}\,\varphi(u)$, and $\varphi(u)$ must be a vertex of (s); that is, $\varphi(u)$ is equal either to $\psi(t_i)$ or to $\psi(t_{i+1})$ as claimed.

Thus, if $u_0 = t_i < u_1 < \cdots < u_r = t_{i+1}$ are the vertices of I' between t_i and t_{i+1}, then $\{\varphi(u_0), \varphi(u_1), \ldots, \varphi(u_r)\}$ are vertices of a common simplex of K. Note that $\varphi(u_0)$ is a vertex of (s) because $\varphi(u_0) = \varphi(t_i) = \psi(t_i)$; similarly $\varphi(u_r)$ is a vertex of (s).

Now consider the parts of ω_ψ and ω_φ arising from the restrictions of ψ and φ to $[t_i, t_{i+1}]$. This part of ω_ψ is just $|\psi(t_i)\psi(t_{i+1})|$. The corresponding part of ω_φ is

$$|\varphi(u_0)\varphi(u_1)||\varphi(u_1)\varphi(u_2)| \cdots |\varphi(u_{r-1})\varphi(u_r)|.$$

Since $\{\varphi(u_0), \ldots, \varphi(u_r)\}$ are vertices of a common simplex of K, this is edge equivalent to $|\varphi(u_0)\varphi(u_r)| = |\varphi(t_i)\varphi(t_{i+1})| = |\psi(t_i)\psi(t_{i+1})|$. Thus these parts of ω_ψ and ω_φ are edge equivalent. Since this is true for each i, $\omega_\psi \overset{\mathrm{E}}{\simeq} \omega_\varphi$. □

Corollary. *Let K be a simplicial complex, let $v_0 \in K^0$, and let $i: K^2 \to K$ be the injection of the 2-skeleton of K into K. Then i induces an isomorphism*

$$i_*: E(K^2, v_0) \to E(K, v_0).$$

Consequently, the induced map

$$i_*: \pi_1([K^2], v_0) \to \pi_1([K], v_0)$$

is an isomorphism.

PROOF. The definition of edge equivalence depends only on K^2. □

Theorem 6. *The n-sphere S^n is simply connected for $n > 1$; that is, $\pi_1(S^n, p) = (e)$ for each $p \in S^n$.*

PROOF. First note that S^n is homeomorphic to the n-skeleton of an $(n + 1)$-simplex. In fact, if s is an $(n + 1)$-simplex in R^{n+1}, the following map $\varphi: [s^n] \to R^{n+1}$ maps $[s^n]$ homeomorphically onto $S^n \subset R^{n+1}$. Let $b = (b_1, \ldots, b_{n+1}) \in R^{n+1}$ be the barycenter of s. For $x = (x_1, \ldots, x_{n+1}) \in [s^n]$, define $\varphi(x) \in R^{n+1}$ by

$$\varphi(x) = \frac{1}{\left[\sum_{i=1}^{n+1} (x_i - b_i)^2\right]^{1/2}} (x_1 - b_1, x_2 - b_2, \ldots, x_{n+1} - b_{n+1})$$

Geometrically, $[s]$ may be regarded as inscribed in S^n, and φ is projection outward from the barycenter of $[s]$.

It suffices then to show that $\pi_1([s^n], v_0) = (e)$. By Theorem 5, every element of $\pi_1([s^n], v_0)$ has a representative α which is a route, and, in particular, its image lies in $[s^1]$. If $n > 1$, then there exists a point $p \in [s^n]$ with $p \notin [s^1]$. But $[s^n] - \{p\}$ is homeomorphic with R^n, which is contractible. Hence $\alpha \simeq e_{v_0}$. □

Definitions. A *graph* is a simplicial complex of dimension less than 2. A *tree* is an arcwise connected graph T such that, for each 1-simplex $s \in T$, $[T] - (s)$ is not connected (see Figure 4.18).

An *end* of a graph is a vertex which is the vertex of at most one 1-simplex.

Remark. Every tree has an end. For otherwise we could build up a route by starting at one vertex, moving to another vertex along a 1-simplex, moving to a third vertex along a different 1-simplex, etc. The route never touches a vertex twice because otherwise the 1-simplex which brings the route back to that vertex could be removed without disconnecting the tree. If the route never reaches an end, we will touch infinitely many vertices in this way (induction). But a complex has only finitely many vertices.

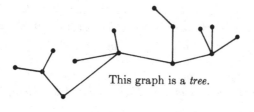

This graph is a *tree.*

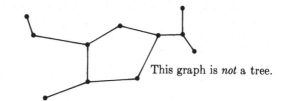

This graph is *not* a tree.

Figure 4.18

Theorem 7. *Every tree is contractible.*

PROOF. By induction on the number of vertices. If T has 1 vertex, the theorem is trivial. Assume the theorem for trees with n vertices. Let T have $n + 1$ vertices, and let v_0 be an end of T. Then there exists a unique 1-simplex $s \in T$ with vertex v_0.

Let $L = T - \{(s), v_0\}$. Then L is a simplicial complex, and $[L] = [T] - (s) \cup \{v_0\}$. L is a tree because if t is a 1-simplex in L such that $[L] - (t)$ is connected, then $[T] - (t)$ would also be connected. Now L has only n vertices, so L is contractible. Moreover, $[L]$ and $[T]$ are of the same homotopy type. (Let $f: [T] \to [L]$ map $(s) \cup \{v_0\}$ into the other vertex of s and map L onto itself. Let $g: [L] \to [T]$ be inclusion. Then $f \circ g \simeq i_{[L]}$ and $g \circ f \simeq i_{[T]}$.) Hence $[T]$ is contractible. $\qquad\square$

Corollary. *Let T be a tree and let v_0 be a vertex of T. Then*

$$\pi_1([T], v_0) = E(T, v_0) = (e).$$

Definition. Let K be a graph. Let α_0 be the number of vertices of K, and let α_1 be the number of 1-simplices. Let $\chi(K) = \alpha_0 - \alpha_1$. The integer $\chi(K)$ is called the *Euler characteristic* of K.

Remark. Note that the integer $\chi(K)$ is invariant under subdivisions, because inserting an extra vertex into K splits some 1-simplex into two 1-simplices, so that α_0 and α_1 both increase by one.

Theorem 8. *Let T be a tree. Then $\chi(T) = 1$.*

PROOF. By induction on the number n of vertices of T. For $n = 1$, the theorem is clear. Assume the theorem for trees with n vertices, and let T have $n + 1$

vertices. Let L be the tree obtained in the proof of Theorem 7. Then L has n vertices so $\chi(L) = 1$. But $\alpha_0(T) = \alpha_0(L) + 1$, and $\alpha_1(T) = \alpha_1(L) + 1$, so $\chi(T) = \chi(L) = 1$. $\qquad\square$

Theorem 9. *Let K be an arcwise connected graph. Let n be the maximum number of open 1-simplices which can be removed from K without disconnecting the space. (n is the number of "basic" circuits in K.) Then $n = 1 - \chi(K)$.*

PROOF. If K is a tree, then $n = 0$, and Theorem 8 applies. If K is not a tree, let (s_1) be an open 1-simplex such that $[K] - (s_1)$ is connected. If $K - (s_1)$ is a tree, stop. Otherwise, let (s_2) be an open 1-simplex such that $[K] - (s_1) \cup (s_2)$ is connected. Continue. Since there are only finitely many 1-simplices in K, the process must stop; that is, for some n, $K - \{(s_1), (s_2), \ldots, (s_n)\}$ is a tree T. Then

$$\chi(K) = \chi(T) - n = 1 - n;$$

that is,

$$n = 1 - \chi(K). \qquad\square$$

(Note that the above formula implies that although the particular 1-simplices which we delete are by no means unique, the number which must be deleted to obtain a tree is independent of the particular method of deletion employed.)

Remark. Recall the definition of the *free group F_n on n generators*. Consider an alphabet consisting of n letters $a_1, a_2, \ldots, a_n$. Consider the symbols

$$a_1{}^{-1}, a_2{}^{-1}, \ldots, a_n{}^{-1} \quad \text{and} \quad e.$$

Let S be the set of all "words" obtained by arranging these symbols in any order in a row of finite length—repetitions are allowed. The "product" $\alpha\beta$ of two words α and β is defined by juxtaposition: β is attached to the end of α. The "inverse" of a word is obtained by reversing the order of the arrangement and at the same time replacing a_j by $a_j{}^{-1}$, $a_j{}^{-1}$ by a_j, and e by e. An equivalence relation $\sim$ is defined on S as follows. We decree that $ee \sim e$ and that for each j,

$$a_j a_j{}^{-1} \sim e, \qquad a_j{}^{-1} a_j \sim e;$$
$$a_j e \sim a_j, \qquad a_j{}^{-1} e \sim a_j{}^{-1};$$
$$e a_j \sim a_j, \qquad e a_j{}^{-1} \sim a_j{}^{-1}.$$

Furthermore, any two words are $\sim$ equivalent if one may be obtained from the other through a sequence of such "elementary" equivalences. The set of equivalence classes forms a group with multiplication and inverse as above, and with identity the equivalence class of e. This group is F_n.

EXAMPLE. The free group on one generator is isomorphic with the integers.

Remark. For $n > 1$, F_n is not commutative. For, in fact, $a_1 a_2 a_1{}^{-1} a_2{}^{-1} \neq e$.

Remark. If F_n is a free group with generators $a_1, a_2, \ldots, a_n$, and G is any group, then any map $h: \{a_1, a_2, \ldots, a_n\} \to G$ can be extended to a homomorphism $\tilde{h}: F_n \to G$. The homomorphism $\tilde{h}$ is defined by

$$\tilde{h}(a_{j_1}^{\pm 1} a_{j_2}^{\pm 1} \cdots a_{j_r}^{\pm 1}) = h(a_{j_1})^{\pm 1} h(a_{j_2})^{\pm 1} \cdots h(a_{j_r})^{\pm 1}.$$

Moreover, this property characterizes the group F_n; that is, if H is a group generated by n elements such that any map from these generators into an arbitrary group extends to a group homomorphism, then H is isomorphic with F_n.

Theorem 10. *Let K be an arcwise connected graph, and let v_0 be a vertex of K. Then $\pi_1([K], v_0)$ is isomorphic with the free group on $n = 1 - \chi(K)$ generators.*

EXAMPLE. Let p_1 and p_2 be distinct points in R^2. Then for $p \in R^2 - \{p_1, p_2\}$, $\pi_1(R^2 - \{p_1, p_2\}, p)$ is the free group on two generators. For consider a graph K as in Figure 4.19. (Note that p_1 and p_2 are not part of the graph.) Then $R^2 - \{p_1, p_2\}$ and $[K]$ are of the same homotopy type. In fact, the map

Figure 4.19

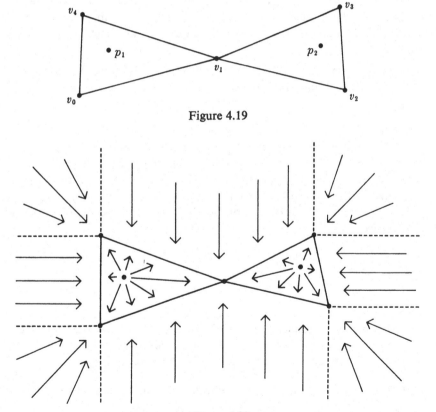

Figure 4.20

$R^2 - \{p_1, p_2\} \to [K]$ defined by projection, as in Figure 4.20, together with the inclusion map $[K] \to R^2 - \{p_1, p_2\}$ give a homotopy equivalence. Hence

$$\pi_1(R^2 - \{p_1, p_2\}, p) = \pi_1([K], v_0).$$

But $\chi(K) = 5 - 6 = -1$, so $n = 1 - (-1) = 2$ and, by Theorem 10, $\pi_1([K], v_0) \cong F_2$.

PROOF OF THEOREM 10. We shall construct homomorphisms

$$h \colon E(K, v_0) \to F_n,$$

$$h_1 \colon F_n \to E(K, v_0)$$

such that $h_1 \circ h$ and $h \circ h_1$ are identity maps. This will show that $E(K, v_0)$ is isomorphic with F_n. The theorem is then a consequence of Theorem 5.

   Construction of h. Let $(s_1), (s_2), \ldots, (s_n)$ $(n = 1 - \chi(K))$ be open 1-simplices of K such that $T = K - \{(s_1), (s_2), \ldots, (s_n)\}$ is a tree (cf. Theorem 9). Let F_n be the free group with generators $(s_1), (s_2), \ldots, (s_n)$. For each $j \in \{1, 2, \ldots, n\}$, let $s_j{}^+$ be the edge $|v_j v_j'|$ in K, where v_j and v_j' are the vertices of s_j. (We are here implicitly choosing some ordering for the vertices of each s_j.) Let $s_j{}^-$ be the edge $|v_j' v_j|$. Then each route ω in K is of the form

$$\omega = \rho_1 s_{j_1}{}^{\pm} \rho_2 s_{j_2}{}^{\pm} \cdots \rho_k s_{j_k}{}^{\pm} \rho_{k+1},$$

where each ρ_i is a route (possibly the trivial route $|v_{j_i} v_{j_i}|$) in the tree T. Now set

$$h(\omega) = (s_{j_1})^{\pm 1}(s_{j_2})^{\pm 1} \cdots (s_{j_k})^{\pm 1}.$$

We must check that h is well defined; that is, that $h(\omega)$ depends only on the edge equivalence class of ω. For this, it suffices to show that if ω_1 and ω_2 are routes in K which differ by an elementary edge equivalence, then $h(\omega_1) = h(\omega_2)$. So suppose

$$\omega_1 = \sigma |v_1 v_2||v_2 v_3| \tau,$$
$$\omega_2 = \sigma |v_1 v_3| \tau,$$

where σ and τ are routes in K, and v_1, v_2, v_3 are vertices of a common simplex in K. Since K is a graph, and hence has no 2-simplices, either $v_1 = v_2 = v_3$, $v_1 = v_2, v_2 = v_3$, or $v_1 = v_3$. In each of the first three cases, at least one of the simplices (v_1, v_2) and (v_2, v_3) is a 0-simplex, hence is not an (s_j), and the other is equal to (v_1, v_3). Thus in each of the first three cases, $h(\omega_1) = h(\omega_2)$. In the fourth case,

$$\omega_1 = \sigma |v_1 v_2||v_2 v_1| \tau$$
$$\omega_2 = \sigma |v_1 v_1| \tau.$$

If (v_1, v_2) is not an (s_j), then clearly $h(\omega_1) = h(\omega_2)$. If $(v_1, v_2) = (s_j)$ for some j, then

$$\omega_1 = \sigma s_j{}^{\pm} s_j{}^{\mp} \tau,$$

$$h(\omega_1) = h(\sigma)(s_j)^{\pm 1}(s_j)^{\mp 1} h(\tau) = h(\sigma) e h(\tau) = h(\sigma) h(\tau) = h(\omega_2).$$

Thus in all cases, $h(\omega_1) = h(\omega_2)$ as required, and h is well defined. Clearly, h is a homomorphism.

Construction of h_1. Since F_n is a free group, it suffices to define h_1 on the generators $(s_j) = (v_j, v_j')$. For this, let σ_j be a route in the tree T from v_0 to v_j, and τ_j be a route in T from v_0 to v_j'. Define $h_1((s_j))$ to be the edge equivalence class of the route $\sigma_j s_j {}^+ \tau_j{}^{-1}$. This definition is independent of σ_j because any other route in T from v_0 to v_j is edge equivalent to σ_j. (T is simply connected by Theorem 7.) (Proceed similarly for τ_j.) Now h_1 extends uniquely to a homomorphism $F_n \twoheadrightarrow E(K, v_0)$.

$h \circ h_1$ is identity because, for each generator (s_j) of F_n,

$$h \circ h_1((s_j)) = h(\sigma_j s_j{}^+ \tau_j) = (s_j).$$

$h_1 \circ h$ is identity because if

$$\omega = \rho_1 s_{j_1}{}^{\pm} \rho_2 s_{j_2}{}^{\pm} \cdots \rho_k s_{j_k}{}^{\pm} \rho_{k+1}$$

is a route in K, then

$$
\begin{aligned}
h_1 \circ h(\omega) &= h_1((s_{j_1})^{\pm 1}(s_{j_2})^{\pm 1} \cdots (s_{j_k})^{\pm 1}) \\
&= (\sigma_{j_1} s_{j_1}{}^+ \tau_{j_1}{}^{-1})^{\pm 1}(\sigma_{j_2} s_{j_2}{}^+ \tau_{j_2}{}^{-1})^{\pm 1} \cdots (\sigma_{j_k} s_{j_k}{}^+ \tau_{j_k}{}^{-1})^{\pm 1} \\
&= \eta_{j_1} s_{j_1}{}^{\pm} \eta_{j_1}' \eta_{j_2} s_{j_2}{}^{\pm} \eta_{j_2}' \cdots \eta_{j_k} s_{j_k}{}^{\pm} \eta_{j_k}',
\end{aligned}
$$

where

$$
\eta_{j_i} = \begin{cases} \sigma_{j_i} & (\text{if } s_{j_i} \text{ appears as } s_{j_i}{}^+) \\ \tau_{j_i} & (\text{if } s_{j_i} \text{ appears as } s_{j_i}{}^-) \end{cases}
$$

and

$$
\eta_{j_i}' = \begin{cases} \tau_{j_i}{}^{-1} & (\text{if } s_{j_i} \text{ appears as } s_{j_i}{}^+) \\ \sigma_{j_i}{}^{-1} & (\text{if } s_{j_i} \text{ appears as } s_{j_i}{}^-). \end{cases}
$$

But ρ_1 and η_{j_1} are both routes in T from v_0 to the origin of $s_{j_1}{}^{\pm}$, hence they are edge equivalent. (T is simply connected.) Similarly, ρ_2 and $\eta_{j_1}' \eta_{j_2}$ are both routes in T from the end of $s_{j_1}{}^{\pm}$ to the origin of $s_{j_2}{}^{\pm}$, hence they are edge equivalent. Continuing by induction, we conclude that $h_1 \circ h(\omega) \overset{\text{E}}{\simeq} \omega$; that is, $h_1 \circ h$ is identity. $\qquad\square$

Corollary. *Let K be a simplicial complex. Then the fundamental group $\pi_1([K], v_0)$ (v_0 a vertex of K) is in a "natural" way a quotient group of a free group.*

PROOF. We may assume that K (and hence K^1 also) is arcwise connected. Let $i: K^1 \hookrightarrow K$ be injection. Then

$$i_*: E(K^1, v_0) \to E(K, v_0)$$

is surjective from the definition of the edge path group. Hence, by Theorem 5,

$$i_*: \pi([K^1], v_0) \to \pi_1([K], v_0)$$

is surjective. Let $n = 1 - \chi(K^1)$. Then $\pi_1([K^1], v_0) = F_n$, the free group on n generators, so

$$i_*: F_n \to \pi_1([K], v_0)$$

is surjective. Let H be the kernel of i_*. Then

$$\pi_1([K], v_0) \simeq F_n/H. \qquad \square$$

Remark. Regarding $E(K^1, v_0)$ as F_n, the subgroup H is the subgroup generated by routes of the form $\rho_1|v_1v_2||v_2v_3||v_3v_1|\rho_2^{-1}$, where ρ_1 and ρ_2 are routes in the tree T from v_0 to v_1, and (v_1, v_2, v_3) is a 2-simplex in K.

Corollary. *Let* $D^2 = \{(x, y) \in R^2; x^2 + y^2 \leq 1\}$ *be the unit disc in* R^2. *There exists no continuous map* $f: D^2 \to S^1$ *such that* $f|_{S^1}$ *is the identity.*

PROOF. Suppose such an f exists. Let $g: S^1 \to D^2$ be inclusion. Then $f \circ g = i_{S^1}$, so that $(f \circ g)_*: \pi_1(S^1, 1) \to \pi_1(S^1, 1)$ is the identity map. But $(f \circ g)_* = f_* \circ g_*$, and $\pi_1(D^2, 1) = (e)$ since D^2 is homotopic to a point. Then

$$\mathrm{Im}\,(f \circ g)_* = \mathrm{Im}\,(f_* \circ g_*) \subset \mathrm{Im}\,f_* = (e).$$

Since $(f \circ g)_*$ is surjective, we conclude $\pi_1(S^1, e) = (e)$, which contradicts Theorem 10. $\qquad \square$

Corollary. *Special case of Brouwer fixed point theorem.) Let* D^2 *be the unit disc in* R^2. *Suppose* $f: D^2 \to D^2$ *is continuous. Then* f *has a fixed point; that is, there exists an* $x \in D^2$ *such that* $f(x) = x$.

PROOF. Suppose there exists no fixed point. Then, for each $x \in D^2$, $f(x) \neq x$, so that $x - f(x) \neq 0$ (vector addition in R^2). Let $g: D^2 \to S^1$ be defined as follows. For each $x \in D^2$, $g(x)$ is the projection of $f(x)$ onto S^1 along the vector $x - f(x)$ (see Figure 4.21). Then g is continuous, and $g|_{S^1}$ is the identity. This contradicts the previous corollary. $\qquad \square$

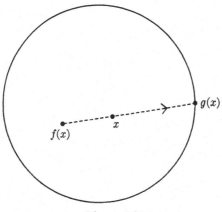

Figure 4.21

Remark. These two corollaries admit the following generalizations to higher dimensions. (1) There exists no continuous map from the n-disc D^n (closed ball in R^n) onto its boundary (an $(n-1)$-sphere S^{n-1}) whose restriction to S^{n-1} is the identity. (2) Every continuous map from the closed n-disc D^n into itself has a fixed point. However, S^{n-1} is simply connected for $n > 2$, so the above proof breaks down. The fundamental group in the proof must be replaced by another topological invariant, the $(n-1)$-th homology group. In the case $n = 1$, the analogues of these corollaries are consequences of the connectedness of $D^1 = I$.

Manifolds 5

5.1 Differentiable manifolds

Definition. A *locally Euclidean space* X of dimension n is a Hausdorff topological space such that, for each $x \in X$, there exists a homeomorphism φ_x mapping some open set containing x onto an open set in R^n.

Remark. We may, if we wish, choose each φ_x so that $\varphi_x(x) = 0$ and so that the image of φ_x is a ball $B_0(\varepsilon)$. Given any φ_x homeomorphically mapping an open set U about x onto an open set in R^n, let $\varepsilon > 0$ be such that $B_{\varphi(x)}(\varepsilon) \subset \varphi_x(U)$. Let

$$\psi \colon B_{\varphi(x)}(\varepsilon) \to B_0(\varepsilon)$$

be translation by $-\varphi(x)$. Then

$$\tilde{\varphi}_x = \psi \circ \varphi_x\big|_{\varphi_x^{-1}(B_{\varphi(x)}(\varepsilon))}$$

maps $\varphi_x^{-1}(B_{\varphi(x)}(\varepsilon))$ homeomorphically onto $B_0(\varepsilon)$.

EXAMPLE 1. R^n is locally Euclidean. For each $x \in R^n$, take φ_x to be the identity map.

EXAMPLE 2. S^n is locally Euclidean. Given $x \in S^n$, let $y \in S^n$, $y \neq x$. Then φ_x = stereographic projection from y maps $S^n - \{y\}$ homeomorphically onto R^n.

EXAMPLE 3. Projective space P^n; that is, the space of all lines through 0 in R^{n+1}, is locally Euclidean. For since P^n is covered by S^n, each $x \in P^n$ is contained in an open set homeomorphic to an open set in S^n that itself contains, about each of its points, an open set homeomorphic to an open set in R^n.

EXAMPLE 4. Each open subset U of a locally Euclidean space X is locally Euclidean. For if $x \in U$, let ψ_x be a homeomorphism mapping an open set about x in X onto an open set in R^n. Take $\varphi_x = \psi_x|_{U \cap \text{domain } \psi_x}$.

EXAMPLE 5. The set of all nonsingular $k \times k$ matrices forms a locally Euclidean space of dimension k^2. Each $k \times k$ matrix may be identified with a k^2-tuple by stringing out the rows in a line. The nonsingular matrices then form an open set of R^{k^2}, namely $\Delta^{-1}(R^1 - \{0\})$ where $\Delta: R^{k^2} \to R^1$ is the determinant function.

Definition. A C^k-*differentiable manifold* of dimension n is a pair (X, Φ) where X is a Hausdorff topological space, and Φ is a collection of maps such that the following conditions hold (see Figure 5.1).

(1) $\{\text{domain } \varphi\}_{\varphi \in \Phi}$ is an open covering of X,
(2) each $\varphi \in \Phi$ maps its domain homeomorphically onto an open set in R^n,
(3) for each $\varphi, \psi \in \Phi$ with $(\text{domain } \varphi) \cap (\text{domain } \psi) \neq \varnothing$, the map $\psi \circ \varphi^{-1}$ is a C^k-map from $\varphi(\text{domain } \varphi \cap \text{domain } \psi) \subset R^n$ into R^n,
(4) Φ is maximal relative to (2) and (3); that is, if ψ is any homeomorphism mapping an open set in X onto an open set in R^n such that, for each $\varphi \in \Phi$ with domain $\varphi \cap$ domain $\psi \neq \varnothing$, $\psi \circ \varphi^{-1}$ and $\varphi \circ \psi^{-1}$ are C^k-maps from

$$\varphi(\text{domain } \varphi \cap \text{domain } \psi) \quad \text{and} \quad \psi(\text{domain } \varphi \cap \text{domain } \psi)$$

into R^n—then $\psi \in \Phi$.

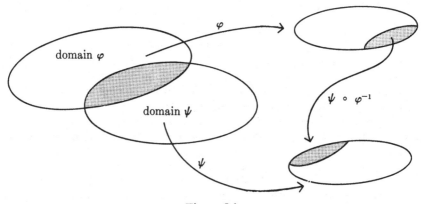

Figure 5.1

Here k may be $0, 1, 2, \ldots, \infty, \omega$. C^0 means continuous. C^k for k finite means all partial derivatives of order less than or equal to k exist and are continuous. C^∞ means all partial derivatives of all orders exist and are continuous. C^ω means real analytic; that is, the function may be expressed as a convergent Taylor series in a neighborhood of each point.

Note that a C^k-manifold is a locally Euclidean space and a locally Euclidean space gives rise to a C^0-manifold.

If $n = 2$ and, in Condition (3), "C^k" is replaced by "complex analytic" (where R^2 is identified with the complex numbers $\mathbb{C}$), (X, Φ) is called a *complex analytic manifold* of complex dimension 1 or a *Riemann surface*. Φ is then called a *complex structure* or *conformal structure* on X.

The maps $\varphi \in \Phi$ are called *coordinate systems*. More precisely, the map $\varphi \in \Phi$ is called a *coordinate system on the open set* (domain φ) $\subset X$. For $x \in X$, a *coordinate system about x* is a coordinate system $\varphi \in \Phi$ such that $x \in$ domain φ.

Remark. Each of the above Examples 1, 2, 3, and 5 of locally Euclidean spaces form the underlying space of a C^∞-manifold. You need only check that the maps φ_x satisfy Condition (3) for a manifold, and then take Φ to be a maximal set containing $\{\varphi_x\}_{x \in X}$. Example 4 above also carries over to manifolds. Namely, if (X, Φ) is a C^k-manifold and U is an open set in X, then $(U, \Phi|_U)$ is a C^k-manifold, where $\Phi|_U = \{\varphi|_U\}_{\varphi \in \Phi}$.

Definitions. Let (X, Φ) be a C^k-manifold. A real-valued function $f: X \to R^1$ is a *C^s-function* ($s \le k$), denoted $f \in C^s(X, R^1)$, if, for each $\varphi \in \Phi$, $f \circ \varphi^{-1}$ is a C^s-function mapping the image of $\varphi \subset R^n$ into R^1.

Let (X, Φ) be a C^k-manifold, and let $x \in X$. A real-valued function f is said to be *of class C^s* ($s \le k$) *in a neighborhood of x*, denoted $f \in C^s(X, x, R^1)$, if

$$U = (\text{domain } f)$$

is an open set in X containing x, and $f \in C^s(U, R^1)$, where U has the C^k-manifold structure as an open set in X.

Remarks. Note that we are able to define C^s-functions on X because (1) X looks locally (via the coordinate systems $\varphi \in \Phi$) like R^n, and we know what it means for a function on R^n to be C^s; and (2) if $U = $ domain φ and $V = $ domain ψ for $\varphi, \psi \in \Phi$, with $U \cap V \ne \varnothing$, the concept of a C^s-function in a neighborhood of x in $U \cap V$ is the same relative to the coordinate system φ as to the coordinate system ψ, because $\psi \circ \varphi^{-1}$ is a C^k-homeomorphism and $k \ge s$.

Note also that if f and g are C^s-functions in a neighborhood of x, then $f + g$ and fg (product) are C^s-functions in a neighborhood of x, where

$$\text{domain } (f + g) = \text{domain } (fg) = (\text{domain } f) \cap (\text{domain } g).$$

Definition. Let (X, Φ) be a C^k-manifold, and let $\varphi \in \Phi$ be a coordinate system on $U = $ domain φ. Let $r_j: R^n \to R^1$ be the jth coordinate function on R^n; that is, $r_j(a_1, a_2, \ldots, a_n) = a_j$ for $(a_1, \ldots, a_n) \in R^n$. The jth *coordinate function* of the coordinate system φ is the function $x_j: U \to R^1$ defined by $x_j = r_j \circ \varphi$.

111

Remark. $x_j: U \rightarrow R^1$ is a C^k-function. The n-tuple of functions $(x_1, \ldots, x_n)$ is sometimes also referred to as a coordinate system.

Definition. Let (X_1, Φ_1) and (X_2, Φ_2) be C^k-manifolds (not necessarily of the same dimension). A mapping $\Psi: X_1 \rightarrow X_2$ is of class C^s ($s \leq k$), denoted $\Psi \in C^s(X_1, X_2)$, if, whenever $f \in C^s(X_2, R^1)$, then $f \circ \Psi \in C^s(X_1, R^1)$.

Exercise 1. Show that, if $\Psi: X_1 \rightarrow X_2$ is of class C^s ($s \geq 0$), then Ψ is continuous.

Remarks. We shall confine our attention to C^∞-manifolds. This will include, in particular, C^ω-manifolds and complex analytic manifolds of dimension 1. We shall use the word "smooth" to denote C^∞.

We now proceed to define the concept of tangent vector on a manifold. Recall that, in Euclidean space, a vector at a point defines a map which sends each smooth function into a real number, namely, the directional derivative with respect to the given vector. Moreover, the vector is determined by its values on all smooth functions. We shall use this property to define tangent vectors on a manifold.

Definition. Let (X, Φ) be a smooth manifold and let $x \in X$. A *tangent vector* at x is a map $v: C^\infty(X, x, R^1) \rightarrow R^1$ such that, if φ is a (fixed) coordinate system with $x \in U = $ domain φ, then there exists an n-tuple $(a_1, a_2, \ldots, a_n)$ of real numbers with the following property. For each $f \in C^\infty(X, x, R^1)$,

$$v(f) = \sum_{i=1}^{n} a_i \frac{\partial}{\partial r_i} (f \circ \varphi^{-1})|_{\varphi(x)}.$$

(Note that if $W = $ domain f, then φ and f are both defined on the open set $U \cap W$ containing x, so that $f \circ \varphi^{-1}$ is a smooth function with domain $\varphi(U \cap W) \subset R^n$ containing $\varphi(x)$.)

Remark. If $v: C^\infty(X, x, R^1) \rightarrow R^1$ has the property required above of a tangent vector with respect to one coordinate system φ about x, then it also has this property with respect to any other coordinate system about x. For, if ψ is another such coordinate system, then, using the chain rule,

$$v(f) = \sum_{i=1}^{n} a_i \frac{\partial}{\partial r_i} (f \circ \varphi^{-1})|_{\varphi(x)}$$

$$= \sum_{i=1}^{n} a_i \frac{\partial}{\partial r_i} (f \circ \psi^{-1} \circ \psi \circ \varphi^{-1})|_{\varphi(x)}$$

$$= \sum_{i=1}^{n} a_i \sum_{j=1}^{n} \frac{\partial}{\partial r_j} (f \circ \psi^{-1})|_{\psi(x)} J_{ji}(\psi \circ \varphi^{-1})|_{\varphi(x)},$$

where $J_{ij}(\psi \circ \varphi^{-1})$ is the Jacobian matrix of the function $\psi \circ \varphi^{-1}$. Hence

$$v(f) = \sum_{j=1}^{n} \left(\sum_{i=1}^{n} a_i J_{ji}(\psi \circ \varphi^{-1})|_{\varphi(x)} \right) \frac{\partial}{\partial r_j} (f \circ \psi^{-1})|_{\psi(x)}.$$

Setting

$$b_j = \sum_{i=1}^{n} a_i J_{ji}(\psi \circ \varphi^{-1})|_{\varphi(x)},$$

we obtain

$$v(f) = \sum_{j=1}^{n} b_j \frac{\partial}{\partial r_j} (f \circ \psi^{-1})|_{\psi(x)}.$$

Thus, to check if v is a tangent vector at x, it suffices to check the required property in any one coordinate system at x.

Notation. Given a coordinate system φ about x, let $x_j = r_j \circ \varphi$ denote the jth coordinate function of φ. By $\partial/\partial x_j$ $(j = 1, \ldots, n)$ is meant the tangent vector at x defined by

$$\frac{\partial}{\partial x_j}(f) = \frac{\partial}{\partial r_j}(f \circ \varphi^{-1})|_{\varphi(x)}$$

for $f \in C^\infty(X, x, R^1)$. Thus $\partial/\partial x_j$ corresponds, relative to the coordinate system φ, to the n-tuple $(0, 0, \ldots, 1, \ldots, 0)$, where the 1 is in the jth spot.

Remark 1. If $x_1, \ldots, x_n$ are the coordinate functions of a coordinate system φ about x, and $y_1, \ldots, y_n$ are those of a coordinate system ψ about x, then the above computation shows that

$$\frac{\partial}{\partial x_j} = \sum_{i=1}^{n} \frac{\partial}{\partial x_j}(y_i)\frac{\partial}{\partial y_i}.$$

Remark 2. A tangent vector v at $x \in X$ has the following properties. For any $f, g \in C^\infty(X, x, R^1)$ and for $\lambda \in R^1$,

(1) $v(f + g) = v(f) + v(g)$
(2) $v(\lambda f) = \lambda v(f)$
(3) $v(fg) = v(f)g(x) + f(x)v(g)$.

These three properties say that the map $v: C^\infty(X, x, R^1) \to R^1$ is a *derivation*. Moreover, these properties characterize tangent vectors; that is, we could have defined a tangent vector to be a map $v: C^\infty(X, x, R^1) \to R^1$ satisfying (1)–(3) above, and then proved that, relative to any coordinate system φ about x, $v = \sum_{i=1}^{n} a_i(\partial/\partial x_i)$ for some n-tuple $(a_1, \ldots, a_n)$ of real numbers, where x_i is the ith coordinate function of φ.

Remark 3. The set X_x of tangent vectors at x forms a vector space under the following rules of addition and scalar multiplication:

$$(v_1 + v_2)(f) = v_1(f) + v_2(f) \qquad (v_1, v_2 \in X_x),$$
$$(\lambda v_1)(f) = \lambda(v_1(f)) \qquad (v_1 \in X_x, \lambda \in R^1).$$

To see that $v_1 + v_2$ and λv_1 are tangent vectors at x, let φ be a coordinate system about x, with coordinate functions $(x_1, \ldots, x_n)$. Then

$$v_1 = \sum_{i=1}^{n} a_i(\partial/\partial x_i) \quad \text{and} \quad v_2 = \sum_{i=1}^{n} b_i(\partial/\partial x_i)$$

for some $(a_1, \ldots, a_n)$ and $(b_1, \ldots, b_n)$. It is then easy to check that

$$v_1 + v_2 = \sum_{i=1}^{n} (a_i + b_i) \frac{\partial}{\partial x_i},$$

$$\lambda v_1 = \sum_{i=1}^{n} (\lambda a_i) \frac{\partial}{\partial x_i}.$$

The map $(a_1, \ldots, a_n) \to \sum_{i=1}^{n} a_i(\partial/\partial x_i)$ gives a vector space isomorphism $R^n \to X_x$, so X_x has dimension n. Moreover, it is clear that $\{\partial/\partial x_i\}_{i \in \{1, \ldots, n\}}$ is a basis for X_x. The space X_x is called the *tangent space* to X at x. It is also denoted by $T(X)_x$ or by $T(X, x)$.

For φ and ψ two coordinate systems at x, with coordinate functions $(x_1, \ldots, x_n)$ and $(y_1, \ldots, y_n)$ respectively, the formula

$$\frac{\partial}{\partial x_j} = \sum_{i=1}^{n} \frac{\partial}{\partial x_j} (y_i) \frac{\partial}{\partial y_i}$$

merely expresses the vector $\partial/\partial x_j$ in terms of the basis $\{\partial/\partial y_i\}_{i \in \{1, \ldots, n\}}$. Thus the change of basis matrix from the basis $\{\partial/\partial y_i\}$ of X_x to the basis $\{\partial/\partial x_i\}$ is precisely the Jacobian matrix $((\partial/\partial x_j)(y_i))$.

Remark 4. The tangent space $T(R^n, a)$ to R^n at a point $a \in R^n$ is naturally isomorphic with R^n itself. The isomorphism $R^n \to T(R^n, a)$ is given by

$$(\lambda_1, \ldots, \lambda_n) \to \sum_{i=1}^{n} \lambda_i \frac{\partial}{\partial r_i}.$$

Notation. We shall henceforth omit the Φ from our notation for a differentiable manifold (X, Φ). To be sure, a locally Euclidean space X may have two or more distinct differentiable structures on it (or it may have none), but we shall denote a manifold (X, Φ) merely by X and shall assume that a definite differentiable structure is given on it.

Definition. Let X and Y be smooth manifolds. Let $\Psi: X \to Y$ be a smooth map. The *differential* of Ψ at $x \in X$ is the map $d\Psi: X_x \to Y_{\Psi(x)}$ defined as follows. For $v \in X_x$ and $g \in C^\infty(Y, \Psi(x), R^1)$, $(d\Psi(v))(g) = v(g \circ \Psi)$.

Remark. It is easily checked that $d\Psi(v)$ is indeed a tangent vector at $\Psi(x)$. For, if φ is a coordinate system about x with coordinate functions $(x_1, \ldots, x_n)$, and τ is a coordinate system about $\Psi(x)$ with coordinate functions

$(y_1, \ldots, y_m)$, then $v = \sum_{i=1}^{n} a_i(\partial/\partial x_i)$ for some real numbers a_i; and if $g \in C^{\infty}(Y, \Psi(x), R^1)$, then

$$[d\Psi(v)](g) = v(g \circ \Psi) = \sum_{i=1}^{n} a_i \frac{\partial}{\partial x_i}(g \circ \Psi)$$

$$= \sum_{i=1}^{n} a_i \frac{\partial}{\partial r_i}(g \circ \tau^{-1} \circ \tau \circ \Psi \circ \varphi^{-1})|_{\varphi(x)}$$

$$= \sum_{i=1}^{n} a_i \sum_{j=1}^{m} \frac{\partial}{\partial s_j}(g \circ \tau^{-1})|_{\tau \circ \Psi(x)} \frac{\partial}{\partial r_i}(s_j \circ \tau \circ \Psi \circ \varphi^{-1})|_{\varphi(x)}$$

$$[(s_1, \ldots, s_m) \text{ coordinates on } R^m]$$

$$= \sum_{i=1}^{n} \sum_{j=1}^{m} a_i \frac{\partial}{\partial y_j}(g) \frac{\partial}{\partial x_i}(y_j \circ \Psi)$$

$$= \left[\sum_{j=1}^{m} v(y_j \circ \Psi) \frac{\partial}{\partial y_j}\right](g).$$

Since this holds for all $g \in C^{\infty}(Y, \Psi(x), R^1)$,

$$\boxed{d\Psi(v) = \sum_{j=1}^{m} v(y_j \circ \Psi) \frac{\partial}{\partial y_j},}$$

and, in particular, $d\Psi(v)$ is a tangent vector. Furthermore, it is clear that $d\Psi$ is a linear transformation $X_x \to Y_{\Psi(x)}$. Since

$$d\Psi\left(\frac{\partial}{\partial x_i}\right) = \sum_{j=1}^{m} \frac{\partial}{\partial x_i}(y_j \circ \Psi) \frac{\partial}{\partial y_j},$$

this linear transformation $d\Psi$ has matrix

$$(d\Psi)_{ij} = \left(\frac{\partial}{\partial x_j}(y_i \circ \Psi)\right)$$

relative to the bases $\{\partial/\partial x_i\}_{i \in \{1,\ldots,n\}}$ and $\{\partial/\partial y_j\}_{j \in \{1,\ldots,m\}}$.

Remark. Let X, Y, and Z be smooth manifolds. Let $\Psi: X \to Y$ and $\Phi: Y \to Z$ be smooth maps. Then $d(\Phi \circ \Psi) = d\Phi \circ d\Psi$.

PROOF. Suppose $v \in X_x$ and $h \in C^{\infty}(Z, \Phi \circ \Psi(x), R^1)$. Then

$$[d(\Phi \circ \Psi)(v)](h) = v(h \circ (\Phi \circ \Psi)) = v((h \circ \Phi) \circ \Psi)$$
$$= d\Psi(v)(h \circ \Phi)$$
$$= [d\Phi(d\Psi(v))](h)$$
$$= [(d\Phi \circ d\Psi)(v)](h). \qquad \square$$

Remark. Let X be a smooth manifold, and let U be open in X. Then U is itself a smooth manifold. Moreover, the inclusion map $i: U \to X$ is a smooth map. Indeed, $f \in C^\infty(X, R^1)$ implies $f|_U \in C^\infty(U, R^1)$. Furthermore, the differential

$$di: T(U, u_0) \to T(X, u_0) \qquad (u_0 \in U)$$

is an isomorphism; we shall identify these two linear spaces.

Exercise 2. If $u_0 \in U$ an open set in X, construct a function $h \in C^\infty(X, R^1)$ such that

$$h(x) = \begin{cases} 1 & (x \in W \text{ an open set containing } u_0), \\ 0 & (x \notin U). \end{cases}$$

[*Hint*: Make use of the smooth function $g: R^1 \to R^1$ defined by

$$g(t) = \begin{cases} e^{-1/t^2} & (t > 0) \\ 0 & (t \leq 0).] \end{cases}$$

If $f_1 \in C^\infty(U, u_0, R^1)$, use Exercise 1 to show that there exists a smaller open set W and $f \in C^\infty(X, R^1)$ such thas $f|_W = f_1|_W$.

Remark. Let X be a smooth manifold, and let $f \in C^\infty(X, R^1)$. Let us compute df. For $v \in T(X, x)$, $df(v) \in T(R^1, f(x))$. Since $T(R^1, f(x))$ is 1-dimensional, $df(v) = \lambda(d/dr)$ for some $\lambda \in R^1$. To determine λ, it suffices to evaluate $df(v)$ on the coordinate function $r: R^1 \to R^1$ as follows.

$$\lambda = \left[\lambda \frac{d}{dr} \right](r) = [df(v)](r) = v(r \circ f) = v(f).$$

Thus $df(v) = v(f) \, (d/dr)$. Now $T(R^1, f(x))$ is naturally isomorphic with R^1 via the isomorphism $\lambda(d/dr) \to \lambda$. Let us identify these two spaces through this isomorphism. Then $df: T(X, x) \to R^1$ is a linear functional on $T(X, x)$; that is, df is a member of the dual space $T^*(X, x)$ and is, as such, given by

$$df(v) = v(f) \qquad (v \in T(X, x)).$$

$T^*(X, x)$ is called the *cotangent space* at x.

Definition. Let X be a smooth manifold. A *smooth curve* in X is a smooth map α from some (open or closed) interval $\subset R^1$ into X. If the domain of α is a closed interval $[a, b]$, smoothness of α means that α admits a smooth extension

$$\tilde{\alpha}: (a - \varepsilon, b + \varepsilon) \to X.$$

(Note that open intervals are open sets in R^1 and hence are smooth manifolds.)

A *broken C^∞-curve* in X is a continuous map $\alpha: [a, b] \to X$ together with a subdivision of $[a, b]$ on whose closed subintervals α is a C^∞ curve.

EXAMPLE.

$$\alpha(t) = \begin{cases} (t, t \sin 1/t) & (t \in (0, 1]) \\ (0, 0) & (t = 0) \end{cases}$$

is *not* a smooth curve in R^2 because it admits no smooth extension past 0.

Definition. Let $\alpha: I \to X$ (I an interval $\subset R^1$) be a smooth curve in X. The *tangent vector* to α at time t ($t \in I$), denoted by $\dot{\alpha}(t)$, is defined by

$$\dot{\alpha}(t) = d\tilde{\alpha}\left(\left(\frac{d}{dr}\right)_t\right).$$

Note that $\dot{\alpha}(t)$ is well defined, even at the endpoints of I.

Remark. Given a tangent vector $v \in X_x$, let $\alpha: I \to X$ be a smooth curve whose tangent vector at time $t = 0$ is v. (Such a curve may be obtained by taking a coordinate system φ about x, finding a curve (for example, the straight line) in R^n whose tangent vector at time 0 is $d\varphi(v)$, and pulling this curve back to X by φ^{-1}.) Then, for $f \in C^\infty(X, x, R^1)$,

$$v(f) = \dot{\alpha}(0)(f) = d\tilde{\alpha}\left(\left(\frac{d}{dr}\right)_0\right)(f) = \frac{d}{dr}(f \circ \tilde{\alpha})|_0.$$

Thus $v(f)$ is the derivative of the "restriction" of f to the curve α. Moreover, two curves α_1 and α_2 have the same tangent vector v at time 0 if and only if $\alpha_1(0) = \alpha_2(0)$ and

$$\frac{d}{dr}(f \circ \tilde{\alpha}_1)|_0 = \frac{d}{dr}(f \circ \tilde{\alpha}_2)|_0$$

for all $f \in C^\infty(X, x, R^1)$ (see Figure 5.2). We may use this equation to define an equivalence relation on the set of all curves α with $\alpha(0) = x$. Then we get a one-to-one correspondence between equivalence classes of curves through x and tangent vectors at x. Thus, we could have defined a tangent vector at x to be such an equivalence class of curves through x.

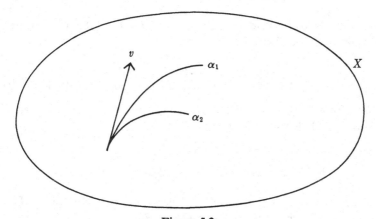

Figure 5.2

5.2 Differential forms

Definitions. Let X be a smooth manifold. Define (see Figure 5.3)

$$T(X) = \bigcup_{x \in X} T(X, x) \quad \text{and} \quad T^*(X) = \bigcup_{x \in X} T^*(X, x).$$

$T(X)$ is called the *tangent bundle* of X. $T^*(X)$ is called the *cotangent bundle* of X.

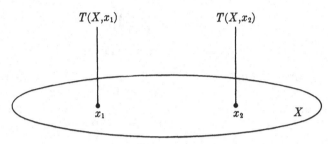

Figure 5.3

A *projection map* $\pi: T(X) \to X$ is defined as follows. If $v \in T(X)$, then $v \in T(X, x)$ for some (unique) $x \in X$; set $\pi(v) = x$. Similarly, there is a projection map from $T^*(X)$ onto X that we shall also denote by π.

A *vector field* on X is a map $V: X \to T(X)$ such that $\pi \circ V = i_X$. A vector field V is *smooth* if for each $f \in C^\infty(X, R^1)$, $Vf \in C^\infty(X, R^1)$. Here Vf is defined by

$$(Vf)(x) = V(x)f.$$

A *differential 1-form* on X is a map $\omega: X \to T^*(X)$ such that $\pi \circ \omega = i_X$. A differential 1-form ω is *smooth* if for each smooth vector field V on X,

$$\omega(V) \in C^\infty(X, R^1).$$

Here $\omega(V)$ is defined by $(\omega(V))(x) = \omega(x)(V(x))$. We shall denote the set of all smooth vector fields on X by $C^\infty(X, T(X))$ and the set of all smooth 1-forms by $C^\infty(X, T^*(X))$.

Exercise. Define a manifold structure on $T(X)$ so that π is a smooth map and so that a vector field V is smooth if and only if it is a smooth map from $X \to T(X)$.

[*Hint*: For $\varphi: U \to R^n$ a local coordinate system on X, with coordinate functions $(x_1, \ldots, x_n)$, define $\tilde{\varphi}: \pi^{-1}(U) \to R^{2n}$ by

$$\tilde{\varphi}(v) = (\varphi \circ \pi(v), b_1, \ldots, b_n),$$

where $b_1, \ldots, b_n \in R^1$ are such that $v = \sum_{i=1}^n b_i \, \partial/\partial x_i$.]

Remark 1. Let $f \in C^\infty(X, R^1)$. Then $df \in C^\infty(X, T^*(X))$. For if $V \in C^\infty(X, T(X))$, then $df(V) = Vf \in C^\infty(X, R^1)$.

Remark 2. $C^\infty(X, T(X))$ and $C^\infty(X, T^*(X))$ are both vector spaces over the reals under the operations of pointwise addition and scalar multiplication. For example, if V_1 and $V_2 \in C^\infty(X, T(X))$, then $V_1 + V_2$ is defined by $(V_1 + V_2)(x) = V_1(x) + V_2(x)$; and if $\lambda \in R^1$, then λV_1 is defined by $(\lambda V_1)(x) = \lambda(V_1(x))$.

Remark 3. Let φ be a coordinate system on X with domain U and coordinate functions $(x_1, x_2, \ldots, x_n)$. Then the following hold.

(1) $(\partial/\partial x_i) \in C^\infty(U, T(U))$ for $i \in \{1, \ldots, n\}$. $\partial/\partial x_i$ is smooth because if

$$f \in C^\infty(U, R^1), \quad \text{then} \quad f \circ \varphi^{-1} \in C^\infty(\varphi(U), R^1),$$

and, for each $x \in U$,

$$\left[\frac{\partial}{\partial x_i}(f)\right](x) = \left[\frac{\partial}{\partial r_i}(f \circ \varphi^{-1})\right](\varphi(x))$$

$$= \left[\left[\frac{\partial}{\partial r_i}(f \circ \varphi^{-1})\right] \circ \varphi\right](x);$$

that is,

$$\frac{\partial}{\partial x_i}(f) = \left[\frac{\partial}{\partial r_i}(f \circ \varphi^{-1})\right] \circ \varphi \in C^\infty(U, R^1).$$

(2) If $V \in C^\infty(U, T(U))$, then there exist functions $a_i \in C^\infty(U, R^1)$ for $i \in \{1, \ldots, n\}$, such that $V = \sum_{i=1}^n a_i(\partial/\partial x_i)$. These functions a_i exist because

$$\{(\partial/\partial x_i)(x)\}_{i \in \{1,\ldots,n\}}$$

is a basis for $T(X, x)$. They are smooth because $(\partial/\partial x_i)(x_j) = \delta_{ij}$, so that

$$a_j = \sum_{i=1}^n a_i \delta_{ij} = \sum_{i=1}^n a_i \frac{\partial}{\partial x_i}(x_j) = V(x_j) \in C^\infty(U, R^1).$$

(3) If $V \in C^\infty(X, T(X))$, then $V|_U \in C^\infty(U, T(U))$ by a previous exercise, and $V|_U = \sum_{i=1}^n a_i(\partial/\partial x_i)$ as in (2) with $a_i \in C^\infty(U, R^1)$.

(4) $dx_j \in C^\infty(U, T^*(U))$ for $j \in \{1, \ldots, n\}$ because $x_j \in C^\infty(U, R^1)$. Furthermore, $\{dx_j\}$ is at each point the dual basis to $\{\partial/\partial x_j\}$ because

$$dx_j\left(\frac{\partial}{\partial x_i}\right) = \frac{\partial}{\partial x_i}(x_j) = \delta_{ij}.$$

(5) If $\omega \in C^\infty(U, T^*(U))$, then there exist $a_i \in C^\infty(U, R^1)$ such that $\omega = \sum_{i=1}^n a_i \, dx_i$. These functions a_i exist because $\{dx_i\}$ is at each point a basis for the cotangent space. They are smooth because

$$a_i = \sum_{j=1}^n a_j \, dx_j\left(\frac{\partial}{\partial x_i}\right) = \omega\left(\frac{\partial}{\partial x_i}\right) \in C^\infty(U, R^1).$$

(6) If $f \in C^\infty(U, R^1)$, then

$$df = \sum_{j=1}^n \frac{\partial}{\partial x_j}(f) \, dx_j$$

because $df = \sum_{j=1}^{n} a_j \, dx_j$ for some a_j, and

$$a_i = \sum_{j=1}^{n} a_j \, dx_j\left(\frac{\partial}{\partial x_i}\right) = df\left(\frac{\partial}{\partial x_i}\right) = \frac{\partial}{\partial x_i}(f).$$

We have just seen that if $f \in C^{\infty}(X, R^1)$, then df is a smooth differential 1-form. We now introduce differential k-forms.

Review of exterior algebra

Let V be an n-dimensional vector space over the reals. Then the following hold.

(1) The vector space $\Lambda^k(V^*)$ is the space of all skew-symmetric k-linear functions on V; that is, each $\tau \in \Lambda^k(V^*)$ is a map $\tau\colon \underbrace{V \oplus \cdots \oplus V}_{k\text{-times}} \to R^1$ such that for all $v_1, \ldots, v_k, v_j' \in V$, $\lambda \in R^1$,

(1) $\tau(v_1, \ldots, v_{j-1}, v_j + v_j', v_{j+1}, \ldots, v_k)$
$\quad = \tau(v_1, \ldots, v_{j-1}, v_j, v_{j+1}, \ldots, v_k) + \tau(v_1, \ldots, v_{j-1}, v_j', v_{j+1}, \ldots, v_k);$

(2) $\tau(v_{\pi(1)}, \ldots, v_{\pi(k)}) = (-1)^\pi \tau(v_1, \ldots, v_k);$ and

(3) $\tau(v_1, \ldots, v_{j-1}, \lambda v_j, v_{j+1}, \ldots, v_k) = \lambda \tau(v_1, \ldots, v_j, \ldots, v_k),$

where π is any element of the permutation group S_k on k letters, and $(-1)^\pi$ is $+1$ if π is an even permutation or -1 if π is an odd permutation. This second condition is equivalent to requiring that if two vectors in the argument of τ are interchanged, then the value of τ on these vectors changes sign. The dimension of $\Lambda^k(V^*)$ is equal to the binomial coefficient $\binom{n}{k}$ for $k \leq n$; it is zero for $k > n$.

(2) If we set $\mathscr{G}(V^*) = \sum_{k=0}^{n} \oplus \Lambda^k(V^*)$, where $\Lambda^0(V^*) = R^1$, a product is defined on $\mathscr{G}(V^*)$ as follows. If $\tau \in \Lambda^k(V^*)$ and $\mu \in \Lambda^l(V^*)$, their product $\tau \wedge \mu$ is the element of $\Lambda^{k+l}(V^*)$ defined by

$$\tau \wedge \mu(v_1, \ldots, v_{k+l})$$
$$= \frac{1}{(k+l)!} \sum_{\pi \in S_{k+l}} (-1)^\pi \tau(v_{\pi(1)}, \ldots, v_{\pi(k)})\mu(v_{\pi(k+1)}, \ldots, v_{\pi(k+l)}).$$

Since $\mathscr{G}(V^*)$ is generated by such μ and τ, this multiplication extends to $\mathscr{G}(V^*)$ by linearity, that is, by requiring that exterior multiplication $\wedge$ be distributive with respect to vector addition. This multiplication is associative and $\mathscr{G}(V^*)$ is an algebra, with unit 1. However, multiplication is not commutative: if $\mu \in \Lambda^k(V^*)$ and $\tau \in \Lambda^l(V^*)$, then

$$\mu \wedge \tau = (-1)^{kl} \tau \wedge \mu.$$

(3) If $\varphi_1, \ldots, \varphi_n$ is a basis for V^*, then

$$[\varphi_{i_1} \wedge \cdots \wedge \varphi_{i_k}; 1 \leq i_1 < i_2 < \cdots < i_k \leq n]$$

is a basis for $\Lambda^k(V^*)$. Hence the union of these sets over $k \in \{1, \ldots, n\}$, together with $1 \in \Lambda^0(V^*)$, is a basis for $\mathscr{G}(V^*)$. It follows that the dimension of $\mathscr{G}(V^*)$ is 2^n.

If $v_1, \ldots, v_k \in V$, the value of $\varphi_{i_1} \wedge \cdots \wedge \varphi_{i_k}$ on these vectors is given by

$$(\varphi_{i_1} \wedge \cdots \wedge \varphi_{i_k})(v_1, \ldots, v_k) = \frac{1}{k!} \sum_{\pi \in S_k} (-1)^\pi \varphi_{i_1}(v_{\pi(1)}) \cdots \varphi_{i_k}(v_{\pi(k)}).$$

(4) $\mathscr{G}(V^*)$ has the following properties:

(1) $1 \in \mathscr{G}(V^*)$, $V^* \subset \mathscr{G}(V^*)$;
(2) $\mathscr{G}(V^*)$ is generated by 1 and V^*;
(3) $\varphi \wedge \varphi = 0$ whenever $\varphi \in V^*$; and
(4) dimension $\mathscr{G}(V^*) = 2^n$.

These properties in fact characterize $\mathscr{G}(V^*)$; that is, if $\tilde{\mathscr{G}}(V^*)$ is any algebra over the reals satisfying properties (1)–(4), then $\tilde{\mathscr{G}}(V^*)$ and $\mathscr{G}(V^*)$ are isomorphic (as algebras).

(Note that Condition (3) is equivalent to the condition that $\varphi_1 \wedge \varphi_2 = -\varphi_2 \wedge \varphi_1$ for all $\varphi_1, \varphi_2 \in V^*$.)

(5) If $L: V^* \to V^*$ is a linear transformation, then L induces a unique algebra homomorphism $\tilde{L}: \mathscr{G}(V^*) \to \mathscr{G}(V^*)$ which extends the map L. $\tilde{L}$ preserves degrees; that is, $\tilde{L}: \Lambda^k(V^*) \to \Lambda^k(V^*)$. In particular, $\tilde{L}: \Lambda^n(V^*) \to \Lambda^n(V^*)$. Hence, since dim $\Lambda^n(V^*) = 1$, there exists a scalar λ such that $\tilde{L}|_{\Lambda^n(V^*)} = \lambda i_{\Lambda^n(V^*)}$. This scalar λ is precisely $\Delta(L)$, the determinant of L.

(6) The algebra $\mathscr{G}(V^*)$ is called the *Grassmann algebra*, or *exterior algebra*, of V^*. Elements of $\mathscr{G}(V^*)$ are called *forms* on V. Forms in $\Lambda^k(V^*)$ are said to be of *degree k*.

Now let X be a smooth manifold. Let

$$\Lambda^k(X) = \bigcup_{x \in X} \Lambda^k(T^*(X, x)),$$

and let

$$\mathscr{G}(X) = \bigcup_{x \in X} \mathscr{G}(T^*(X, x)).$$

As usual, we shall denote the projection maps from these spaces onto X by π. These spaces can each be given the structure of a smooth manifold such that π is a smooth map.

Definition. A *k-form* on X is a mapping $\mu: X \to \Lambda^k(X)$ such that $\pi \circ \mu = i_X$. A k-form μ on X is *smooth* if whenever $V_1, \ldots, V_k$ are smooth vector fields on X, then

$$\mu(V_1, \ldots, V_k) \in C^\infty(X, R^1),$$

where

$$\mu(V_1, \ldots, V_k)(x) = \mu(x)(V_1(x), \ldots, V_k(x)).$$

A *differential form* on X is a mapping $\omega: X \to \mathscr{G}(X)$ such that $\pi \circ \omega = i_X$; it is *smooth* if its component in $\Lambda^k(X)$ is smooth for each k. The set of smooth k-forms on X is denoted by $C^\infty(X, \Lambda^k(X))$. The set of all smooth

121

differential forms is denoted by $C^\infty(X, \mathscr{G}(X))$. Note that $C^\infty(X, \Lambda^k(X))$ is a vector space under pointwise addition and scalar multiplication, and that $C^\infty(X, \mathscr{G}(X))$ is an algebra under the additional operation of pointwise exterior multiplication.

Remark 1. A 0-form on X is just a real-valued function on X; it is a smooth 0-form if and only if it is a smooth function.

Remark 2. Let φ be a local coordinate system on X, with domain U and coordinate functions $(x_1, \ldots, x_n)$. Then $\{dx_1, \ldots, dx_n\}$ is a basis for $T^*(X, x)$ for each $x \in U$. Hence

$$[dx_{i_1} \wedge \cdots \wedge dx_{i_k}; i_1 < \cdots < i_k]$$

is a basis for $\Lambda^k(T^*(X, x))$ for each $x \in U$. Thus, the restriction to U of each k-form μ on X can be expressed as

$$\mu = \sum_{i_1 < \cdots < i_k} a_{i_1 \cdots i_k} \, dx_{i_1} \wedge \cdots \wedge dx_{i_k},$$

where each $a_{i_1 \cdots i_k}$ is a real-valued function on U. Furthermore, μ is smooth if and only if, for each (φ, U), $a_{i_1 \cdots i_k} \in C^\infty(U, R^1)$. This is because

$$a_{i_1 \cdots i_k} = k! \, \mu\left(\frac{\partial}{\partial x_{i_1}}, \ldots, \frac{\partial}{\partial x_{i_k}}\right).$$

Theorem 1. *Let X be a smooth manifold. There exists a unique linear map $d: C^\infty(X, \mathscr{G}(X)) \to C^\infty(X, \mathscr{G}(X))$, called the* exterior differential, *such that the following properties hold.*

(1) $d: C^\infty(X, \Lambda^k(X)) \to C^\infty(X, \Lambda^{k+1}(X))$;
(2) $d(f) = df$ *(ordinary differential) for* $f \in C^\infty(X, \Lambda^0(X))$;
(3) *if* $\mu \in C^\infty(X, \Lambda^k(X))$ *and* $\tau \in C^\infty(X, \mathscr{G}(X))$, *then*
$\quad d(\mu \wedge \tau) = (d\mu) \wedge \tau + (-1)^k \mu \wedge d\tau$; *and*
(4) $d^2 = 0$.

Remark. For the proof we need the following lemma, which asserts that for any exterior differentiation operator d, $(d\omega)(x)$ depends only on the behavior of ω in a small neighborhood of x.

Lemma. *Let $d: C^\infty(X, \mathscr{G}(X)) \to C^\infty(X, \mathscr{G}(X))$ be linear and satisfy the conditions of the theorem. Suppose $\omega \in C^\infty(X, \mathscr{G}(X))$ is such that $\omega|_W = 0$ for some open set $W \subset X$. Then $(d\omega)|_W = 0$. Hence, if $\omega, \tau \in C^\infty(X, \mathscr{G}(X))$ are such that $\omega|_W = \tau|_W$ for some open set W, then $(d\omega)|_W = (d\tau)|_W$.*

PROOF. Suppose $\omega|_W = 0$. Let $x_0 \in W$. Let $f \in C^\infty(X, R^1)$ be such that $f(x_0) = 1$ and $f(x) = 0$ for all $x \notin W$. Then $f\omega$ is identically zero on X, so that

$$0 = d(f\omega) = (df) \wedge \omega + f \, d\omega.$$

Evaluating at x_0 gives $(d\omega)(x_0) = 0$. Since this holds for all $x_0 \in W$, $d\omega|_W = 0$. If $\omega|_W = \tau|_W$, then $(\omega - \tau)|_W = 0$, so that

$$0 = [d(\omega - \tau)]|_W = [d\omega - d\tau]|_W \quad \text{and} \quad d\omega|_W = d\tau|_W. \qquad \square$$

PROOF OF THEOREM 1

Uniqueness. Suppose $d: C^\infty(X, \mathscr{G}(X)) \to C^\infty(X, \mathscr{G}(X))$ satisfies the conditions of the theorem. Let $x \in X$, and let φ be a local coordinate system about x with domain U and coordinate functions $(x_1, \ldots, x_n)$. Let $\omega \in C^\infty(X, \Lambda^k(X))$. Then the restriction of ω to U can be expressed as

$$\omega|_U = \sum a_{i_1 \cdots i_k} \, dx_{i_1} \wedge \cdots \wedge dx_{i_k}$$

for some $a_{i_1 \cdots i_k} \in C^\infty(U, R^1)$. Now the right-hand side of this equation is not a differential form on X, so we cannot apply d to it. However, let U_1 be an open ball containing x with $\overline{U}_1$ compact and $\subset U$, and let $g \in C^\infty(X, R^1)$ be such that $g(x) = 1$ for $x \in U_1$ and $g(x) = 0$ for $x \notin U$. Then $\tilde{\omega} \in C^\infty(X, \Lambda^k(X))$, where

$$\tilde{\omega} = \sum (ga_{i_1 \cdots i_k}) \, d(gx_{i_1}) \wedge \cdots \wedge d(gx_{i_k}).$$

Here, by gh, for $h \in C^\infty(U, R^1)$, is meant the smooth function on X defined by

$$(gh)(x) = \begin{cases} g(x)h(x) & (x \in U) \\ 0 & (x \notin U). \end{cases}$$

Furthermore, $\tilde{\omega}|_{U_1} = \omega|_{U_1}$. By the lemma, $(d\omega)|_{U_1} = (d\tilde{\omega})|_{U_1}$. Now

$$\begin{aligned} d\tilde{\omega} &= \sum d[ga_{i_1 \cdots i_k} \, d(gx_{i_1}) \wedge \cdots \wedge d(gx_{i_k})] &\text{(by linearity)} \\ &= \sum d(ga_{i_1 \cdots i_k}) \wedge d(gx_{i_1}) \wedge \cdots \wedge d(gx_{i_k}) \\ &\quad + \sum ga_{i_1 \cdots i_k} \, d(d(gx_{i_1}) \wedge \cdots \wedge d(gx_{i_k})) &\text{(by Property (3))} \\ &= \sum d(ga_{i_1 \cdots i_k}) \wedge d(gx_{i_1}) \wedge \cdots \wedge d(gx_{i_k}), \end{aligned}$$

since each term of the second sum is zero by Properties (3) and (4). In particular, since g is identically 1 on U_1, and since $(d\omega)|_{U_1} = (d\tilde{\omega})|_{U_1}$,

$$(d\omega)|_{U_1} = \sum_{i_1 < \cdots < i_k} \sum_{j=1}^{n} \frac{\partial}{\partial x_j} (a_{i_1 \cdots i_k}) \, dx_j \wedge dx_{i_1} \wedge \cdots \wedge dx_{i_k}.$$

Thus if d exists, its value at x on k-forms must be given by this formula. Since x was arbitrary in X, and since every differential form is a sum of k-forms, $k \in \{0, 1, \ldots, n\}$, uniqueness is established.

Existence. We first define d locally. Let φ be a local coordinate system on X with domain U and coordinate functions $(x_1, \ldots, x_n)$. (Note that U is itself a smooth manifold.) Define $d_U: C^\infty(U, \mathscr{G}(U)) \to C^\infty(U, \mathscr{G}(U))$ as follows. For

$$\omega = \sum a_{i_1 \cdots i_k} \, dx_{i_1} \wedge \cdots \wedge dx_{i_k} \in C^\infty(U, \Lambda^k(U)),$$

define

$$d_U\omega = \sum \sum_{j=1}^{n} \frac{\partial}{\partial x_j} (a_{i_1\cdots i_k}) \, dx_j \wedge \, dx_{i_1} \wedge \cdots \wedge \, dx_{i_k}.$$

Extend d_U to $C^\infty(U, \mathcal{G}(U))$ by linearity. Then Properties (1) and (2) are clearly satisfied. To verify (3) and (4), note first that each form in $C^\infty(U, \mathcal{G}(U))$ is a sum of forms of the type $a_{i_1\cdots i_k} \, dx_{i_1} \wedge \cdots \wedge \, dx_{i_k}$. By linearity of d_U, together with distributivity of exterior multiplication with respect to addition, it suffices to check (3) and (4) on forms of this type.

Property (3). Suppose

$$\mu = a_{i_1\cdots i_k} \, dx_{i_1} \wedge \cdots \wedge \, dx_{i_k} \quad \text{and} \quad \tau = b_{j_1\cdots j_l} \, dx_{j_1} \wedge \cdots \wedge \, dx_{j_l}.$$

Then

$$d_U(\mu \wedge \tau) = d_U[a_{i_1\cdots i_k} b_{j_1\cdots j_l} \, dx_{i_1} \wedge \cdots \wedge \, dx_{i_k} \wedge \, dx_{j_1} \wedge \cdots \wedge \, dx_{j_l}]$$

$$= \sum_{r=1}^{n} \left[\frac{\partial}{\partial x_r} (a_{i_1\cdots i_k}) b_{j_1\cdots j_l} + a_{i_1\cdots i_k} \frac{\partial}{\partial x_r} (b_{j_1\cdots j_l}) \right]$$

$$dx_r \wedge \, dx_{i_1} \wedge \cdots \wedge \, dx_{j_l}$$

$$= \left(\sum_{r=1}^{n} \frac{\partial}{\partial x_r} (a_{i_1\cdots i_k}) \, dx_r \wedge \, dx_{i_1} \wedge \cdots \wedge \, dx_{i_k} \right)$$

$$\wedge (b_{j_1\cdots j_l} \, dx_{j_1} \wedge \cdots \wedge \, dx_{j_l})$$

$$+ (-1)^k (a_{i_1\cdots i_k} \, dx_{i_1} \wedge \cdots \wedge \, dx_{i_k})$$

$$\wedge \left(\sum_{r=1}^{n} \frac{\partial}{\partial x_r} (b_{j_1\cdots j_l}) \, dx_r \wedge \, dx_{j_1} \wedge \cdots \wedge \, dx_{j_l} \right)$$

$$= (d_U\mu) \wedge \tau + (-1)^k \mu \wedge d_U\tau.$$

Property (4). For $\mu = a_{i_1\cdots i_k} \, dx_{i_1} \wedge \cdots \wedge \, dx_{i_k}$,

$$d_U{}^2\mu = d_U\left[\sum_{r=1}^{n} \frac{\partial}{\partial x_r} (a_{i_1\cdots i_k}) \, dx_r \wedge \, dx_{i_1} \wedge \cdots \wedge \, dx_{i_k} \right]$$

$$= \sum_{r,s=1}^{n} \frac{\partial}{\partial x_s} \left[\frac{\partial}{\partial x_r} (a_{i_1\cdots i_k}) \right] dx_s \wedge \, dx_r \wedge \, dx_{i_1} \wedge \cdots \wedge \, dx_{i_k}.$$

But certainly the terms in this expression with $r = s$ are zero, since $dx_r \wedge dx_r = 0$. Moreover, for $r \neq s$, the equality of mixed partial derivatives on R^n implies that

$$\frac{\partial}{\partial x_s} \frac{\partial}{\partial x_r} (a_{i_1\cdots i_k}) = \frac{\partial}{\partial x_r} \frac{\partial}{\partial x_s} (a_{i_1\cdots i_k}),$$

so that

$$\frac{\partial}{\partial x_s} \frac{\partial}{\partial x_r} (a_{i_1\cdots i_k}) \, dx_s \wedge \, dx_r = -\frac{\partial}{\partial x_r} \frac{\partial}{\partial x_s} (a_{i_1\cdots i_k}) \, dx_r \wedge \, dx_s;$$

thus the remaining terms match up in pairs which cancel each other.

Thus the operator d_U has Properties (1)–(4). By uniqueness, every linear operator on $C^\infty(U, \mathscr{G}(U))$ having these properties must be given by the above boxed formula. In particular, if U_1 is any open subset of U, then $\varphi|_{U_1}$ is a coordinate system, and $d_{U_1}: C^\infty(U_1, \mathscr{G}(U_1)) \to C^\infty(U_1, \mathscr{G}(U_1))$ is given in the coordinate system $\varphi|_{U_1}$ by the same formula. Thus, if $\omega \in C^\infty(X, \mathscr{G}(X))$, then

$$d_{U_1}(\omega|_{U_1}) = (d_U(\omega|_U))|_{U_1}.$$

This relation enables us to define d globally by $(d\omega)|_U = d_U(\omega|_U)$ for all

$$\omega \in C^\infty(X, \mathscr{G}(X))$$

and any coordinate neighborhood U. This d is well defined because if U and V are overlapping coordinate neighborhoods, then

$$(d_U(\omega|_U))|_{U \cap V} = d_{U \cap V}(\omega|_{U \cap V}) = (d_V(\omega|_V))|_{U \cap V}.$$

Clearly, d has the required properties, since d_U has them for each U. $\qquad\square$

Digression on vector analysis

The multilinear algebra developed above is particularly simple in the case $n = 3$. We want to show how the classical approach of vector analysis fits into the scheme of differential forms.

In order to develop the connection, we consider first the general situation in an n-dimensional vector space T.

Definition. A *volume element* of T is a choice of basis in $\Lambda^n(T^*)$; since $\Lambda^n(T^*)$ is 1-dimensional, a volume element is a choice of a nonzero element in $\Lambda^n(T^*)$.

EXAMPLE. If T is the tangent space to a manifold and $\{dx_1, \ldots, dx_n\}$ is a basis for T^*, then $dx_1 \wedge \cdots \wedge dx_n$ is a volume element of T. (Note that a volume element ω determines an isomorphism $\Lambda^n(T^*) \cong R^1$, where $r\omega$ corresponds to r. Conversely, such an isomorphism defines a volume element ω corresponding to 1.)

Remark. Given a volume element ω of T, there exists a natural isomorphism $m: \Lambda^{n-1}(T^*) \to T$ defined as follows. Recall that T is naturally isomorphic to its double dual T^{**}. Identifying T^{**} with T through this isomorphism, m will have values in T^{**}. For $\varphi \in \Lambda^{n-1}(T^*)$, $m(\varphi)$ is then defined by $[m(\varphi)](\psi) = \lambda$, where, for $\psi \in T^*$, λ is the real number such that $\varphi \wedge \psi = \lambda\omega$. To show that m is an isomorphism, let $\{\varphi_1, \ldots, \varphi_n\}$ be a basis for T^* such that $\omega = \varphi_1 \wedge \cdots \wedge \varphi_n$. Then the set $\{\varphi_1 \wedge \cdots \wedge \varphi_{j-1} \wedge \varphi_{j+1} \wedge \cdots \wedge \varphi_n\}$ is a basis for $\Lambda^{n-1}(T^*)$. The value of m on these basis vectors is then given by

$$m(\varphi_1 \wedge \cdots \wedge \varphi_{j-1} \wedge \varphi_{j+1} \wedge \cdots \wedge \varphi_n) = (-1)^{n+j} e_j,$$

where $\{e_1, \ldots, e_n\}$ is the basis for T dual to $\{\varphi_1, \ldots, \varphi_n\}$.

Remark. Given an inner product $\langle\,,\,\rangle$ on a finite dimensional vector space T, there exists a natural isomorphism $g: T \to T^*$ defined by

$$[g(v)](w) = \langle v, w\rangle \qquad (v, w \in T).$$

If $\{e_1, \ldots, e_n\}$ is a basis for T, let $g_{ij} = \langle e_i, e_j\rangle$, $(i, j \in \{1, \ldots, n\})$. Then in terms of the dual basis $\{\varphi_1, \ldots, \varphi_n\}$ for T^*,

$$g(e_i) = \sum_{j=1}^{n} g_{ij}\varphi_j \qquad (i \in \{1, \ldots, n\}).$$

In particular, if $\{e_1, \ldots, e_n\}$ is orthonormal, then $g_{ij} = \delta_{ij}$, and

$$g(e_i) = \varphi_i.$$

Applications. Take $T = R^n$. Then T has an inner product and a natural volume element $\omega = \varphi_1 \wedge \cdots \wedge \varphi_n$, where $\{\varphi_i\}$ is the dual basis to the natural basis $\{e_i\}$ for R^n. Thus the isomorphisms m and g are defined. Also, we have natural identifications $T(R^n, x) \leftrightarrow R^n$ for each $x \in R^n$.

(1) Let $f \in C^\infty(R^n, R^1)$. Then the *gradient* of f is the vector field on R^n given by

$$\boxed{\operatorname{grad} f = g^{-1} \circ (df).}$$

Relative to the usual coordinates $(x_1, \ldots, x_n) = (r_1, \ldots, r_n)$ on R^n,

$$\operatorname{grad} f = g^{-1} \circ (df) = g^{-1}\left(\sum_{j=1}^{n} \frac{\partial f}{\partial x_j}\, dx_j\right) = \sum_{j=1}^{n} \frac{\partial f}{\partial x_j} \frac{\partial}{\partial x_j} \leftrightarrow \left(\frac{\partial f}{\partial x_1}, \ldots, \frac{\partial f}{\partial x_n}\right).$$

(2) Let V be a vector field on R^3. Then $g \circ V$ is a 1-form and $d(g \circ V)$ is a 2-form. Now for dimension $T = 3$, $\Lambda^2(T^*) = \Lambda^{n-1}(T^*)$, so the isomorphism m maps $\Lambda^2(T^*) \to T$. Thus $m(d(g \circ V))$ is a vector field on R^3. It is called the *curl* of V.

$$\boxed{\operatorname{curl} V = (m \circ d \circ g)(V).}$$

Exercise. Compute the coordinate expression for curl V.

(3) Let v_1 and v_2 be vectors in R^3. Then $g(v_1)$ and $g(v_2)$ are 1-forms. Their exterior product is a 2-form; its image under m is a vector, called the *cross product* of v_1 and v_2.

$$\boxed{v_1 \times v_2 = m(g(v_1) \wedge g(v_2)).}$$

(4) Let V be a vector field on R^n. Then $m^{-1}(V)$ is an $(n-1)$-form on R^n. Its differential is an n-form; that is, a multiple of the volume element ω. This multiple is (up to sign) the *divergence* of V:

$$\boxed{(-1)^{n-1}d \circ m^{-1}(V) = (\operatorname{div} V)\omega.}$$

Remark. Using these formulas, certain important formulas of vector analysis become trivial consequences of $d^2 = 0$.

A. curl grad $f = 0$ because

$$\begin{aligned} \text{curl grad } f &= m \circ d \circ g(g^{-1} \circ d(f)) \\ &= m(d^2 f) \\ &= 0. \end{aligned}$$

B. div curl $V = 0$ because

$$\begin{aligned} d \circ m^{-1}(\text{curl } V) &= d \circ m^{-1}(m \circ d \circ g(V)) \\ &= d^2(g(V)) \\ &= 0. \end{aligned}$$

Definition. Let X and Y be smooth manifolds, and let $\Psi \colon X \to Y$ be a smooth map. Then an *induced map* $\Psi^* \colon C^\infty(Y, \mathscr{G}(Y)) \to C^\infty(X, \mathscr{G}(X))$ is defined as follows. For $f \in C^\infty(Y, \Lambda^0(Y))$, $\Psi^*(f) = f \circ \Psi$; for $\omega \in C^\infty(Y, \Lambda^k(Y))$ $[k > 0]$,

$$(\Psi^*\omega)(x)(v_1, \ldots, v_k)$$
$$= \omega(\Psi(x))(d\Psi(v_1), \ldots, d\Psi(v_k)) \ [v_1, \ldots, v_k \in T(X, x), x \in X];$$

Ψ^* is extended to $C^\infty(Y, \mathscr{G}(Y))$ by linearity.

Remarks. It is easy to check that, if ω is a smooth differential form, then so is $\Psi^*\omega$. It is clear that Ψ^* maps k-forms into k-forms. In fact, it is easily checked that Ψ^* is an algebra homomorphism; i.e., Ψ^* is linear and

$$\Psi^*(\omega \wedge \tau) = (\Psi^*\omega) \wedge (\Psi^*\tau)$$

for all ω, τ.

Theorem 2. *Let X and Y be smooth and let $\Psi \colon X \to Y$ be a smooth map. Then*

$$d \circ \Psi^* = \Psi^* \circ d.$$

PROOF

(1) If $f \in C^\infty(Y, \Lambda^0(Y))$, then for $v \in T(X, x)$,

$$\begin{aligned} [d \circ \Psi^*(f)](v) &= [d(f \circ \Psi)](v) \\ &= [df \circ d\Psi](v) \quad \text{(since d on functions is ordinary differential)} \\ &= [\Psi^*(df)](v) \\ &= [(\Psi^* \circ d)(f)](v). \end{aligned}$$

(2) For ω a 1-form on Y of the type $\omega = df$,

$$\begin{aligned} (d \circ \Psi^*)(\omega) &= d(\Psi^*(df)) \\ &= d(\Psi^* \circ d(f)) \\ &= d(d \circ \Psi^*(f)) \quad \text{(by (1))} \\ &= 0, \end{aligned}$$

and

$$(\Psi^* \circ d)(\omega) = \Psi^*(d\omega) = \Psi^*(ddf) = \Psi^*(0) = 0.$$

(3) Using (1) and (2), together with the fact that Ψ^* is an algebra homomorphism, the result is established in general by checking it locally on k-forms ω restricted to local coordinate neighborhoods:

$$\omega|_U = \sum a_{i_1 \cdots i_k} \, dx_{i_1} \wedge \cdots \wedge dx_{i_k}.$$

(Details are left to the reader.) □

Definitions. Let X be a smooth manifold. A smooth differential form ω on X is *closed* if $d\omega = 0$. A form ω is *exact* if it is the differential of another form on X; that is, ω is exact if $\omega = d\tau$ for some smooth form τ. (Note that every exact form is closed, since $d^2 = 0$. The converse question is fundamental to our subject.)

Let $Z^k(X, d)$ denote the vector space of closed k-forms on X. Let $B^k(X, d)$ denote the space of exact k-forms on X. Then $B^k(X, d) \subset Z^k(X, d)$ because $d^2 = 0$. Let $H^k(X, d) = Z^k(X, d)/B^k(X, d)$. $H^k(X, d)$ is called the kth *De Rham cohomology group* of X. Its dimension, which we shall see is finite for compact X, is called the kth *Betti number* of X.

Remark. Although these cohomology groups are defined in terms of the manifold structure of X, they are topological invariants; that is, if two manifolds are homeomorphic (by a not necessarily smooth homeomorphism), then they have isomorphic cohomology groups. In fact, these groups can be defined directly using only the topological structure of X.

EXAMPLE 1. $H^0(X, d) \cong R^1$ if X is connected. For since there are no forms of degree less than 0, $B^0(X, d) = 0$. Thus

$$H^0(X, d) = Z^0(X, d) = [f \in C^\infty(X, R^1); df = 0].$$

If U is any connected coordinate neighborhood of X, with coordinate functions $(x_1, \ldots, x_n)$, then $df = 0$ on U means

$$0 = df = \sum_{i=1}^{n} \frac{\partial}{\partial x_i}(f) \, dx_i;$$

that is, $(\partial/\partial x_i)(f) = 0$ for all i. But this implies that f is constant on U. Since X is connected, and since f is constant on each connected coordinate neighborhood in X, then f must be constant on X; that is, $Z^0(X, d) = $ [constant functions on X] $\cong R^1$.

EXAMPLE 2. $H^1(S^1, d) \cong R^1$, where S^1 is the circle. For since there are no nonzero k-forms on S^1 for $k > 1$, $Z^1(S^1, d) = C^\infty(S^1, \Lambda^1(S^1))$. Moreover,

$$B^1(S^1, d) = [df; f \in C^\infty(S^1, R^1)].$$

Now, if θ denotes the polar coordinate on S^1, then $\partial/\partial\theta$ is a nonzero vector

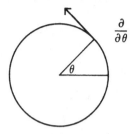

Figure 5.4

field on S^1 and its dual 1-form $d\theta$ is a nonzero 1-form on S^1 (see Figure 5.4). Furthermore, $d\theta$ is not exact (in spite of the notation!)—but, given any 1-form $\omega = g(\theta)\, d\theta$ on S^1, $\omega - (c\, d\theta)$ *is* exact for some $c \in R^1$. Thus

$$Z^1(S^1, d)/B^1(S^1, d) \cong [c\, d\theta; c \in R^1] \cong R^1.$$

Exercise. Verify the above facts. Take $c = (1/2\pi) \int_0^{2\pi} g(\theta)\, d\theta$.

Remarks. Let $\psi\colon X \to Y$ be smooth. Then

$$\psi^*\colon Z^k(Y, d) \to Z^k(X, d) \quad \text{and} \quad \psi^*\colon B^k(Y, d) \to B^k(X, d).$$

For if ω is a closed k-form on Y, then $d(\psi^*\omega) = \psi^*(d\omega) = \psi^*(0) = 0$. If $\omega = d\tau$ is an exact k-form on Y, then $\psi^*(\omega) = \psi^*(d\tau) = d(\psi^*(\tau))$. Thus ψ^* induces a linear map $\tilde{\psi}$ on cohomology, such that

$$\tilde{\psi}\colon Z^k(Y, d)/B^k(Y, d) \to Z^k(X, d)/B^k(X, d);$$

that is,

$$\tilde{\psi}\colon H^k(Y, d) \to H^k(X, d).$$

If $S\colon W \to X$ and $T\colon X \to Y$ are smooth, it is easy to check that $(T \circ S)^* = S^* \circ T^*$, and hence $\widetilde{(T \circ S)} = \tilde{S} \circ \tilde{T}$:

$$W \xrightarrow{S} X \xrightarrow{T} Y,$$

$$H^k(W, d) \xleftarrow{\tilde{S}} H^k(X, d) \xleftarrow{\tilde{T}} H^k(Y, d).$$

Thus we have attached to each smooth manifold X new algebraic invariants $H^k(X, d)$ such that given smooth maps between manifolds, there are induced algebraic maps between these algebraic objects. As in the case of the fundamental group, we are thus able to solve certain difficult topological problems by studying their algebraic counterparts.

Now let us show that $H^k(R^n, d) = 0$ for all $k > 0$. Since R^n is *diffeomorphic* (isomorphic as a smooth manifold) with the unit ball $B_0(1)$ about 0 in R^n, we may as well show that $H^k(B_0(1), d) = 0$ for all $k > 0$. For this we need the following technical lemma.

Lemma. *Let X be a smooth manifold. Then, for each k, consider the maps*

$$C^\infty(X, \Lambda^{k-1}(X)) \xrightarrow{\;d\;} C^\infty(X, \Lambda^k(X)) \xrightarrow{\;d\;} C^\infty(X, \Lambda^{k+1}(X)).$$

$$\underset{h_{k-1}}{\longleftarrow} \qquad \underset{h_k}{\longleftarrow}$$

Suppose there exist linear maps

$$h_j : C^\infty(X, \Lambda^{j+1}(X)) \to C^\infty(X, \Lambda^j(X)) \qquad (j = k-1 \text{ or } k)$$

such that $h_k \circ d + d \circ h_{k-1}$ is the identity map on $C^\infty(X, \Lambda^k(X))$. Then $H^k(X, d) = 0$; that is, every closed k-form is exact.

PROOF. Suppose $\omega \in C^\infty(X, \Lambda^k(X))$ is closed. Then

$$\omega = (h_k \circ d + d \circ h_{k-1})(\omega) = h_k(d\omega) + d(h_{k-1}\omega) = d(h_{k-1}\omega). \qquad \square$$

Remark. If a sequence of such linear maps h_j is defined for all $j \geq 0$, the sequence h_j is called a *homotopy operator.*

Theorem 3 (Poincaré's lemma). *Let $U = B_0(1) \subset R^n$. Then $H^k(U, d) = 0$ for all $k > 0$.*

PROOF. We shall construct maps h_{k-1}, h_k satisfying the conditions of the lemma. This is done through an integration process. Since these maps are to be linear, it suffices to define h_{k-1} on forms $\omega = g\, dx_{i_1} \wedge \cdots \wedge dx_{i_k}$; similarly for h_k. For such ω, set

$$h_{k-1}(\omega)(x) = \left(\int_0^1 t^{k-1} g(tx)\, dt \right) \mu,$$

where

$$\mu = x_{i_1}\, dx_{i_2} \wedge \cdots \wedge dx_{i_k} - x_{i_2}\, dx_{i_1} \wedge dx_{i_3} \wedge \cdots \wedge dx_{i_k}$$
$$+ \cdots + (-1)^{k-1} x_{i_k}\, dx_{i_1} \wedge \cdots \wedge dx_{i_{k-1}}.$$

(Note that $d\mu = k\, dx_{i_1} \wedge \cdots \wedge dx_{i_k}$.)

The map h_k is defined similarly by replacing k everywhere by $k+1$.

Now, for $\omega = g\, dx_{i_1} \wedge \cdots \wedge dx_{i_k} \in C^\infty(U, \Lambda^k(U))$ and $x \in U$,

$(d \circ h_{k-1})(\omega)(x)$

$$= d\left[\left(\int_0^1 t^{k-1} g(tx)\, dt \right) \mu \right]$$

$$= \sum_{j=1}^n \frac{\partial}{\partial x_j} \left(\int_0^1 t^{k-1} g(tx)\, dt \right) dx_j \wedge \mu + \left(\int_0^1 t^{k-1} g(tx)\, dt \right) d\mu$$

$$= \sum_{j=1}^n \left(\int_0^1 t^{k-1} \frac{\partial}{\partial x_j}(g(tx))\, dt \right) dx_j \wedge \mu + \left(\int_0^1 t^{k-1} g(tx)\, dt \right) d\mu$$

$$= \sum_{j=1}^n \left(\int_0^1 t^k \frac{\partial g}{\partial x_j}(tx)\, dt \right) dx_j \wedge \mu + k\left(\int_0^1 t^{k-1} g(tx)\, dt \right) dx_{i_1} \wedge \cdots \wedge dx_{i_k},$$

and

$$(h_k \circ d)(\omega)(x) = h_k\left(\sum_{j=1}^{n} \frac{\partial g}{\partial x_j} dx_j \wedge dx_{i_1} \wedge \cdots \wedge dx_{i_k}\right)$$

$$= \sum_{j=1}^{n} \left(\int_0^1 t^k \frac{\partial g}{\partial x_j}(tx) dt\right)[x_j dx_{i_1} \wedge \cdots \wedge dx_{i_k} - dx_j \wedge \mu].$$

Thus,

$$(d \circ h_{k-1} + h_k \circ d)(\omega)(x)$$

$$= \left[k\left(\int_0^1 t^{k-1} g(tx) dt\right) + \sum_{j=1}^{n}\left(\int_0^1 t^k \frac{\partial g}{\partial x_j}(tx)x_j dt\right)\right] dx_{i_1} \wedge \cdots \wedge dx_{i_k}$$

$$= \left\{\int_0^1 \left[kt^{k-1}g(tx) + t^k \frac{d}{dt}(g(tx))\right] dt\right\} dx_{i_1} \wedge \cdots \wedge dx_{i_k}$$

$$= \left\{\int_0^1 \frac{d}{dt}[t^k g(tx)] dt\right\} dx_{i_1} \wedge \cdots \wedge dx_{i_k}$$

$$= t^k g(tx)|_0^1 dx_{i_1} \wedge \cdots \wedge dx_{i_k}$$

$$= g(x) dx_{i_1} \wedge \cdots \wedge dx_{i_k}$$

$$= \omega(x) \qquad \text{(for all } x \in U\text{)}.$$

Since $d \circ h_{k-1} + h_k \circ d$ acts as identity on such ω, it acts by linearity as identity on all k-forms. $\qquad \square$

Remark 1. The maps h_{k-1} and h_k used in this proof were not just picked out of the air. They were constructed as follows. Given a vector space T and $v \in T$, v defines a map $i(v): \Lambda^k(T^*) \to \Lambda^{k-1}(T^*)$ by

$$[i(v)(\omega)](v_1, \ldots, v_{k-1}) = \omega(v, v_1, \ldots, v_{k-1}).$$

Note that i is a bilinear map $T \otimes \Lambda^k(T^*) \to \Lambda^{k-1}(T^*)$. This map i is called *interior multiplication.* The map h_{k-1} was obtained by applying $i(x)$ to ω and averaging over the line through the origin in the direction x.

Remark 2. Theorem 3 is a special case of a more general result. Let U be a smooth manifold. Suppose there exists a smooth map $\Psi: U \times I_\varepsilon \to U$, where $I_\varepsilon = [r \in R^1; -\varepsilon < r < 1 + \varepsilon]$, such that $\Psi(u, 1) = u$ for all $u \in U$, and $\Psi(u, 0) = u_0$ for all $u \in U$; some $u_0 \in U$ (see Figure 5.5). Then $H^k(U, d) = 0$ for all $k > 0$. The map Ψ is a *smooth homotopy.* This theorem says that if U is smoothly homotopic to a point, then the cohomology of U is that of a point.

In the case covered by Theorem 3, a smooth homotopy is given by

$$\Psi(x, t) = tx \qquad (t \in I_\varepsilon; x \in B_0(1)).$$

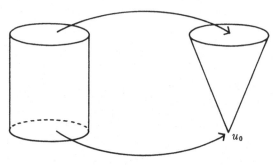

Figure 5.5

Note that the above proof of Poincaré's lemma works equally well for a *star-shaped* region, that is, an open set U such that for some $x_0 \in U$, the line segment joining x_0 to any other point in U lies completely in U.

5.3 Miscellaneous facts

Theorem 1. *Let X and Y be smooth manifolds, with X connected, and let $\psi: X \to Y$ be smooth. Assume $d\psi \equiv 0$. Then ψ is a constant map; that is, $\psi(x) = y_0$ for some $y_0 \in Y$ and for all $x \in X$.*

PROOF. Let $y_0 \in \psi(X)$. Then $\psi^{-1}(y_0)$ is a closed set in X. We shall show this set is also open, hence $\psi^{-1}(y_0) = X$ since X is connected.

Suppose $x_0 \in \psi^{-1}(y_0)$. It is sufficient to find an open set U in X such that $x_0 \in U$ and $U \subset \psi^{-1}(y_0)$. Let V be a coordinate neighborhood of y_0, with coordinate functions $(y_1, \ldots, y_m)$. Take U to be any connected coordinate neighborhood of x_0 such that $U \subset \psi^{-1}(V)$. Let $(x_1, \ldots, x_n)$ denote the coordinate functions in U. Then, for each $x \in U$, the matrix for $d\psi(x)$ relative to the bases $\{\partial/\partial x_i\}$ for $T(X, x)$ and $\{\partial/\partial y_i\}$ for $T(Y, \psi(x))$ is

$$\left(\frac{\partial}{\partial x_j}(y_i \circ \psi)\right).$$

Now, $d\psi \equiv 0$ implies $(\partial/\partial x_j)(y_i \circ \psi) \equiv 0$ on U for all i, j. But this implies that $y_i \circ \psi$ is constant on U for all i. Hence $y_i \circ \psi(x) = y_i \circ \psi(x_0)$ for all i and all $x \in U$; that is, $\psi(x) = \psi(x_0) = y_0$ for all $x \in U$, and $U \subset \psi^{-1}(y_0)$ as required. □

Definition. Let X be a smooth manifold, and let V and W be smooth vector fields on X. The *bracket* $[V, W]$ of V and W is the smooth vector field on X defined by

$$[V, W](f) = V(Wf) - W(Vf) \qquad (f \in C^\infty(X, R^1)).$$

Remark. $[V, W]$ is a vector field, because if φ is a local coordinate system with domain U and coordinate functions $(x_1, \ldots, x_n)$, then

$$V|_U = \sum_{i=1}^{n} a_i \left(\frac{\partial}{\partial x_i} \right) \quad \text{and} \quad W|_U = \sum_{i=1}^{n} b_i \left(\frac{\partial}{\partial x_i} \right)$$

for some $a_i, b_i \in C^\infty(U, R^1)$. Since $[V, W]$ is clearly bilinear, it suffices to check that $[V, W]$ is a vector field when $V = a(\partial/\partial x_i)$ and $W = b(\partial/\partial x_j)$. Then, since mixed partials are equal,

$$[V, W](f) = a \frac{\partial}{\partial x_i} \left(b \frac{\partial}{\partial x_j} (f) \right) - b \frac{\partial}{\partial x_j} \left(a \frac{\partial}{\partial x_i} (f) \right)$$

$$= a \frac{\partial}{\partial x_i} (b) \frac{\partial}{\partial x_j} (f) + ab \frac{\partial}{\partial x_i} \frac{\partial}{\partial x_j} (f)$$

$$- b \frac{\partial}{\partial x_j} (a) \frac{\partial}{\partial x_i} (f) - ab \frac{\partial}{\partial x_j} \frac{\partial}{\partial x_i} (f)$$

$$= \left[a \frac{\partial}{\partial x_i} (b) \frac{\partial}{\partial x_j} - b \frac{\partial}{\partial x_j} (a) \frac{\partial}{\partial x_i} \right] (f)$$

Since $a(\partial/\partial x_i)(b)$ and $b(\partial/\partial x_j)(a) \in C^\infty(X, R^1)$, $[V, W]$ is indeed a smooth vector field.

Remark. The bracket of vector fields has the following properties, each of which is easily verified.

(1) $[V, W] = -[W, V]$
(2) $[V_1 + V_2, W] = [V_1, W] + [V_2, W]$
(3) $[cV, W] = c[V, W]$ for $c \in R^1$
(4) $[[V, W], Z] + [[W, Z], V] + [[Z, V], W] = 0$.

Property (4) is called the *Jacobi identity*. These four properties say that $C^\infty(X, T(X))$ is a *Lie algebra* under bracket multiplication. Note that such an algebra is non-associative.

Theorem 2. *Let ω be a smooth 1-form, and let V and W be smooth vector fields on X. Then*

$$d\omega(V, W) = \tfrac{1}{2}\{V(\omega(W)) - W(\omega(V)) - \omega([V, W])\}.$$

Remark. In some texts, the fraction $\frac{1}{2}$ is missing from this formula. This is due to a slightly different definition of exterior multiplication.

PROOF OF THEOREM 2. It suffices to verify this formula in a local coordinate neighborhood. Furthermore, since both sides are linear in ω, we need only

133

check it on forms of the type $\omega = f\,dg$ (since every 1-form is locally a sum $\sum a_i\,dx_i$). For $\omega = f\,dg$,

$$
\begin{aligned}
d\omega(V, W) &= (df \wedge dg)(V, W)\\
&= \tfrac{1}{2}\{df(V)\,dg(W) - df(W)\,dg(V)\}\\
&= \tfrac{1}{2}\{(Vf)(Wg) - (Wf)(Vg)\}.
\end{aligned}
$$

On the other hand, we also have

$$
\begin{aligned}
\tfrac{1}{2}\{V(\omega(W)) &- W(\omega(V)) - \omega([V, W])\}\\
&= \tfrac{1}{2}\{V(f\,dg(W)) - W(f\,dg(V)) - f\,dg([V, W])\}\\
&= \tfrac{1}{2}\{V(f(Wg)) - W(f(Vg)) - f([V, W]g)\}\\
&= \tfrac{1}{2}[(Vf)(Wg) + fV(Wg) - (Wf)(Vg) - fW(Vg) - fV(Wg) + fW(Vg)\}\\
&= \tfrac{1}{2}\{(Vf)(Wg) - (Wf)(Vg)\}. \qquad\qquad\qquad\qquad\qquad \square
\end{aligned}
$$

Theorem 3 (Inverse function theorem). *Let X and Y be smooth manifolds of dimension n. Let $\psi\colon X \to Y$ be a smooth map. Suppose $x_0 \in X$ is such that*

$$
d\psi(x_0)\colon T(X, x_0) \to T(Y, \psi(x_0))
$$

is an isomorphism. Then there exists a neighborhood U_0 of x_0 such that

(1) $\psi|_{U_0}$ *is injective,*
(2) $\psi(U_0)$ *is open in Y, and*
(3) $\psi^{-1}\colon \psi(U_0) \to U_0$ *is smooth.*

PROOF. Let φ_2 be a coordinate system about $\psi(x_0)$ with domain V and coordinate functions $(y_1, \ldots, y_n)$. Let φ_1 be a coordinate system about x_0 such that

$$
U = \text{domain } \varphi_1 \subset \psi^{-1}(V).
$$

Let $(x_1, \ldots, x_n)$ denote the coordinate functions of φ_1. Then, relative to the bases $\{\partial/\partial x_i\}$ for $T(X, x_0)$ and $\{\partial/\partial y_i\}$ for $T(Y, \psi(x_0))$, $d\psi(x_0)$ has the matrix

$$
\left(\frac{\partial}{\partial x_j}\,(y_i \circ \psi)|_{x_0}\right),
$$

which is nonsingular since $d\psi(x_0)$ is an isomorphism.

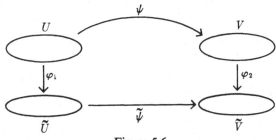

Figure 5.6

Now transfer everything to R^n via φ_1 and φ_2. Let $\tilde{U} = \varphi_1(U)$, $\tilde{V} = \varphi_2(V)$, and $\tilde{\psi}: \tilde{U} \to \tilde{V}$ be defined by $\tilde{\psi} = \varphi_2 \circ \psi \circ \varphi_1^{-1}$ (see Figure 5.6). Then $\tilde{\psi}(x) = (\tilde{\psi}_1(x), \ldots, \tilde{\psi}_n(x))$ for $x \in \tilde{U}$, where $\tilde{\psi}_i = r_i \circ \tilde{\psi}$. The Jacobian of $\tilde{\psi}$ at $\tilde{x}_0 = \varphi_1(x_0)$ is

$$\left(\frac{\partial \tilde{\psi}_i}{\partial r_j} \bigg|_{\tilde{x}_0} \right) = \left(\frac{\partial}{\partial x_j} (y_i \circ \psi) \big|_{x_0} \right),$$

which is nonsingular. Hence, by the classical inverse function theorem, there exists an open set $\tilde{U}_0 \subset \tilde{U}$ containing $\tilde{x}_0$ such that $\tilde{V}_0 = \tilde{\psi}(\tilde{U}_0)$ is open, and such that the equations

$$\tilde{\psi}_i(r_1, \ldots, r_n) = s_i \qquad (i \in \{1, \ldots, n\})$$

have a unique solution in $\tilde{U}_0$ for each $(s_1, \ldots, s_n) \in \tilde{V}_0$. Moreover, this solution depends smoothly on $(s_1, \ldots, s_n)$. In other words, there exist smooth functions

$$h_j: \tilde{V}_0 \to R^1 \qquad (j \in \{1, \ldots, n\})$$

such that for each $s = (s_1, \ldots, s_n) \in \tilde{V}_0$,

$$\tilde{\psi}_i(h_1(s), \ldots, h_n(s)) = s_i.$$

Setting $h(s) = (h_1(s), \ldots, h_n(s))$ for $s \in \tilde{V}_0$, this says that $h = \tilde{\psi}^{-1}$. Transferring back to X and Y, we find the conditions of the theorem are satisfied, with

$$U_0 = \varphi_1^{-1}(\tilde{U}_0). \qquad \square$$

Theorem 4 (Implicit function theorem). *Let X and Y be smooth manifolds, with $\dim X > \dim Y$. Let $\psi: X \to Y$ be a smooth map. Let $y_0 \in \psi(X)$ and let*

$$X_0 = \psi^{-1}(y_0) = [x \in X; \psi(x) = y_0].$$

Assume that for each $x \in X_0$, $d\psi(x): T(X, x) \to T(Y, \psi(x))$ is surjective. Then X_0 has a manifold structure, whose underlying topology is the relative topology of X_0 in X, and in which the inclusion map $X_0 \to X$ is smooth. Furthermore, $\dim X_0 = \dim X - \dim Y$.

We give some applications before proving Theorem 4.

Applications
(1) The n-sphere S^n is a smooth manifold whose topology is the induced topology in R^{n+1}. For let $\psi: R^{n+1} \to R^1$ be defined by

$$\psi(r_1, \ldots, r_{n+1}) = \sum_{i=1}^{n+1} r_i^2.$$

Then $S^n = \psi^{-1}(1)$. Since $\dim R^1 = 1$, we need only check that $d\psi \neq 0$ at each point of $\psi^{-1}(1)$. But $d\psi = 2 \sum_{i=1}^{n+1} r_i \, dr_i$. Since $\{dr_1, \ldots, dr_{n+1}\}$ is linearly independent, $d\psi \neq 0$ unless $r_i = 0$ for all i. In particular, $d\psi \neq 0$ on $\psi^{-1}(1)$. Note that $\dim S^n = \dim R^{n+1} - \dim R^1 = n$, as expected.
(2) Let $X = R^{n^2}$, viewed as the space of all real $n \times n$ matrices. Let $Y = R^1$, and let $\psi: X \to Y$ be the determinant function. Then $\psi^{-1}(1)$ is the

135

group of all $n \times n$ matrices of determinant 1. It is called the *unimodular group*. To verify that this group has a manifold structure, we need only show that $d\psi \neq 0$ at each point of $\psi^{-1}(1)$. Now, for (r_{ij}) coordinate functions on R^{n^2},

$$\psi \circ r_{ij} = \sum_{\pi \in S_n} (-1)^\pi r_{1\pi(1)} \cdots r_{n\pi(n)}.$$

Hence

$$d\psi = \sum_{j=1}^{n} \sum_{\pi \in S_n} (-1)^\pi r_{1\pi(1)} \cdots r_{j-1\pi(j-1)} r_{j+1\pi(j+1)} \cdots r_{n\pi(n)} \, dr_{j\pi(j)}.$$

For each (i, j), the coefficient of dr_{ij} in this sum is, up to sign, the determinant of the cofactor of r_{ij} in (r_{ij}). These cannot all be zero at any point of $\psi^{-1}(1)$ since $\det (r_{ij}) \neq 0$ at such points. Since $\{dr_{ij}\}$ is a linearly independent set, we are done.

Note that this unimodular group has dimension $n^2 - 1$.

(3) Let $X = R^{n^2}$ as in (2). Let Y be the set of all symmetric $n \times n$ real matrices. Y is a manifold, for it can be naturally identified with $R^{n(n+1)/2}$: merely string out in a row the entries on and below the main diagonal. Let $\psi: X \to Y$ be defined by $\psi(x) = xx^t$ where, for each $x \in X$, x^t denotes the transpose of x. Note that ψ is smooth, since each entry of $\psi(x)$ is a polynomial in the entries of x. Let $X_0 = \psi^{-1}(1)$. Thus X_0 is the group of orthogonal $n \times n$ matrices; that is, X_0 is the *orthogonal group*.

To verify that X_0 is a manifold, we must show that $d\psi(x)$ is surjective for each $x \in X_0$. For this, it suffices to show that $d\psi(e)$ is surjective, where $e = (\delta_{ij})$ is the identity matrix. For assuming that $d\psi(e)$ is surjective, let $x \in X_0$. Then the map $R_x: X \to X$, defined by $R_x(y) = yx$ (matrix multiplication), is a smooth map with a smooth inverse, namely $R_{x^{-1}}$, and hence dR_x is everywhere an isomorphism. Moreover, $\psi \circ R_x = \psi$ for all $x \in X_0$. For if $y \in X$, then

$$\psi \circ R_x(y) = \psi(yx) = (yx)(yx)^t = yxx^t y^t = yey^t = yy^t = \psi(y).$$

Hence,

$$d\psi|_x = d(\psi \circ R_{x^{-1}})|_x = d\psi|_{R_{x^{-1}(x)}} \circ dR_{x^{-1}}|_x = d\psi|_e \circ dR_{x^{-1}}|_x,$$

so $d\psi(x)$ is a composition of surjective maps, hence is surjective.

We still must check that $d\psi(e)$ is surjective. But

$$(r_{ij} \circ \psi)(x) = \sum_{h=1}^{n} r_{ih}(x) r_{jh}(x) \qquad (1 \leq i \leq j \leq n);$$

hence the entries in the matrix for $d\psi(x)$, where $1 \leq k, l \leq n$, and $1 \leq i \leq j \leq n$, are

$$\frac{\partial}{\partial r_{kl}} (r_{ij} \circ \psi)|_x = \begin{cases} r_{jl}(x) & \text{(if } k = i \neq j\text{)}, \\ r_{il}(x) & \text{(if } k = j \neq i\text{)}, \\ 2r_{il}(x) & \text{(if } k = i = j\text{)}, \\ 0 & \text{(otherwise)}. \end{cases}$$

In particular, the entries in the matrix for $d\psi(e)$, where $1 \le k, l \le n, 1 \le i \le j \le n$, are

$$\frac{\partial}{\partial r_{kl}}(r_{ij} \circ \psi)\big|_e = \begin{cases} 1 & (\text{if } (k, l) = (i, j); i \ne j) \\ 1 & (\text{if } (k, l) = (j, i); i \ne j) \\ 2 & (\text{if } (k, l) = (i, j); i = j) \\ 0 & (\text{otherwise}). \end{cases}$$

Thus the square submatrix, consisting of those entries with $k \le l$, is a diagonal matrix with diagonal entries 1 and 2, and so $d\psi(e)$ has rank $n(n + 1)/2$; that is, $d\psi(e)$ is surjective.

Note that $\dim X_0 = \dim X - \dim Y = n(n - 1)/2$.

(4) Let $X = $ the set of all complex $n \times n$ matrices $= R^{2n^2}$. Let $Y = [x \in X; \bar{x}^t = x]$. Let $\psi: X \to Y$ be defined by $\psi(x) = x\bar{x}^t$. Then, as in (3), the set $\psi^{-1}(e)$ is a manifold. $\psi^{-1}(e)$ is the *unitary group*. Its dimension is $2n^2 - n^2 = n^2$.

Remark. Examples (2), (3), and (4) are examples of *Lie groups*; namely, they are groups whose underlying spaces are C^ω-manifolds and are such that the group operations are analytic.

PROOF OF THEOREM 4. Let V be a coordinate neighborhood of y_0 in Y, with coordinate functions $(y_1, \ldots, y_m)$. For $x_0 \in X_0$, let U be a coordinate neighborhood of x_0 in X such that $U \subset \psi^{-1}(V)$. Let $(x_1, \ldots, x_n)$ denote the coordinate functions on U. We may assume that this coordinate system is chosen so that $x_i(x_0) = 0$ $(1 \le i \le n)$. Now $d\psi$ surjective at x_0 means that the $m \times n$ matrix $((\partial/\partial x_j)(y_i \circ \psi)|_{x_0})$ has rank m. By renumbering the coordinate functions on U if necessary, we may assume that the last m columns of this matrix are independent, that is, that this matrix has the form

$$(* \vdots J),$$

where J is a nonsingular $m \times m$ matrix. Let $\tilde{\psi}: U \to R^{n-m} \times V$ be defined by

$$\tilde{\psi}(x) = (x_1(x), \ldots, x_{n-m}(x), \psi(x)) \qquad (x \in U).$$

Then $d\tilde{\psi}(x_0)$ has matrix

$$\begin{pmatrix} I & 0 \\ * & J \end{pmatrix},$$

where I is the identity $(n - m) \times (n - m)$ matrix. Hence $d\tilde{\psi}(x_0)$ is an isomorphism. By the inverse function theorem, there exists a neighborhood U_0 of x_0 such that $\tilde{\psi}|_{U_0}$ is injective, $\tilde{\psi}(U_0)$ is open in $R^{n-m} \times V$, and $\tilde{\psi}^{-1}: \tilde{\psi}(U_0) \to U_0$ is smooth. We may assume that $\tilde{\psi}(U_0)$ is of the form $W_0 \times V_0$, where $0 \in W_0$ and $y_0 \in V_0$, since open sets of this type form a basis for the topology on $R^{n-m} \times V$ (see Figure 5.7). Now note that $\tilde{\psi}^{-1}(W_0 \times \{y_0\}) = X_0 \cap U_0$. Since $\tilde{\psi}|_{U_0}$ is a homeomorphism, $\tilde{\psi}|_{X_0 \cap U_0}$ maps $X_0 \cap U_0$ homeomorphically onto $W_0 \times \{y_0\} \cong W_0 \subset R^{n-m}$. Thus $\tilde{\psi}|_{X_0 \cap U_0}$ is a coordinate system about x_0 in X_0.

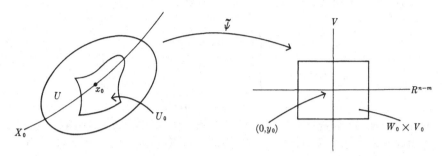

Figure 5.7

To see that such coordinate systems actually define a smooth manifold structure on X_0, we must check that they behave properly on overlaps. So suppose

$$\tilde{\psi}: U_0 \to W_0 \times V_0 \quad \text{and} \quad \tilde{\varphi}: U_1 \to W_1 \times V_1$$

are such that $(X_0 \cap U_0) \cap (X_0 \cap U_1) \neq \varnothing$ (see Figure 5.8). Since $\tilde{\psi}^{-1}$ is smooth, so is

$$\tilde{\varphi} \circ \tilde{\psi}^{-1}|_{\tilde{\psi}(U_0 \cap U_1)}.$$

Restricting to $\tilde{\psi}(X_0 \cap U_0 \cap U_1) = \tilde{\psi}(U_0 \cap U_1) \cap (R^{n-m} \times \{y_0\})$, it follows that

$$\tilde{\varphi} \circ \tilde{\psi}^{-1}|_{\tilde{\psi}(X_0 \cap U_0 \cap U_1)}: \tilde{\psi}(X_0 \cap U_0 \cap U_1) \to \tilde{\varphi}(X_0 \cap U_0 \cap U_1)$$

is smooth. Hence X_0 is a smooth manifold, of dimension $n - m$. $\qquad\square$

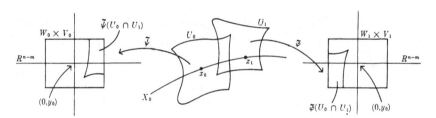

Figure 5.8

Definition. A *submanifold* of a smooth manifold Y is a pair (X, ψ), where X is a smooth manifold and $\psi: X \to Y$ is an injective smooth map such that $d\psi$ is injective at each point of X.

EXAMPLES. The manifold X_0 of the previous theorem, together with the inclusion map $X_0 \to X$, is a submanifold of X. In particular, S^n is a submanifold of R^{n+1}, and each of the Lie groups discussed above are submanifolds of the space of all $n \times n$ real (complex in the case of the unitary group) matrices.

Remark. Note that $\psi: X \to Y$ being injective does not imply that $d\psi$ is injective at each point. For example, the smooth map $\psi: R^1 \to R^1$ defined by $\psi(x) = x^3$ is injective, and yet $d\psi(0) = 0$. Note also that (X, ψ) being a submanifold of Y does *not* imply that ψ is a homeomorphism of X onto $\psi(X)$ with the relative topology.

EXAMPLE. Consider the torus

$$S^1 \times S^1 = [(z_1, z_2); z_1, z_2 \text{ complex, with } |z_1| = |z_2| = 1].$$

Define $\psi: R^1 \to S^1 \times S^1$ by $\psi(t) = (e^{2\pi i t}, e^{2\pi i \alpha t})$, where α is an irrational number. Then (R^1, ψ) is a submanifold of $S^1 \times S^1$. However, $\psi(R^1)$ is dense in $S^1 \times S^1$, so ψ is not a homeomorphism. This submanifold is called the *skew line* on the torus. Representing the torus as a square with opposite sides identified, ψ maps R^1 as in Figure 5.9.

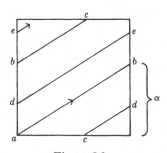

Figure 5.9

Theorem 5. *Let (X, ψ) be a submanifold of Y, with X compact, Suppose X has dimension m and Y has dimension n, where $m \leq n$. Then, for each $x_0 \in X$, there exists a coordinate system $\varphi_Y: V \to R^n$ about $\psi(x_0)$ with coordinate functions $(y_1, \ldots, y_n)$, such that*

$$\psi(X) \cap V = [y \in V; y_{m+1}(y) = \cdots = y_n(y) = 0].$$

Furthermore, a coordinate system $\varphi_X: U \to R^m$ can be chosen about x_0, with coordinate functions $(x_1, \ldots, x_m)$, such that $U \subset \psi^{-1}(V)$ and such that $x_j = y_j \circ \psi$ for all $j \leq m$. Thus, on U,

$$y_j \circ \psi = \begin{cases} x_j & (j \leq m) \\ 0 & (j > m). \end{cases}$$

PROOF. The proof of Theorem 5 is left as an exercise. $\square$

139

Remark. When a coordinate system φ_Y is chosen as in Theorem 5, $\psi(X) \cap V$ is said to be a *slice* in φ_Y. Note that the coordinate systems obtained in the proof of Theorem 4 are of this type.

Corollary. *If* (X, ψ) *is a submanifold of* Y, *and* X *is compact, then*

$$\psi: X \to \psi(X)$$

is a homeomorphism. Moreover, for each submanifold obtained by applying the implicit function theorem, the inclusion map is a homeomorphism.

PROOF. Since $\psi(X)$ is Hausdorff in the relative topology, the first statement is proved. $\qquad\qquad\qquad\qquad\qquad\qquad\qquad\qquad\qquad\qquad\qquad\qquad\qquad\square$

Definition. Let X be a smooth manifold, and let V be a smooth vector field on X. An *integral curve* of V is a smooth curve $\alpha: (a, b) \to X$ (Figure 5.10), such that the tangent vector to α at each point is equal to the value of V at that point; that is,

$$\dot\alpha(t) = V(\alpha(t)) \qquad \text{(for all } t \in (a, b)\text{)}.$$

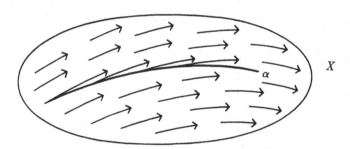

Figure 5.10

Remark. Let $\varphi: U \to R^n$ be a local coordinate system on X, with coordinate functions $(x_1, \ldots, x_n)$. Let $\alpha: (a, b) \to U$ be a smooth curve in U. Then, by definition, $\dot\alpha = d\alpha(d/dt)$. Hence, the ith component of $\dot\alpha$ relative to the basis $\{\partial/\partial x_j\}$ is

$$dx_i(\dot\alpha) = dx_i(d\alpha(d/dt)) = d(x_i \circ \alpha)(d/dt) = d/dt(x_i \circ \alpha),$$

so that

$$\dot\alpha = \sum_{i=1}^n \frac{d}{dt}(x_i \circ \alpha)\frac{\partial}{\partial x_i}.$$

Thus α is an integral curve of a vector field $V = \sum a_i(\partial/\partial x_i)$ if and only if

$$(*) \qquad\qquad \frac{d}{dt}(x_i \circ \alpha) = a_i \qquad (i \in \{1, \ldots, n\}).$$

140

Thus, to find integral curves of a given vector field V on a coordinate neighborhood U, we need solve the system (*) of differential equations. Solutions are guaranteed by the following classical theorem.

Theorem 6. *Let W be an open set in R^n, let $w_0 \in W$, and let $a_i \in C^\infty(W, R^1)$, $(1 \leq i \leq n)$. Then there exists an open set $W_0 \subset W$ about w_0, an interval $(-\varepsilon, \varepsilon) \subset R^1$, and a smooth map $\psi: (-\varepsilon, \varepsilon) \times W_0 \to W$ such that, for each $w \in W_0$, $\psi|_{(-\varepsilon,\varepsilon)\times\{w\}}$ is a solution of the equations*

$$\frac{df_i}{dt} = a_i(f_1(t), \ldots, f_n(t)) \qquad (1 \leq i \leq n)$$

subject to the initial conditions $f_i(0) = w_i$; that is, if $\alpha_w: (-\varepsilon, \varepsilon) \to W$ is defined by

$$\alpha_w(t) = \psi(t, w),$$

then for $1 \leq i \leq n$,

(A) $\qquad \dfrac{d}{dt}(r_i \circ \alpha_w)(t) = a_i(r_1 \circ \alpha_w(t), r_2 \circ \alpha_w(t), \ldots, r_n \circ \alpha_w(t))$

for all $t \in (-\varepsilon, \varepsilon)$, and

(B) $\qquad (r_i \circ \alpha_w)(0) = r_i(w) \qquad (1 \leq i \leq n)$.

Furthermore, α_w is the unique function $\alpha_w: (-\varepsilon, \varepsilon) \to W$ satisfying (A) and (B).

Reinterpreting Theorem 6 in terms of vector fields, we obtain Theorem 7.

Theorem 7. *Let X be a smooth manifold and let V be a smooth vector field on X. Let $x_0 \in X$. Then there exist an open set U about x_0, an interval $(-\varepsilon, \varepsilon) \subset R^1$, and a smooth map $\psi: (-\varepsilon, \varepsilon) \times U \to X$, such that for each $u \in U$, the curve*

$$\alpha_u: (-\varepsilon, \varepsilon) \to X$$

defined by $\alpha_u(t) = \psi(t, u)$ is the unique integral curve: $(-\varepsilon, \varepsilon) \to X$ of V, with $\alpha_u(0) = u$.

Furthermore, the smooth maps $\psi_t: U \to X$, defined for each $t \in (-\varepsilon, \varepsilon)$ by $\psi_t(u) = \psi(t, u)$, have the properties

(1) $\psi_{t_1 + t_2} = \psi_{t_1} \circ \psi_{t_2}$ *on* $\psi_{t_2}^{-1}(U)$ *whenever t_1, t_2 and $t_1 + t_2 \in (-\varepsilon, \varepsilon)$*
(2) $\psi_{-t} = \psi_t^{-1}$ *on* $\psi_t(U) \cap U$ *for each $t \in (-\varepsilon, \varepsilon)$.*

PROOF. Let W be a coordinate system about x_0, with coordinate functions $(x_1, \ldots, x_n)$. Then, on W, $V = \sum_{i=1}^n a_i(\partial/\partial x_i)$ for some smooth functions

$$a_i \in C^\infty(W, R^1).$$

By Theorem 6, there exist $U \subset W$, $(-\varepsilon, \varepsilon) \subset R^1$, and $\psi: (-\varepsilon, \varepsilon) \times U \to W \subset X$ with the required properties. The last statement is a consequence of the uniqueness of the solution; namely, it is easy to check that $t_1 \rightarrow \psi(t_1 + t_2, u)$ and $t_1 \rightarrow \psi(t_1, \psi_{t_2}(u))$ are both integral curves of V which send 0 into $\psi_{t_2}(u)$, and hence they are equal; that is, $\psi_{t_1 + t_2} = \psi_{t_1} \circ \psi_{t_2}$. Similarly, $\psi_{-t} = \psi_t^{-1}$. ☐

Remark. Properties (1) and (2) of Theorem 7 express the fact that ψ_t is a *local one-parameter group of transformations.*

Remark. The previous theorem guarantees the existence locally of integral curves for vector fields. However, it is not always possible to obtain integral curves globally; that is, it is not possible in general to find a curve $\alpha: R^1 \rightarrow X$ through x_0 such that α is an integral curve of a given vector field V. For example, let $X = R^2 - \{0\}$ and let $V = \partial/\partial r_1$. Then the integral curve of V through $(-1, 0)$ cannot be extended to values of $t \geq 1$ (see Figure 5.11).

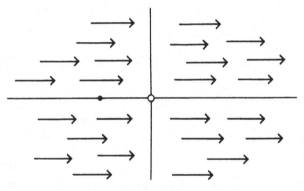

Figure 5.11

However, if X is compact, then every vector field admits through each point integral curves defined on all of R^1.

Remark. In studying the motion of a particle in R^3 under the influence of a force field F, Newton's law tells us that the path of motion is a curve $(x_i(t))$ such that

$$m \frac{d^2 x_i(t)}{dt^2} = F_i \quad (1 \leq i \leq 3),$$

where m is the mass of the particle. Setting $p_i = m(dx_i/dt)$, we have

$$\frac{dx_i}{dt} = \frac{p_i}{m}, \qquad \frac{dp_i}{dt} = F_i \quad (1 \leq i \leq 3).$$

But $(x_1, x_2, x_3, p_1, p_2, p_3)$ may be regarded as coordinate functions on the cotangent bundle of R^3. Hence the orbit of the particle is just the projection onto R^3 of the integral curve of a vector field on the cotangent bundle. In fact, the cotangent bundle is the natural domain for the study of mechanics on a manifold.

Remark. The use of integral curves provides a geometric interpretation of the bracket of two vector fields. Let V and W be smooth vector fields on X, and let $x_0 \in X$. Suppose we move along the integral curve of V through x_0 until the parameter has moved from 0 to $\sqrt{s}$; then move along an integral curve of W from 0 to $\sqrt{s}$; then move back along an integral curve of V, the parameter now varying from 0 to $-\sqrt{s}$; and finally move back along an integral curve of W from 0 to $-\sqrt{s}$, as in Figure 5.12. We will not in general

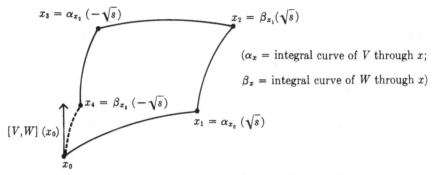

Figure 5.12

return to our starting point. As $s \to 0$, our end point will trace out a curve through x_0. The bracket $[V, W](x_0)$ is precisely the tangent vector to this curve.

Definition. Let V be an n-dimensional real vector space. Then $\Lambda^n(V^*)$ has dimension 1, so it is isomorphic to R^1. Thus $\Lambda^n(V^*) - \{0\}$ is disconnected; it is the union of two connected components. An *orientation* of V is a choice of one of these components. An *oriented vector space* is a pair $(V, \mathscr{A})$ where $\mathscr{A}$ is an orientation of V.

Remarks. Thus each vector space V has two possible orientations. An ordered basis $\{\varphi_1, \ldots, \varphi_n\}$ of V^* determines an orientation of V; namely, the component of $\Lambda^n(V^*)$ in which $\varphi_1 \wedge \cdots \wedge \varphi_n$ lies. Given two ordered bases $\{\varphi_1, \ldots, \varphi_n\}$ and $\{\varphi'_n, \ldots, \varphi'_n\}$ of V^*, with $\varphi'_i = \sum c_{ij}\varphi_j$, then $\varphi'_1 \wedge \cdots \wedge \varphi'_n = \det(c_{ij})\varphi_1 \wedge \cdots \wedge \varphi_n$. Hence two ordered bases determine the same orientation if and only if the determinant of the change of basis matrix is positive. In particular, if $\{\varphi_1, \ldots, \varphi_n\}$ is an ordered basis for V^*, then the orientation determined by the basis

$$\{\varphi_2, \varphi_1, \varphi_3, \ldots, \varphi_n\}$$

is different from the one determined by $\{\varphi_1, \varphi_2, \ldots, \varphi_n\}$.

In R^2, an orientation amounts to a sense of rotation. The orientation determined by $\{dr_1, dr_2\}$ gives the usual sense of positive rotation on R^2; namely, so that the rotation sending $\partial/\partial r_1$ into $\partial/\partial r_2$ is one of $+\pi/2$. The

143

orientation determined by $\{dr_2, dr_1\}$ defines the opposite sense of rotation, so that $\partial/\partial r_2 \to \partial/\partial r_1$ is a rotation of $+\pi/2$ (see Figure 5.13). Similarly, an orientation of R^3 amounts to choosing either the right-handed rule or the left-handed rule for cross products.

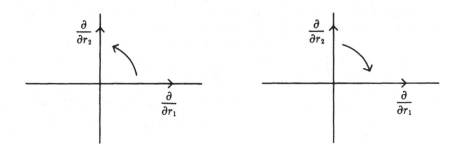

Orientation of $dr_1 \wedge dr_2$ Orientation of $dr_2 \wedge dr_1$

Figure 5.13

Definition. A smooth manifold (X, Φ) is *orientable* if there exists a subset $\Phi' \subset \Phi$ such that

(1) $\{\text{domain } \varphi\}_{\varphi \in \Phi'}$ is a covering of X, and
(2) If φ_1 and φ_2 are coordinate systems in Φ', with domains U and V and coordinate functions $(x_1, \ldots, x_n)$ and $(y_1, \ldots, y_n)$ respectively, then the function $\lambda: U \cap V \to R^1$ determined by

$$dx_1 \wedge \cdots \wedge dx_n = \lambda \, dy_1 \wedge \cdots \wedge dy_n$$

is everywhere positive.

An *orientation* of an orientable manifold (X, Φ) is a choice of subset $\Phi' \subset \Phi$ satisfying (1) and (2) and maximal with respect to (2). An *oriented manifold* is a triple (X, Φ, Φ'), where (X, Φ) is an orientable manifold and Φ' is an orientation of (X, Φ).

Remark. The function λ such that

$$dx_1 \wedge \cdots \wedge dx_n = \lambda \, dy_1 \wedge \cdots \wedge dy_n$$

is just the Jacobian determinant of $\varphi_1 \circ \varphi_2^{-1}$; that is,

$$\lambda = \det\left(\frac{\partial}{\partial y_j}(x_i)\right) = \det d(\varphi_1 \circ \varphi_2^{-1}).$$

In view of this, it is easy to check that a connected orientable manifold (X, Φ) has exactly two orientations Φ' and Φ'', and that Φ is the disjoint union $\Phi' \cup \Phi''$.

Remark. A more sophisticated approach to orientation of manifolds is to consider the set $\Lambda^n(X)$. This set can be given the structure of an $(n + 1)$-dimensional manifold as follows. Let $\varphi: U \to R^n$ be a local coordinate system on X, with coordinate functions $(x_1, \ldots, x_n)$. Then a coordinate system $\tilde{\varphi} \colon \pi^{-1}(U) \to R^{n+1}$ is defined on $\pi^{-1}(U)$ by

$$\tilde{\varphi}(\omega) = (\varphi(\pi(\omega)), \lambda(\omega)) \qquad (\omega \in \pi^{-1}(U)),$$

where $\lambda \colon \pi^{-1}(U) \to R^1$ is the function such that

$$\lambda(\omega) \, dx_1 \wedge \cdots \wedge dx_n = \omega \qquad (\omega \in \pi^{-1}(U)).$$

In terms of $\Lambda^n(X)$, we have the following characterization of orientability.

Theorem 8. *Let X be a connected smooth manifold (see Figure 5.14). Let*

$$O = \bigcup_{x \in X} \{0 \text{ element in } \Lambda^n(T^*(X, x))\} \subset \Lambda^n(X).$$

Then either $\Lambda^n(X) - O$ is connected, in which case X is not orientable, or $\Lambda^n(X) - O$ breaks up into exactly two connected components, in which case X is orientable. An orientation of an orientable manifold X amounts to a choice of one of these two components.

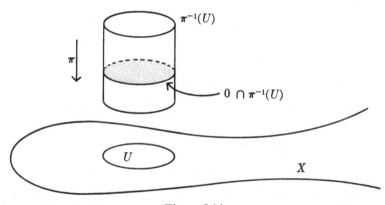

Figure 5.14

PROOF. We omit the proof. □

Theorem 9. *Let (X, Φ) be a smooth manifold of dimension n. Suppose there exists a smooth n-form ω on X which is nowhere zero. Then X is orientable.*

PROOF. Let $\varphi \in \Phi$ be a local coordinate system on X, with connected domain U and coordinate functions $(x_1, \ldots, x_n)$. Then, on U,

$$\omega = f_\varphi \, dx_1 \wedge \cdots \wedge dx_n$$

for some smooth function $f_\varphi \colon U \to R^1$. Since ω is never zero, neither is f_φ. Thus either $f_\varphi > 0$ everywhere, or $f_\varphi < 0$ everywhere. Let

$$\Phi' = [\varphi \in \Phi; f_\varphi > 0].$$

145

Then Φ' is an orientation of X. Φ' covers X because if $x \in X$ and φ is a co-ordinate system about x with $f_\varphi < 0$, then the new coordinate system $\bar{\varphi}$ about x, obtained by changing the sign of one of the coordinate functions of φ, has $f_{\bar{\varphi}} > 0$. Furthermore, if φ, $\psi \in \Phi'$ have domains U and V and coordinate functions

$$(x_1, \ldots, x_n) \quad \text{and} \quad (y_1, \ldots, y_n)$$

respectively, then on $U \cap V$

$$dy_1 \wedge \cdots \wedge dy_n = \frac{1}{f_\psi} \omega = \frac{f_\varphi}{f_\psi} dx_1 \wedge \cdots \wedge dx_n$$

and $f_\varphi / f_\psi > 0$. Maximality is clear. $\qquad \square$

Theorem 10. *Let (X, ψ) be an n-dimensional submanifold of R^{n+1}. Suppose (X, ψ) admits a nonzero "normal vector field"; that is, suppose there exists a smooth map $V: X \to T(R^{n+1})$ such that for each $x \in X$, $V(x)$ is a nonzero vector in $T(R^{n+1}, \psi(x))$ perpendicular to $d\psi(T(X, x))$ (see Figure 5.15). Then X is orientable.*

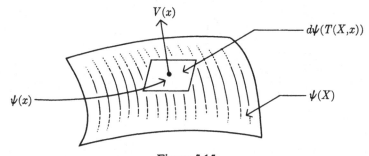

Figure 5.15

Remark. Perpendicularity in $T(R^{n+1}, \psi(x))$ means with respect to the inner product $\langle \, , \, \rangle$ given by

$$\left\langle \frac{\partial}{\partial r_i}, \frac{\partial}{\partial r_j} \right\rangle = \delta_{ij}.$$

PROOF OF THEOREM 10. Given a normal vector field V, consider the n-form μ defined at points of $\psi(X)$ by

$$\mu = i(V) \, dr_1 \wedge \cdots \wedge dr_{n+1}.$$

Let $\omega = \psi^* \mu$. Then ω is a smooth n-form on X. By Theorem 9, it suffices to show ω is never zero on X. Suppose it were; that is, suppose $\omega(x) = 0$ for some $x \in X$. Then

$$\begin{aligned}
0 &= \omega(x)(v_1, \ldots, v_n) \\
&= \psi^* \mu(x)(v_1, \ldots, v_n) \\
&= \mu(\psi(x))(d\psi(v_1), \ldots, d\psi(v_n))
\end{aligned}$$

for all $v_1, \ldots, v_n \in T(X, x)$. Now each vector $w \in T(R^{n+1}, \psi(x))$ is of the form $w = d\psi(v) + cV(x)$ for some $v \in T(X, x)$, $(c \in R^1)$. Thus, for arbitrary vectors

$$w_i = d\psi(v_i) + c_i V(x) \in T(R^{n+1}, \psi(x)) \qquad (1 \le i \le n),$$

we have

$$\begin{aligned}
\mu(\psi(x))(w_1, \ldots, w_n) &= \mu(\psi(x))(d\psi(v_1) + c_1 V(x), \ldots, d\psi(v_n) + c_n V(x)) \\
&= \mu(\psi(x))(d\psi(v_1), \ldots, d\psi(v_n)) \\
&\quad + \sum_{j=1}^{n} c_j \mu(\psi(x))(d\psi(v_1), \ldots, d\psi(v_{j-1}), V(x), \\
&\quad\qquad d\psi(v_{j+1}), \ldots, d\psi(v_n)).
\end{aligned}$$

All other terms are zero since $V(x)$ appears twice as an argument, and μ is skew symmetric. Moreover, the first term vanishes by the above discussion, and each term of the sum is zero because

$$\begin{aligned}
\mu(\psi(x))(\ldots, V, \ldots) &= i(V)\, dr_1 \wedge \cdots \wedge dr_{n+1}(\ldots, V, \ldots) \\
&= dr_1 \wedge \cdots \wedge dr_{n+1}(V, \ldots, V, \ldots) \\
&= 0.
\end{aligned}$$

Since $w_1, \ldots, w_n \in T(R^{n+1}, \psi(x))$ were arbitrary, this shows that $\mu(\psi(x)) = 0$. But

$$\begin{aligned}
\mu &= i(V)\, dr_1 \wedge \cdots \wedge dr_{n+1} \\
&= \sum_{j=1}^{n} (-1)^{j-1}(Vr_j)\, dr_1 \wedge \cdots \wedge dr_{j-1} \wedge dr_{j+1} \wedge \cdots \wedge dr_{n+1}.
\end{aligned}$$

Since $V(x)r_j \ne 0$ for some j, $\mu(\psi(x)) \ne 0$. This contradiction proves the theorem. $\qquad\square$

Corollary. *The unit sphere S^n is orientable.*

PROOF. S^n admits a nonzero normal vector field, namely, the restriction to S^n of the unit vector field on $R^{n+1} - \{0\}$ pointing radially outward. $\qquad\square$

Remark. It can be shown that every compact connected n-dimensional submanifold of R^{n+1} separates R^{n+1} into two connected pieces, one bounded and one unbounded. Thus every such submanifold admits a unit normal vector field (for example, the one pointing into the unbounded component), hence is orientable.

Remark. A nonorientable 2-dimensional manifold is called a one-sided surface.

EXAMPLE 1. The *Möbius strip S*, obtained from an open rectangular strip by giving the strip a half twist and glueing the ends, is nonorientable. Note that a nonzero normal vector field cannot exist on S, for if such a field varies continuously along the center line, it would have to point in the opposite direction after a full circuit.

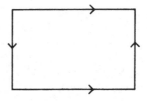

Figure 5.16

EXAMPLE 2. The *Klein bottle K*, obtained from $I \times I$ by identifying opposite sides (to get a cylinder) and then identifying the other pair of sides with a twist (Figure 5.16), is nonorientable. This surface cannot be represented as a submanifold of R^3. However, there does exist a map $\psi: K \to R^3$ with $d\psi$ injective at each point, and such that ψ is one-to-one except along a circle in R^3 (see Figure 5.17).

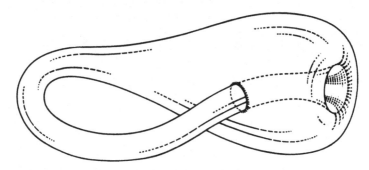

Figure 5.17

Definition. Let X be a topological space, and let $\mathcal{U}$ be an open covering of X. The covering $\mathcal{U}$ is *locally finite* if, for each $x \in X$, there exists an open set W_x containing x such that

$$[U \in \mathcal{U}; U \cap W_x \neq \varnothing]$$

is a finite set.

Definition. A topological space X is *paracompact* if every open covering of X has a locally finite refinement; that is, if for every open covering $\mathcal{U}$, there exists a locally finite open covering $\mathcal{V}$ such that for each $V \in \mathcal{V}$ there exists a $U \in \mathcal{U}$ with $V \subset U$.

Remark. It can be shown that all metric spaces are paracompact. Also, every regular topological space whose topology has a countable basis is paracompact.

Definition. Let X be a smooth manifold. A smooth *partition of unity* on X is a pair $(\mathcal{V}, \mathcal{F})$, where $\mathcal{V}$ is a locally finite covering of X and $\mathcal{F} = \{f_V\}_{V \in \mathcal{V}}$ is a collection of smooth real-valued functions on X such that

(1) each $f_V \geq 0$,

(2) for each $V \in \mathcal{V}$, the support of f_V = the closure of the set

$$[x \in X; f_V(x) \neq 0]$$

is contained in V, and

(3) $\sum_{V \in \mathcal{V}} f_V = 1$.

(Note that this sum makes sense since for each $x \in X$, $f_V(x) = 0$ for all but finitely many $V \in \mathcal{V}$.

Theorem 11. *Let X be a paracompact manifold. Then, given any open covering $\mathcal{U}$ of X, there exists a smooth partition of unity $(\mathcal{V}, \mathcal{F})$ on X such that $\mathcal{V}$ is a refinement of $\mathcal{U}$.*

PROOF. Since X is a manifold, there is a refinement $\mathcal{W}$ of $\mathcal{U}$ such that each open set $W \in \mathcal{W}$ is a coordinate neighborhood, and $\overline{W}$ is compact. Since X is paracompact, there is a locally finite refinement $\mathcal{V}$ of the open covering $\mathcal{W}$. Note that $\mathcal{V}$ is a refinement of $\mathcal{U}$, and if $V \in \mathcal{V}$, then $\overline{V}$ is compact, and V is a coordinate neighborhood.

Suppose we can "shrink the covering $\mathcal{V}$ slightly" and still get a covering. That is, suppose for each $V \in \mathcal{V}$, we can choose an open set $\alpha(V)$ such that $\overline{\alpha(V)} \subset V$ and $\{\alpha(V)\}_{V \in \mathcal{V}}$ is a covering. We then proceed as follows. Since $V \in \mathcal{V}$ is a coordinate neighborhood, and $\overline{\alpha(V)}$ is a compact set in V, we can find a smooth nonnegative function $g_V: X \to R^1$ such that $g_V(x) = 1$ for $x \in \alpha(V)$ and $g_V(x) = 0$ for $x \notin V$. Let $g = \sum_{V \in \mathcal{V}} g_V$. Then g is well defined and in $C^\infty(X, R^1)$ because $\mathcal{V}$ is locally finite. Furthermore, g never vanishes on X because $\{\alpha(V)\}_{V \in \mathcal{V}}$ is a covering; hence $f_V = g_V/g \in C^\infty(X, R^1)$. Let $\mathcal{F} = \{f_V\}_{V \in \mathcal{V}}$; then $(\mathcal{V}, \mathcal{F})$ is a smooth partition of unity.

To "shrink the covering $\mathcal{V}$ slightly," proceed as follows. Consider the family β of all functions β such that

(1) domain of β is a subset $\mathcal{D}_\beta$ of $\mathcal{V}$;

(2) if $V \in \mathcal{D}_\beta$, then $\beta(V)$ is an open set in V such that $\overline{\beta(V)} \subset V$; and

(3) the collection of open sets $[\beta(V); V \in \mathcal{D}_\beta] \cup [V; V \notin \mathcal{D}_\beta]$ is an open covering of X.

The family β is partially ordered: $\beta < \gamma$ if $\mathcal{D}_\beta \subset \mathcal{D}_\gamma$ and $V \in \mathcal{D}_\beta \Rightarrow \beta(V) = \gamma(V)$. We leave the following point set argument to the reader: since $\mathcal{V}$ is locally finite, the maximum principle implies that β has a maximal element α and $\mathcal{D}_\alpha = \mathcal{V}$, so that α is the required shrinkage. $\square$

Theorem 12. *Let X be a paracompact manifold that is orientable. Then there exists a smooth n-form ω on X such that ω never vanishes.*

PROOF. Let Φ' be an orientation of X. Let $\mathcal{U} = \{\text{domain } \varphi\}_{\varphi \in \Phi'}$. Then $\mathcal{U}$ is an open covering of X. Let $(\mathcal{V}, \mathcal{F})$ be a smooth partition of unity such that $\mathcal{V}$ is a refinement of $\mathcal{U}$. For each $V \in \mathcal{V}$, let $\varphi_V \in \Phi'$ be such that $V \subset \text{domain } \varphi_V$. Then the restriction of φ_V to V is also an element of Φ'. Let $(x_1^V, \ldots, x_n^V)$ denote the coordinate functions on V. Then the n-form $\omega^V = dx_1^V \wedge \cdots \wedge dx_n^V \in C^\infty(V, \Lambda^n(V))$ is nowhere zero on V. Let $\omega = \sum_{V \in \mathcal{V}} f_V \omega^V$, where $f_V \omega^V$ is by definition zero outside V. Then $\omega \in C^\infty(X, \Lambda^n(X))$.

We must show that ω is nowhere zero. For $x \in X$, let $\varphi \in \Phi'$ be a coordinate system about x, with domain U and coordinate functions $(y_1, \ldots, y_n)$. Then, for each $V \in \mathcal{V}$ with $U \cap V \neq \varnothing$,

$$\omega^V = dx_1^V \wedge \cdots \wedge dx_n^V = g_V \, dy_1 \wedge \cdots \wedge dy_n \quad \text{on} \quad U \cap V,$$

and $g_V > 0$ on $U \cap V$ since both φ_V and φ are members of Φ'. Thus,

$$\omega|_U = \sum_{V \in \mathcal{V}} (f_V \omega^V)|_U = \left(\sum_{V \in \mathcal{V}} f_V g_V \right) dy_1 \wedge \cdots \wedge dy_n.$$

Since $\sum_{V \in \mathcal{V}} f_V = 1$, there exists $V_0 \in \mathcal{V}$ such that $f_{V_0}(x) > 0$. Since $g_{V_0}(x) \neq 0$ and each $f_V g_V \geq 0$, $(\sum_{V \in \mathcal{V}} f_V g_V)(x) \neq 0$ and $\omega(x) \neq 0$. $\square$

Remark. Theorems 9 and 12 completely characterize orientability of paracompact manifolds by the existence or nonexistence of a nonzero n-form. This characterization can be applied to show that the projective space P^n is orientable if and only if n is odd. This is done by considering the sphere S^n as a covering space of P^n with covering map p. Let ω be the nonzero n-form on S^n constructed in the proof of Theorem 10 and its corollary. Then one can show that for n odd, ω defines an n-form $\tilde{\omega}$ on P^n such that $\omega = p^*\tilde{\omega}$. If P^n were orientable for n even, then there would exist a nonzero n-form $\tilde{\omega}$ on P^n and then $p^*\tilde{\omega} = g\omega$ for some $g \neq 0$. On the other hand, one can check that if $x_1 \neq x_2 \in S^n$ are such that $p(x_1) = p(x_2)$, then $g(x_1) > 0 \Leftrightarrow g(x_2) < 0$, contradicting the fact that g is never zero.

Remark. A nonzero smooth n-form on a smooth n-manifold is called a *volume element.* Thus every orientable paracompact manifold admits a volume element. The form $i(V) \, dr_1 \wedge \cdots \wedge dr_{n+1}$ on S^n discussed in Theorem 10 and its corollary is the usual volume element on the n-sphere.

Definition. A *Riemannian manifold* is a smooth manifold X, together with a map

$$\langle,\rangle \colon X \to \bigcup_{x \in X} \{\text{inner products on } T(X, x)\}$$

such that for each $x \in X$, $\langle,\rangle(x)$ (usually denoted $\langle,\rangle_x$) is an inner product on $T(X, x)$, and such that $\langle,\rangle$ is smooth; that is, for each pair V_1, V_2 of smooth vector fields on X, $\langle V_1, V_2 \rangle$ is a smooth function, where

$$\langle V_1, V_2 \rangle(x) = \langle V_1(x), V_2(x) \rangle_x.$$

The map $\langle,\rangle$ is called a *Riemannian structure* on X.

Theorem 13. *Let X be a paracompact smooth manifold. Then there exists a Riemannian structure on X.*

PROOF. Let $(\mathscr{V}, \mathscr{F})$ be a smooth partition of unity on X such that each $V \in \mathscr{V}$ is a coordinate neighborhood. Define a Riemannian structure $\langle,\rangle_V$ on each $V \in \mathscr{V}$ by

$$\left\langle \frac{\partial}{\partial x_i}, \frac{\partial}{\partial x_j} \right\rangle_V = \delta_{ij},$$

where $(x_1, \ldots, x_n)$ are the coordinate functions on V. Then define $\langle,\rangle$ on X by

$$\langle,\rangle = \sum_{V \in \mathscr{V}} f_V \langle,\rangle_V. \qquad \square$$

Remark. The converse of Theorem 13 also holds; namely, every Riemannian manifold is paracompact.

EXAMPLE 1. R^n is a Riemannian manifold: take $\{\partial/\partial r_i\}$ as an orthonormal basis for the tangent space at each point.

EXAMPLE 2. Let X be a Riemannian manifold, and let (Y, i) be a submanifold of X. Then a Riemannian structure is given on Y by

$$\langle v_1, v_2 \rangle_y = \langle di(v_1), di(v_2) \rangle_{i(y)} \qquad (v_1, v_2 \in T(Y, y)).$$

EXAMPLE 3. In view of Example 2, every submanifold of R^n has a Riemannian structure.

EXAMPLE 4. Let X and Y be Riemannian manifolds. Then the manifold $X \times Y$ has a Riemannian structure given as follows. For $(x, y) \in X \times Y$, the tangent space $T(X \times Y, (x, y))$ is naturally isomorphic to the direct sum of the vector spaces $T(X, x)$ and $T(Y, y)$. An inner product on $T(X \times Y, (x, y))$ is then given by requiring that this isomorphism be an isometry with the orthogonal direct sum $T(X, x) \oplus T(Y, y)$.

Definition. Let X and Y be Riemannian manifolds. A map $\varphi \colon X \to Y$ is an *isometry* if it is smooth, injective, surjective, has a smooth inverse, and is such that $d\varphi$ is an isometry at each point; that is,

$$\langle d\varphi(v_1), d\varphi(v_2) \rangle_{\varphi(x)} = \langle v_1, v_2 \rangle_x$$

for all $v_1, v_2 \in T(X, x)$ and $x \in X$.

Remark. Thus an isometry preserves all the structure of a Riemannian manifold. Two manifolds are equivalent from the viewpoint of Riemannian geometry if there exists an isometry between them. Such manifolds are said to be *isometric*. Note that two Riemannian manifolds as smooth manifolds can be the same; yet as Riemannian manifolds, be distinct.

151

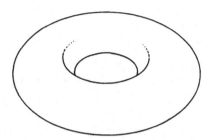

Figure 5.18

EXAMPLE 5. Consider the torus $S^1 \times S^1$. It has a Riemannian structure as a submanifold of R^3 (see Figure 5.18). On the other hand, it has a Riemannian structure as a product $S^1 \times S^1$, where S^1 is given a Riemannian structure by way of its usual imbedding into R^2. These two structures are distinct. In fact, the product structure on $S^1 \times S^1$ cannot be obtained by representing $S^1 \times S^1$ as a submanifold of R^3 (see Chapter 8). However, it can be obtained as a submanifold of R^4 since

$$S^1 \times S^1 \subset R^2 \times R^2 = R^4.$$

Homology theory and the De Rham theory

6

6.1 Simplicial homology

We have defined the De Rham cohomology groups $H^l(X, d)$ for a smooth manifold X. These groups came from a sequence of maps

$$C^\infty(X, \Lambda^{l-1}(X)) \xrightarrow{d} C^\infty(X, \Lambda^l(X)) \xrightarrow{d} C^\infty(X, \Lambda^{l+1}(X))$$

and $H^l(X, d) = \operatorname{Ker} d/\operatorname{Im} d$. We saw that the dimension of $H^0(X, d)$ measured the number of connected components of X, and we saw, at least for the circle $X = S^1$, that the dimension of $H^1(X, d)$ measured the number of "holes" in X. We shall now develop similar groups for simplicial complexes. We shall study a sequence of maps

$$C_{l-1} \xleftarrow{\partial} C_l \xleftarrow{\partial} C_{l+1},$$

where each C_k is an abelian group and where $\partial^2 = 0$. Then homology groups H_l will be defined by $H_l = Z_l/B_l$, where $Z_l = \operatorname{Ker} \partial \colon C_l \to C_{l-1}$ and $B_l = \operatorname{Im} \partial \colon C_{l+1} \to C_l$. An element of Z_l will geometrically be a "chain" of l-simplices without boundary. An element of B_l will geometrically be a boundary of a chain of $(l + 1)$-simplices. The boundary of a 1-simplex (v_0, v_1) will be the sum of the 0-simplices v_0 and v_1 with appropriate signs attached. Similarly, the boundary of a 2-simplex (v_0, v_1, v_2) will be an appropriate linear combination of its edges (v_0, v_1), (v_1, v_2), and (v_2, v_0).

Definition. Let s be an l-simplex, with vertices $v_0, v_1, \ldots, v_l$. Two orderings $(v_{j_1}, v_{j_2}, \ldots, v_{j_l})$ and $(v_{k_1}, v_{k_2}, \ldots, v_{k_l})$ of the vertices of s are *equivalent* if $(k_1, \ldots, k_l)$ is an even permutation of $(j_1, \ldots, j_l)$. This is clearly an equivalence relation, and for $l > 1$, it partitions the orderings of $v_0, \ldots, v_l$ into two equivalence classes. An *oriented simplex* is a simplex s together with a choice of one of these equivalence classes. If $v_0, v_1, \ldots, v_l$ are the vertices of s, the oriented simplex determined by the ordering $(v_0, \ldots, v_l)$ will be denoted by $\langle v_0, v_1, \ldots, v_l \rangle$.

Remark. Note that an oriented 1-simplex has a sense of direction attached to it, an oriented 2-simplex has a sense of rotation attached to it, and so on (see Figure 6.1). In fact, each l-simplex s lies in an l-dimensional plane in

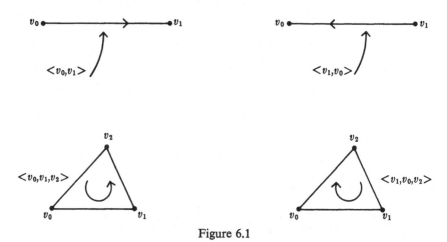

Figure 6.1

some R^m. Orienting s by $\langle v_0, v_1, \ldots, v_l \rangle$ is the same as orienting the l-plane containing s by means of the ordered basis $\{v_1 - v_0, v_2 - v_0, \ldots, v_l - v_0\}$.

Definition. Let K be a simplicial complex, and let $\mathscr{I}$ denote the group of integers. Let $C_l(K, \mathscr{I})$ denote the factor group of the free abelian group generated by all oriented simplices of K, modulo the subgroup generated by all elements of the form $\langle v_0, v_1, v_2, \ldots, v_l \rangle + \langle v_1, v_0, v_2, \ldots, v_l \rangle$. Thus $C_l(K, \mathscr{I})$ is an abelian group called the *group of l-chains* of K with integer coefficients. A typical element of this group is of the form

$$\sum_{s \text{ an } l\text{-simplex}} n_s \langle s \rangle \qquad (n_s \in \mathscr{I}),$$

where, for each l-simplex s, $\langle s \rangle$ is some fixed orientation of s, and where s with the opposite orientation is identified with $-\langle s \rangle$.

Remark. Given an arbitrary abelian group $\mathscr{G}$, the group $C_l(K, \mathscr{G})$ of l-chains of K with coefficients in $\mathscr{G}$ can be defined as the set of all formal linear combinations

$$\sum_s g_s \langle s \rangle \qquad (g_s \in \mathscr{G})$$

subject to the identifications $-g_s \langle v_0, v_1, \ldots, v_l \rangle = g_s \langle v_1, v_0, \ldots, v_l \rangle$. (We are writing the group operation in $\mathscr{G}$ additively.) In particular, $C_l(K, \mathscr{I})$ is defined for any field $\mathscr{F}$, in which case $C_i(K, \mathscr{F})$ is a vector space over $\mathscr{F}$ whose dimension equals the number of l-simplices of K. We shall only be interested in the cases where $\mathscr{G}$ constitutes the integers $\mathscr{I}$, the reals R, the complexes $\mathbf{C}$, or the integers $\mathscr{I}_2$ modulo 2; that is, the group of order 2.

Definition. Let $\langle s \rangle = \langle v_0, v_1, \ldots, v_{l+1} \rangle$ be an oriented $(l+1)$-simplex. The *boundary* $\partial \langle s \rangle$ of $\langle s \rangle$ is the l-chain defined by

$$\partial \langle s \rangle = \sum_{j=0}^{l+1} (-1)^j \langle v_0, v_1, \ldots, \hat{v}_j, \ldots, v_{l+1} \rangle,$$

where $\hat{\ }$ over a symbol means that symbol is deleted.

Remark. Note that $\partial \langle s \rangle$ is well defined and that $\bigcup_{j=0}^{l+1} [v_0, v_1, \ldots, \hat{v}_j, \ldots, v_{l+1}]$, the union of the faces occurring in $\partial \langle s \rangle$, is the topological boundary of $[s]$.

EXAMPLES

(1) $\partial \langle v_0, v_1 \rangle = \langle v_1 \rangle - \langle v_0 \rangle$.
(2) $\partial \langle v_0, v_1, v_2 \rangle = \langle v_1, v_2 \rangle - \langle v_0, v_2 \rangle + \langle v_0, v_1 \rangle = \langle v_0, v_1 \rangle + \langle v_1, v_2 \rangle + \langle v_2, v_0 \rangle$ (see Figure 6.2).

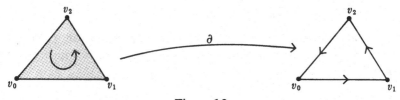

Figure 6.2

Definition. Let K be a simplicial complex, and let $\mathscr{G}$ be an abelian group. The *boundary map*

$$C_i(K, \mathscr{G}) \xleftarrow{\ \partial\ } C_{l+1}(K, \mathscr{G})$$

is the group homomorphism defined by

$$\partial\left(\sum g_s \langle s \rangle\right) = \sum g_s \, \partial \langle s \rangle.$$

Lemma. *The maps*

$$C_{l-1}(K, \mathscr{G}) \xleftarrow{\ \partial\ } C_l(K, \mathscr{G}) \xleftarrow{\ \partial\ } C_{l+1}(K, \mathscr{G})$$

satisfy $\partial^2 = \partial \circ \partial = 0$.

PROOF Since $\partial \circ \partial$ is linear, it suffices to check this on generators

$$\langle v_0, v_1, \ldots, v_{l+1} \rangle$$

155

as follows:

$$\partial(\partial\langle v_0,\ldots,v_{l+1}\rangle) = \partial\left[\sum_{j=0}^{l+1}(-1)^j\langle v_0,\ldots,\hat{v}_j,\ldots,v_{l+1}\rangle\right]$$

$$= \sum_{j=0}^{l+1}(-1)^j\,\partial\langle v_0,\ldots,\hat{v}_j,\ldots,v_{l+1}\rangle$$

$$= \sum_{j=0}^{l+1}(-1)^j\left[\sum_{i=0}^{j-1}(-1)^i\langle v_0,\ldots,\hat{v}_i,\ldots,\hat{v}_j,\ldots,v_{l+1}\rangle\right.$$

$$\left. + \sum_{i=j+1}^{l+1}(-1)^{i-1}\langle v_0,\ldots,\hat{v}_j,\ldots,\hat{v}_i,\ldots,v_{l+1}\rangle\right]$$

$$= \sum_{i<j}(-1)^{i+j}\langle v_0,\ldots,\hat{v}_i,\ldots,\hat{v}_j,\ldots,v_{l+1}\rangle$$

$$+ \sum_{i>j}(-1)^{i+j-1}\langle v_0,\ldots,\hat{v}_j,\ldots,\hat{v}_i,\ldots,v_{l+1}\rangle$$

$$= \sum_{i<j}[(-1)^{i+j}+(-1)^{i+j-1}]\langle v_0,\ldots,\hat{v}_i,\ldots,\hat{v}_j,\ldots,v_{l+1}\rangle$$

$$= 0. \qquad\qquad\qquad \square$$

Definition. Given K and $\mathscr{G}$, let

$$Z_l(K,\mathscr{G}) = [c \in C_l(K,\mathscr{G}); \partial c = 0],$$
$$B_l(K,\mathscr{G}) = [\partial c; c \in C_{l+1}(K,\mathscr{G})],$$
$$H_l(K,\mathscr{G}) = Z_l(K,\mathscr{G})/B_l(K,\mathscr{G}).$$

Elements of $Z_l(K,\mathscr{G})$ are called *cycles*, and of $B_l(K,\mathscr{G})$ are called *boundaries*. The group $H_l(K,\mathscr{G})$ is called the *l*th *homology group* of K with coefficients in $\mathscr{G}$.

Remark. It turns out that the groups $H_l(K,\mathscr{G})$ depend only on the topology of $[K]$. If $f\colon [K]\to[L]$ is a homeomorphism, then there is induced an isomorphism

$$f_*\colon H_l(K,\mathscr{G})\to H_l(L,\mathscr{G}).$$

In particular, if K_1 and K_2 are simplicial complexes with $[K_1] = [K_2]$, then they have the same homology groups.

Exercise. Show that the vector space $H_0(K,R)$ has dimension equal to the number of connected components in $[K]$.

EXAMPLE 1. Let K be the 1-skeleton of a 2-simplex; so K consists of three vertices v_0, v_1, v_2 and three 1-simplices (v_0, v_1), (v_1, v_2), and (v_2, v_0) (see Figure 6.3). Then both $C_0(K,\mathscr{I})$ and $C_1(K,\mathscr{I})$ are isomorphic to $\mathscr{I}\oplus\mathscr{I}\oplus\mathscr{I}$. $C_l(K,\mathscr{I}) = 0$ for $l > 1$. A typical element c_1 of $C_1(K,\mathscr{I})$ is of the form

$$c_1 = m_1\langle v_0, v_1\rangle + m_2\langle v_1, v_2\rangle + m_3\langle v_2, v_0\rangle \qquad (m_1, m_2, m_3 \in \mathscr{I}).$$

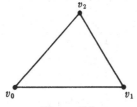

Figure 6.3

Its boundary ∂c_1 is given by

$$\partial c_1 = m_1(\langle v_1 \rangle - \langle v_0 \rangle) + m_2(\langle v_2 \rangle - \langle v_1 \rangle) + m_3(\langle v_0 \rangle - \langle v_2 \rangle)$$

$$= (m_3 - m_1)\langle v_0 \rangle + (m_1 - m_2)\langle v_1 \rangle + (m_2 - m_3)\langle v_2 \rangle.$$

Thus $c_1 \in Z_1(K, \mathscr{I})$ if and only if

$$m_3 - m_1 = 0, \qquad m_1 - m_2 = 0, \qquad m_2 - m_3 = 0,$$

that is, if and only if $m_1 = m_2 = m_3$, so

$$Z_1(K, \mathscr{I}) = [n(\langle v_0, v_1 \rangle + \langle v_1, v_2 \rangle + \langle v_2, v_0 \rangle); n \in \mathscr{I}] \cong \mathscr{I}.$$

Furthermore, $B_1(K, \mathscr{I}) = 0$ because $C_2(K, \mathscr{I}) = 0$. Hence

$$\boxed{H_1(K, \mathscr{I}) = Z_1(K, \mathscr{I})/B_1(K, \mathscr{I}) \cong \mathscr{I}.}$$

To compute $H_0(K, \mathscr{I})$, note that a typical cycle $c_0 \in Z_0(K, \mathscr{I})$ is of the form

$$c_0 = n_1\langle v_0 \rangle + n_2\langle v_1 \rangle + n_3\langle v_2 \rangle \qquad (n_1, n_2, n_3 \in \mathscr{I}).$$

Then $c_0 = \partial c_1$ for some

$$c_1 = m_1\langle v_0, v_1 \rangle + m_2\langle v_1, v_2 \rangle + m_3\langle v_2, v_0 \rangle \in C_1(K, \mathscr{I})$$

if and only if there exist (integer) solutions to the equations

$$m_3 - m_1 = n_1,$$
$$m_1 - m_2 = n_2,$$
$$m_2 - m_3 = n_3.$$

It is easy to check that such a solution exists if and only if $n_1 + n_2 + n_3 = 0$. Thus

$$B_0(K, \mathscr{I}) = [n_1\langle v_0 \rangle + n_2\langle v_1 \rangle + n_3\langle v_2 \rangle; n_1 + n_2 + n_3 = 0].$$

Let $\varphi: Z_0(K, \mathscr{I}) \to \mathscr{I}$ be the homomorphism defined by

$$\varphi(n_1\langle v_0 \rangle + n_2\langle v_1 \rangle + n_3\langle v_2 \rangle) = n_1 + n_2 + n_3.$$

Then the kernel of φ is just $B_0(K, \mathscr{I})$; thus

$$\boxed{H_0(K, \mathscr{I}) = Z_0(K, \mathscr{I})/B_0(K, \mathscr{I}) \cong \mathscr{I}.}$$

EXAMPLE 2. Let K be the complex consisting of all the faces of a 2-simplex (v_0, v_1, v_2). Then, as in Example 1,

$$\boxed{H_0(K, \mathcal{I}) \cong \mathcal{I}.}$$

Moreover, as before,

$$Z_1(K, \mathcal{I}) = [n(\langle v_0, v_1 \rangle + \langle v_1, v_2 \rangle + \langle v_2, v_0 \rangle); n \in \mathcal{I}].$$

Now, however,

$$C_2(K, \mathcal{I}) = [n\langle v_0, v_1, v_2 \rangle; n \in \mathcal{I}],$$

so that

$$\begin{aligned}
B_1(K, \mathcal{I}) &= [\partial(n\langle v_0, v_1, v_2 \rangle); n \in \mathcal{I}] \\
&= [n(\langle v_1, v_2 \rangle - \langle v_0, v_2 \rangle + \langle v_0, v_1 \rangle); n \in \mathcal{I}] \\
&= [n(\langle v_0, v_1 \rangle + \langle v_1, v_2 \rangle + \langle v_2, v_0 \rangle); n \in \mathcal{I}] \\
&= Z_1(K, \mathcal{I}).
\end{aligned}$$

Hence

$$\boxed{H_1(K, \mathcal{I}) = Z_1(K, \mathcal{I})/B_1(K, \mathcal{I}) = 0.}$$

Finally, since $\partial(n\langle v_0, v_1, v_2 \rangle) = 0$ if and only if $n = 0$, $Z_2(K, \mathcal{I}) = 0$, and hence

$$\boxed{H_2(K, \mathcal{I}) = 0.}$$

Definitions. Let K be a simplicial complex. The lth *Betti number* β_l of K is the integer

$$\beta_l = \dim H_l(K, R).$$

The *Euler characteristic* $\chi(K)$ of K is the integer

$$\chi(K) = \sum_{l=0}^{\dim K} (-1)^l \beta_l.$$

Theorem. *Let K be a simplicial complex. For each l with $0 \leq l \leq \dim K$, let α_l denote the number of l-simplices in K. Then*

$$\chi(K) = \sum_{l=0}^{\dim K} (-1)^l \alpha_l;$$

that is, $\chi(K)$ is equal to the number of vertices $-$ the number of edges $+$ the number of 2-faces $- \cdots$.

PROOF. For each l, $0 \leq l \leq \dim K$, consider the linear map

$$C_{l-1}(K, R) \xleftarrow{\ \partial\ } C_l(K, R),$$

where C_{-1} is by definition the zero space. Then, by the rank and nullity theorem of linear algebra,

$$\alpha_l = \dim C_l(K, R) = \dim \text{Ker } \partial + \dim \text{Im } \partial$$
$$= \dim Z_l(K, R) + \dim B_{l-1}(K, R) \qquad (l = 0, 1, \ldots, \dim K).$$

Moreover,

$$\beta_l = \dim H_l(K, R) = \dim [Z_l(K, R)/B_l(K, R)]$$
$$= \dim Z_l(K, R) - \dim B_l(K, R).$$

Thus

$$\chi(K) = \sum_{l=0}^{\dim K} (-1)^l \beta_l$$

$$= \sum_{l=0}^{\dim K} (-1)^l [\dim Z_l(K, R) - \dim B_l(K, R)]$$

$$= \sum_{l=0}^{\dim K} (-1)^l \dim Z_l(K, R) + \sum_{l=0}^{\dim K} (-1)^{l+1} \dim B_l(K, R)$$

$$= \sum_{l=0}^{\dim K} (-1)^l \dim Z_l(K, R) + \sum_{l=1}^{\dim K} (-1)^l \dim B_{l-1}(K, R)$$
$$\text{(since } \dim B_l = 0 \text{ for } l = \dim K)$$

$$= \sum_{l=0}^{\dim K} (-1)^l [\dim Z_l(K, R) + \dim B_{l-1}(K, R)]$$
$$\text{(since } \dim B_{-1} = 0)$$

$$= \sum_{l=0}^{\dim K} (-1)^l \alpha_l. \qquad \square$$

Remark. If $[K]$ is homeomorphic to a connected compact orientable 2-dimensional manifold, then it turns out that $\beta_0 = 1$ and $\beta_2 = 1$, so that

$$\chi(K) = \beta_0 - \beta_1 + \beta_2 = 2 - \beta_1$$

or

$$\beta_1 = 2 - \chi(K).$$

Furthermore, β_1 is always even for such K. It can be shown that any such surface is homeomorphic to a sphere with a certain number of "handles" attached; $\frac{1}{2}\beta_1$ is just the number of handles (see Figure 6.4).

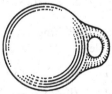
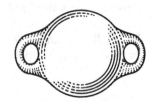

Sphere: $\beta_1 = 0$ Torus: $\beta_1 = 2$ Sphere with two handles: $\beta_1 = 4$

Figure 6.4

Thus the homology groups completely determine the homeomorphism class of connected compact orientable surfaces. However, for higher dimensional manifolds, the homology groups contain comparatively little information.

Remark. We have been discussing a *homology* theory for simplicial complexes, that is, a theory arising from a sequence of groups and homomorphisms

$$\cdots \xleftarrow{\partial} C_{l-1}(K, R) \xleftarrow{\partial} C_l(K, R) \xleftarrow{\partial} C_{l+1}(K, R) \xleftarrow{} \cdots,$$

where the map ∂ lowers the dimension of chains. On the other hand, in studying De Rham *cohomology*, we used a sequence

$$\cdots \xrightarrow{d} C^\infty(X, \Lambda^{l-1}(X)) \xrightarrow{d} C^\infty(X, \Lambda^l(X)) \xrightarrow{d} C^\infty(X, \Lambda^{l+1}(X)) \xrightarrow{} \cdots,$$

where the map d raised dimension (degree). In order to compare these two theories, it is convenient to define a *simplicial cohomology* theory. This is done by passing to dual spaces.

Definition. Let K be a simplicial complex. For $0 \le l \le \dim K$, let

$$C^l(K) = [C_l(K, R)]^*.$$

Let $\partial^*: C^l(K) \to C^{l+1}(K)$ be the adjoint of the map $\partial: C_{l+1}(K, R) \to C_l(K, R)$. Thus ∂^* is defined by

$$[\partial^*(\varphi)](c) = \varphi(\partial c) \qquad (\varphi \in C^l(K); c \in C_{l+1}(K, R)).$$

Then we get a sequence

$$\cdots \longrightarrow C^{l-1}(K) \xrightarrow{\partial^*} C^l(K) \xrightarrow{\partial^*} C^{l+1}(K) \longrightarrow \cdots.$$

Moreover, $\partial^* \circ \partial^* = 0$ since $\partial \circ \partial = 0$. Let

$$Z^l(K) = [\varphi \in C^l(K); \partial^*\varphi = 0],$$
$$B^l(K) = [\partial^*\varphi; \varphi \in C^{l-1}(K)],$$
$$H^l(K) = Z^l(K)/B^l(K).$$

Elements of $C^l(K)$ are called *cochains*; elements of $Z^l(K)$ are *cocycles*; elements of $B^l(K)$ are *coboundaries*. The map ∂^* is the *coboundary operator*. $H^l(K)$ is the *l*th *cohomology group* of K.

Exercise. Verify that $H^l(K)$ is isomorphic to $[H_l(K, R)]^*$.

We shall need an explicit formula exhibiting the effect of the coboundary operator ∂^*. For each oriented *l*-simplex $\langle s \rangle$ of K, let $\varphi_{\langle s \rangle} \in C^l(K)$ be defined by

$$\varphi_{\langle s \rangle}\langle t \rangle = \begin{cases} 1 & (\text{if } \langle t \rangle = \langle s \rangle) \\ -1 & (\text{if } \langle t \rangle = -\langle s \rangle) \\ 0 & (\text{if } t \ne s). \end{cases}$$

Thus, if $\{\langle s_1 \rangle, \ldots, \langle s_m \rangle\}$ is a basis for $C_l(K, R)$ (so that $\{s_1, \ldots, s_m\}$ is the set of all l-simplices of K), then $\{\varphi_{\langle s_1 \rangle}, \ldots, \varphi_{\langle s_m \rangle}\}$ is the dual basis for $C^l(K)$. Since ∂^* is linear, we need only compute the effect of ∂^* on these generators $\varphi_{\langle s \rangle}$.

Lemma

$$\partial^* \varphi_{\langle v_0, \ldots, v_l \rangle} = \sum_v{}' \varphi_{\langle v, v_0, \ldots, v_l \rangle},$$

where $\sum_v'$ denotes the sum over all vertices $v \in K$ such that $(v, v_0, v_1, \ldots, v_l)$ is an $(l + 1)$-simplex of K.

PROOF. We need only check this formula on oriented $(l + 1)$-simplices

$$\langle t \rangle = \langle w_0, w_1, \ldots, w_{l+1} \rangle$$

of K. If we set $\langle s \rangle = \langle v_0, v_1, \ldots, v_l \rangle$, the left side yields

$$(\partial^* \varphi_{\langle s \rangle})(\langle t \rangle) = \varphi_{\langle s \rangle}(\partial \langle t \rangle)$$
$$= \varphi_{\langle s \rangle}\left(\sum_{i=0}^{l+1} (-1)^i \langle w_0, \ldots, \hat{w}_i, \ldots, w_{l+1} \rangle\right)$$
$$= \sum_{i=0}^{l+1} (-1)^i \varphi_{\langle s \rangle}(\langle w_0, \ldots, \hat{w}_i, \ldots, w_{l+1} \rangle).$$

But each term of this sum is zero unless, for some j, $(w_0, \ldots, \hat{w}_j, \ldots, w_{l+1}) = (s)$; that is, unless (s) is a face of (t). If (s) is a face of (t), then $(t) = (v, v_0, \ldots, v_l)$ for some vertex $v \in K$, in which case either

(1) $\langle t \rangle = \langle v, v_0, \ldots, v_l \rangle$ and $(\partial^* \varphi_{\langle s \rangle})(\langle t \rangle) = 1$; or
(2) $\langle t \rangle = -\langle v, v_0, \ldots, v_l \rangle$ and $(\partial^* \varphi_{\langle s \rangle})(\langle t \rangle) = -1$.

Thus

$$(\partial^* \varphi_{\langle s \rangle})(\langle t \rangle) = \begin{cases} 1 & (\text{if } \langle t \rangle = \langle v, v_0, \ldots, v_l \rangle \text{ for some } v) \\ -1 & (\text{if } \langle t \rangle = -\langle v, v_0, \ldots, v_l \rangle \text{ for some } v) \\ 0 & (\text{in all other cases}) \end{cases}$$
$$= \left(\sum_v{}' \varphi_{\langle v, v_0, \ldots, v_l \rangle}\right)(\langle t \rangle).$$

Since this holds for arbitrary $\langle t \rangle$, the formula is established. $\square$

6.2 De Rham's theorem

Definition. A *smoothly triangulated manifold* is a triple (X, K, h), where X is a C^∞ manifold, K is a simplicial complex, and $h: [K] \to X$ is a homeomorphism such that for each simplex s of K, the map $h|_{[s]}: [s] \to X$ has an extension h_s to a neighborhood U of $[s]$ in the plane of $[s]$ such that $h_s: U \to X$ is a smooth submanifold.

Remark. If dim $X = n$, we need only require that this last condition be satisfied for each n-simplex of K, since every simplex of K is a face of an n-simplex and since restrictions of smooth maps to submanifolds are smooth.

EXAMPLE. Let $X = S^n$. Let K be the n-skeleton of an $(n + 1)$-simplex circumscribed about S^n. Let $h: [K] \to S^n$ be radial projection. Then (X, K, h) is a smoothly triangulated manifold (Figure 6.5).

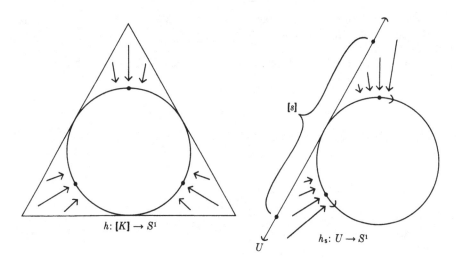

$h: [K] \to S^1$

$h_s: U \to S^1$

Figure 6.5

Remark. It can be shown that every compact smooth manifold can be smoothly triangulated. The proof is difficult and will not be presented here. Note that smoothly triangulated manifolds are compact because $[K]$ is compact for each (finite) simplicial complex K.

The goal of this section is to show that for smoothly triangulated manifolds (X, K, h), the De Rham cohomology of X is isomorphic to the simplicial cohomology of K. For this, we shall need the following facts about barycentric coordinates. Recall that we have previously discussed the barycentric coordinates of a point relative to the vertices of a simplex containing it. We now extend this concept.

Definition. Let K be a simplicial complex and let $v_1, \ldots, v_m$ denote the vertices of K. Suppose $x \in [K]$. For $j \in \{1, \ldots, m\}$, the jth *barycentric coordinate* $b_j(x)$ of x is defined as follows. If $x \notin \mathrm{St}(v_j)$, then $b_j(x) = 0$; if $x \in \mathrm{St}(v_j)$, then $x \in (s)$ for some simplex s having v_j as a vertex, and $b_j(x)$ is equal to the barycentric coordinate of x in s relative to the vertex v_j.

Remark. The following facts are easily verified.

(1) $b_j: [K] \to R$ is a continuous function.
(2) $b_j(x) \geq 0$ and $\sum_{j=1}^m b_j(x) = 1$ for each $x \in [K]$.
(3) $x = \sum_{j=1}^m b_j(x)v_j$.
(4) $b_{j_0}(x) \neq 0, b_{j_1}(x) \neq 0, \ldots, b_{j_l}(x) \neq 0$ for some $x \in [K]$ if and only if $v_{j_0}, \ldots, v_{j_l}$ are the vertices of an l-simplex of K.

Definition. Let K be a simplicial complex, and let s be a simplex of K. The *star* of s is the union of all the open simplices (t) of K such that (s) is a face of (t).

Remarks

(1) For $s = v$ a 0-simplex (i.e., a vertex) of K, $\mathrm{St}(s) = \mathrm{St}(v)$, as defined above.
(2) $\mathrm{St}(s)$ is an open set in $[K]$. (This is an elementary consequence of (3).)
(3) If $(s) = (v_{j_0}, \ldots, v_{j_l})$ and $x \in [K]$, then $x \in \mathrm{St}(s)$ if and only if $b_{j_i}(x) \neq 0$ for all $i \in \{0, \ldots, l\}$.
(4) If $(s) = (v_{j_0}, \ldots, v_{j_l})$, then

$$[K] - \mathrm{St}(s) = [x \in [K]; b_{j_i}(x) = 0 \text{ for some } i \in \{0, \ldots, l\}].$$

(5) If s_1 and s are l-simplices of K with $s_1 \neq s$, then $[s_1] \subset [K] - \mathrm{St}(s)$.

Given a smoothly triangulated manifold (X, K, h), we want to define, for each l, an isomorphism from $H^l(X, d)$ onto $H^l(K)$. To do this, note that homomorphisms $\tilde{f}_l: H^l(X, d) \to H^l(K)$ are defined whenever there is given a sequence of linear maps $f_l: C^\infty(X, \Lambda^l(X)) \to C^l(K)$ such that $\partial^* \circ f_l = f_{l+1} \circ d$ for all l.

$$
\begin{array}{ccccc}
\cdots \to C^\infty(X, \Lambda^l(X)) & \xrightarrow{\ d\ } & C^\infty(X, \Lambda^{l+1}(X)) & \to & \cdots \\
\downarrow{\scriptstyle f_l} & & \downarrow{\scriptstyle f_{l+1}} & & \\
\cdots \to C^l(K) & \xrightarrow{\ \partial^*\ } & C^{l+1}(K) & \to & \cdots
\end{array}
$$

For then $f_l(Z^l(X, d)) \subset Z^l(K)$, because $d\omega = 0$ $(\omega \in C^l(X, d))$ implies that

$$\partial^*(f_l(\omega)) = f_{l+1}(d\omega) = f_{l+1}(0) = 0.$$

Also $f_l(B^l(X, d)) \subset B^l(K)$, because $\omega = d\tau$ $(\tau \in C^{l-1}(X, d))$ implies that

$$f_l(\omega) = f_l(d\tau) = \partial^*(f_{l-1}\tau) \in \mathrm{Im}\, \partial^*.$$

Thus f_l induces

$$\tilde{f}_l: H^l(X, d) = Z^l(X, d)/B^l(X, d) \to Z^l(K)/B^l(K) = H^l(K).$$

We now proceed to define such a sequence of linear maps

$$\int_l : C^\infty(X, \Lambda^l(X)) \to C^l(K).$$

For $\omega \in C^\infty(X, \Lambda^l(X))$, $\int_l (\omega)$ will be a linear functional on $C_l(K)$. Thus it suffices to specify the values of $\int_l (\omega)$ on basis elements of $C_l(K)$, that is, on

163

oriented l-simplices $\langle s \rangle$. To do this, consider the smooth map $h_s: U \to X$. Then $h_s^*(\omega)$ is a smooth l-form on U, an open set in the plane of $[s]$; that is, in an l-dimensional Euclidean space. We define $\int_l (\omega)(\langle s \rangle)$ to be the integral of this l-form over $\langle s \rangle$:

$$\int_l (\omega)(\langle s \rangle) = \int_{\langle s \rangle} h_s^*(\omega).$$

In other words, let $(r_1, \ldots, r_l)$ denote coordinates in the plane of $[s]$ consistent with the orientation of $\langle s \rangle$; so if $\langle s \rangle = \langle v_0, \ldots, v_l \rangle$, let $(r_1, \ldots, r_l)$ be coordinates relative to the ordered basis $\{v_1 - v_0, \ldots, v_l - v_0\}$. Then

$$h_s^*(\omega) = g \, dr_1 \wedge \cdots \wedge dr_l$$

for some continuous function g on U, and

$$\int_l (\omega)(\langle s \rangle) = \int_{[s]} g \, dr_1 \cdots dr_l \qquad \text{(Riemann integral)}.$$

Note that this integral is independent of the homeomorphism h; that is, it depends only on the point set $h([s])$ and its orientation by the change of variables theorem for integrals.

$$\textit{Claim:} \qquad \boxed{ \partial^* \circ \int_l = \int_{l+1} \circ \, d. }$$

This is just Stokes's theorem. For given any smooth l-form ω and oriented $(l + 1)$-simplex $\langle s \rangle$,

$$\left[\int_{l+1} \circ \, d(\omega) \right](\langle s \rangle) = \int_{\langle s \rangle} (h_s)^*(d\omega)$$

$$= \int_{\langle s \rangle} d(h_s^*(\omega))$$

$$= \int_{\partial \langle s \rangle} h_s^*(\omega) \qquad \text{(by Stokes's theorem)}$$

$$= \int_l (\omega)(\partial \langle s \rangle)$$

$$= \left[\partial^* \circ \int_l (\omega) \right] \langle s \rangle.$$

Thus $\int_l$ induces a homomorphism $\tilde{\int}_l: H^l(X, d) \to H^l(K)$.

Theorem (De Rham's Theorem). *Let (X, K, h) be a smoothly triangulated manifold. Then*

$$\tilde{\int}_l : H^l(X, d) \to H^l(K)$$

is an isomorphism for each l ($0 \le l \le \dim X$).

This theorem is a consequence of the following two lemmas.

Lemma 1. *There exists a sequence of linear maps*

$$\alpha_l: C^l(K) \to C^\infty(X, \Lambda^l(X)) \qquad (0 \le l \le \dim X)$$

with the following properties.

(1) $d \circ \alpha_l = \alpha_{l+1} \circ \partial^*$.

(2) $\int_l \circ \alpha_l = $ *identity.*

(3) *If c^0 denotes the 0-cochain such that $c^0(v) = 1$ for each vertex v in K, then $\alpha_0(c^0) = 1$; that is, $\alpha_0(c^0)$ is the 0-form equal to the constant function 1.*

(4) *If $\langle s \rangle$ is an oriented l-simplex of K, then the l-form $\alpha_l(\varphi_{\langle s \rangle})$ is identically zero in a neighborhood of $X - \mathrm{St}\,(s)$.*

Lemma 2. *Let ω be a closed l-form on X. Suppose $\int_l (\omega) = \partial^* c$ for some $c \in C^{l-1}(K)$. Then there exists an $(l - 1)$-form τ on X such that $\int_{l-1} (\tau) = c$ and $d\tau = \omega$.*

Remark. Lemma 1 shows that $\tilde{\int}_l$ is surjective. For given $z \in Z^l(K)$, let $\omega = \alpha_l(z)$. Then $\omega \in Z^l(X, d)$ because

$$d\omega = d \circ \alpha_l(z) = \alpha_{l+1} \circ \partial^*(z) = \alpha_{l+1}(0) = 0.$$

Furthermore, $\int_l (\omega) = \int_l \circ \alpha_l(z) = z$. Thus $\int_l: Z^l(X, d) \to Z^l(K)$ is surjective; hence so is $\tilde{\int}_l$. (Note that Property (1) says that the map α_l induces a homomorphism $\tilde{\alpha}_l: H^l(K) \to H^l(X, d)$. Property (2) says that this map is a right inverse to $\tilde{\int}_l$.)

Lemma 2 shows that $\tilde{\int}_l$ is injective. For if $\omega \in Z^l(X, d)$ and $\int_l (\omega) \in B^l(K)$, then $\omega \in B^l(X, d)$ by Lemma 2.

Thus Lemmas 1 and 2 together do, as claimed, imply De Rham's theorem.

PROOF OF LEMMA 1. For notational convenience, we shall identity $[K]$ and X through the homeomorphism h; that is, we shall assume that $[K] = X$ and that $h = $ identity.

Step 1. We first construct a special partition of unity, subordinate to the open covering

$$\{\mathrm{St}\,(v); v \text{ is a vertex of } K\}$$

of X. Let $v_1, \ldots, v_m$ denote the vertices of K. For each $j \in \{1, \ldots, m\}$, let b_j denote the jth barycentric coordinate function on $[K] = X$ and let

$$F_j = \left[x \in X; b_j(x) \ge \frac{1}{n + 1} \right] \qquad (n = \dim X),$$

$$G_j = \left[x \in X; b_j(x) \le \frac{1}{n + 2} \right].$$

Then F_j and G_j are closed sets in X with the following properties (see Figure 6.6).

(1) $F_j \subset \text{St}(v_j)$.
(2) $X - \text{St}(v_j) \subset G_j$.
(3) $F_j \cap G_j = \varnothing$; that is, $F_j \subset G_j'$.

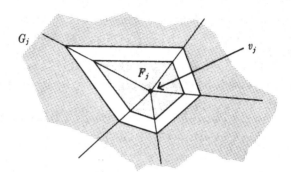

Figure 6.6

(4) There exists a smooth function $f_j \geq 0$ such that $f_j > 0$ on F_j, and $f_j = 0$ on G_j. (F_j is a closed set in the compact space X, hence is compact; thus an $f_j \geq 0$ can be found which is greater than 0 on F_j and equal to 0 outside the open set $G_j' \supset F_j$.)

(5) The closed sets F_j cover X. (Given $x \in X$, then $x \in (s)$ for some simplex $(s) = (v_{j_0}, \ldots, v_{j_l})$ of dimension $l \leq n$. Now $b_j(x) = 0$ for $j \notin \{j_0, \ldots, j_l\}$ and $\sum_{i=0}^{l} b_{j_i}(x) = 1$. Since $l + 1 \leq n + 1$, $b_j(x) \geq 1/(n + 1)$ for some $j \in \{j_0, \ldots, j_l\}$. Thus $x \in F_j$ for this j.) In particular, for each $x \in X$, $f_j(x) \neq 0$ for some j. Furthermore, G_j' is an open covering of X.

(6) From (5), $\sum_{j=1}^{m} f_j > 0$, so that $g_j = f_j / \sum_{k=1}^{m} f_k$ is defined and smooth on X. Furthermore, $\{g_j\}$ is a smooth partition of unity on X subordinate to $\{G_j'\}$; that is, $\sum_{j=1}^{m} g_j = 1$, and g_j vanishes outside G_j'. Since $G_j' \subset \text{St}(v_j)$, the partition of unity $\{g_j\}$ is also subordinate to the open covering $\{\text{St}(v_j)\}$.

Step 2. We shall now define α_l in terms of the smooth functions $\{g_j\}$ defined above. Since α_l is to be linear, it suffices to specify the values of α_l on the generators $\varphi_{\langle s \rangle}$ of $C^l(K)$. For $\langle s \rangle = \langle v_{j_0}, \ldots, v_{j_l} \rangle$ an oriented l-simplex, we define $\alpha_l(\varphi_{\langle s \rangle})$ to be the l-form

$$\alpha_l(\varphi_{\langle s \rangle}) = l! \sum_{i=0}^{l} (-1)^i g_{j_i} \, dg_{j_0} \wedge \cdots \wedge \widehat{dg_{j_i}} \wedge \cdots \wedge dg_{j_l}.$$

Verification of properties (1)–(4)

PROPERTY (1). Clearly,

$$d \circ \alpha_l(\varphi_{\langle s \rangle}) = (l + 1)! \, dg_{j_0} \wedge \cdots \wedge dg_{j_l}.$$

On the other hand,

$$\alpha_{l+1} \circ \partial^*(\varphi_{\langle s \rangle}) = \alpha_{l+1}\left(\sum_{v_k}' \varphi_{\langle v_k, v_{j_0}, \dots, v_{j_l}\rangle}\right)$$

$$= (l+1)! \sum_k' \left[g_k \, dg_{j_0} \wedge \cdots \wedge dg_{j_l} \right.$$

$$\left. - \sum_{i=0}^{l} (-1)^i g_{j_i} \, dg_k \wedge dg_{j_0} \wedge \cdots \wedge \widehat{dg_{j_i}} \wedge \cdots \wedge dg_{j_l} \right].$$

Claim. If the vertices $v_k, v_{j_0}, \dots, v_{j_l}$ are distinct and yet are not the vertices of an $(l+1)$-simplex of K, then

$$g_k \, dg_{j_0} \wedge \cdots \wedge dg_{j_l} \equiv 0.$$

For, if $x \notin \text{St}(v_k)$, then $g_k(x) = 0$. If $x \in \text{St}(v_k)$, then $b_k(x) \neq 0$. But now $b_{j_i}(x) = 0$ for some $i \in \{0, \dots, l\}$, for otherwise $b_k(x) \neq 0, b_{j_0}(x) \neq 0, \dots,$ $b_{j_l}(x) \neq 0$, hence $(v_k, v_{j_0}, \dots, v_{j_l})$ is an $(l+1)$-simplex. But this is a contradiction. Using this i, let

$$U = \left[y \in X; b_{j_i}(y) < \frac{1}{n+2} \right].$$

Then U is an open set in X containing x, and g_{j_i} is identically zero on U because $U \subset G_{j_i}$. Hence $dg_{j_i} \equiv 0$ on U, and, in particular, $dg_{j_i}(x) = 0$. This completes the proof of the claim.

Applying this result to the terms of the above expression for $\alpha_{l+1} \circ \partial^*(\varphi_{\langle s \rangle})$ yields

(A) $$\sum_k' g_k \, dg_{j_0} \wedge \cdots \wedge dg_{j_l} = \sum_{\{k \neq j_0, \dots, j_l\}} g_k \, dg_{j_0} \wedge \cdots \wedge dg_{j_l},$$

since those terms on the right-hand side which do not appear on the left are identically zero; and

(B) $$\sum_k' \sum_{i=0}^{l} (-1)^i g_{j_i} \, dg_k \wedge dg_{j_0} \wedge \cdots \wedge \widehat{dg_{j_i}} \wedge \cdots \wedge dg_{j_l}$$

$$= \sum_{i=0}^{l} (-1)^i \sum_k' g_{j_i} \, dg_k \wedge \cdots \wedge \widehat{dg_{j_i}} \wedge \cdots \wedge dg_{j_l}$$

$$= \sum_{i=0}^{l} (-1)^i \sum_{k \notin \{j_0, \dots, j_l\}} g_{j_i} \, dg_k \wedge dg_{j_0} \wedge \cdots \wedge \widehat{dg_{j_i}} \wedge \cdots \wedge dg_{j_l}$$

$$= \sum_{i=0}^{l} (-1)^i \sum_{k \neq j_i} g_{j_i} \, dg_k \wedge dg_{j_0} \wedge \cdots \wedge \widehat{dg_{j_i}} \wedge \cdots \wedge dg_{j_l}$$

$$= \sum_{i=0}^{l} (-1)^i g_{j_i}\left(\sum_{k \neq j_i} dg_k\right) \wedge dg_{j_0} \wedge \cdots \wedge \widehat{dg_{j_i}} \wedge \cdots \wedge dg_{j_l}$$

$$= \sum_{i=0}^{l} (-1)^i g_{j_i}(-dg_{j_i}) \wedge dg_{j_0} \wedge \cdots \wedge \widehat{dg_{j_i}} \wedge \cdots \wedge dg_{j_l}$$

$$= -\sum_{i=0}^{l} g_{j_i} \, dg_{j_0} \wedge \cdots \wedge dg_{j_l} \qquad \left(\text{since } \sum_{k=1}^{m} g_k = 1 \Rightarrow \sum_{k=1}^{m} dg_k = 0\right).$$

Hence, subtracting (B) from (A),

$$\alpha_{l+1} \circ \partial^*(\varphi_{\langle s \rangle}) = (l+1)! \left(\sum_{k=1}^{m} g_k \right) dg_{j_0} \wedge \cdots \wedge dg_{j_1}$$
$$= (l+1)! \, dg_{j_0} \wedge \cdots \wedge dg_{j_1}$$
$$= d \circ \alpha_l(\varphi_{\langle s \rangle}).$$

PROPERTY (3). Since $\alpha_0(\varphi_{\langle v_j \rangle}) = g_j$,

$$\alpha_0(c^0) = \alpha_0 \left(\sum_{j=1}^{m} \varphi_{\langle v_j \rangle} \right) = \sum_{j=1}^{m} g_j = 1.$$

PROPERTY (4). Suppose $\langle s \rangle = \langle v_{j_0}, \ldots, v_{j_1} \rangle$. Then

$$\alpha_l(\varphi_{\langle s \rangle}) = l! \sum_{j=1}^{l} (-1)^i g_{j_i} \, dg_{j_0} \wedge \cdots \wedge \widehat{dg_{j_i}} \wedge \cdots \wedge dg_{j_1}.$$

Note that if $x \in X$ is such that $b_{j_k}(x) < 1/(n+2)$ for some $k \in \{0, \ldots, l\}$, then $x \in G_{j_k}$, so that g_{j_k} and dg_{j_k}, and hence $\alpha_l(\varphi_{\langle s \rangle})$, are zero at x. Thus $\alpha_l(\varphi_{\langle s \rangle})$ is identically zero on

$$\left[x \in X; b_{j_k}(x) < \frac{1}{n+2} \text{ for some } k \in \{0, \ldots, l\} \right],$$

which is an open set containing $X - \mathrm{St}\,(s)$.

PROPERTY (2). PROOF BY INDUCTION. For $l = 0$, $\int_0 \circ \alpha_0(\varphi_{\langle v_j \rangle})$ $(j \in \{1, \ldots, m\})$ is the 0-cochain given by

$$\left[\int_0 \circ \alpha_0(\varphi_{\langle v_j \rangle}) \right] (\langle v_k \rangle) = \left[\int_0 (g_j) \right] \langle v_k \rangle = g_j(v_k).$$

But note that $g_j(v_k) = 0$ for $k \neq j$ since $v_k \notin \mathrm{St}\,(v_j)$ and $g_j = 0$ outside $\mathrm{St}\,(v_j)$. Furthermore,

$$1 = \sum_{j=1}^{m} g_j(v_k) = g_k(v_k) \qquad \text{(for each } k\text{)}.$$

Hence

$$\left[\int_0 \circ \alpha_0(\varphi_{\langle v_j \rangle}) \right] (\langle v_k \rangle) = \begin{cases} 1 & \text{(if } k = j) \\ 0 & \text{(if } k \neq j) \end{cases}$$
$$= \varphi_{\langle v_j \rangle}(\langle v_k \rangle).$$

Since this holds for all j and k, $\int_0 \circ \alpha_0 = $ identity as required.

Now assume Property (2) for dimension $l - 1$. For $\langle s \rangle$ and $\langle t \rangle$ oriented l-simplices of K,

$$\left[\int_l \circ \alpha_l(\varphi_{\langle s \rangle}) \right] \langle t \rangle = \int_{\langle t \rangle} \alpha_l(\varphi_{\langle s \rangle}).$$

We must show that this equals 1 if $\langle s \rangle = \langle t \rangle$, and 0 if $s \neq t$. That this is zero for $s \neq t$ is a consequence of Property (4) since $\alpha_l(\varphi_{\langle s \rangle})$ is identically zero in a neighborhood of $X - \text{St}(s) \supset [t]$. So we need only check that $\int_{\langle s \rangle} \alpha_l(\varphi_{\langle s \rangle}) = 1$. For this, let $\langle r \rangle = \langle v_{j_1}, \ldots, v_{j_l} \rangle$ and $s = \langle v_{j_0}, v_{j_1}, \ldots, v_{j_l} \rangle$. Then

$$\int_{\langle s \rangle} \alpha_l(\partial^* \varphi_{\langle r \rangle}) = \int_{\langle s \rangle} d[\alpha_{l-1}(\varphi_{\langle r \rangle})]$$

$$= \int_{\partial \langle s \rangle} \alpha_{l-1}(\varphi_{\langle r \rangle}).$$

But $\partial \langle s \rangle = \langle r \rangle$ plus an alternating sum of other oriented $(l-1)$-simplices, so

$$\int_{\partial \langle s \rangle} \alpha_{l-1}(\varphi_{\langle r \rangle}) = \int_{\langle r \rangle} \alpha_{l-1}(\varphi_{\langle r \rangle}) = 1$$

by induction. Hence

$$1 = \int_{\langle s \rangle} \alpha_l(\partial^* \varphi_{\langle r \rangle}) = \int_{\langle s \rangle} \alpha_l(\varphi_{\langle s \rangle} + \text{terms of type } \varphi_{\langle t \rangle} \ (t \neq s))$$

$$= \int_{\langle s \rangle} \alpha_l(\varphi_{\langle s \rangle}). \qquad \square$$

In order to prove Lemma 2, we shall need the following lemma.

Lemma 3. *Let s be a k-simplex in R^n.*

(a$_r$) *Suppose $r \geq 0$ and $k \geq 1$. Let ω be a smooth closed r-form defined "near" $[s^{k-1}]$; that is, defined in a neighborhood of $[s^{k-1}]$. If $k = r + 1$, assume further that $\int_{\partial \langle s \rangle} \omega = 0$. Then there exists a smooth closed r-form τ defined near $[s]$ such that $\tau = \omega$ near $[s^{k-1}]$.*

(b$_r$) *Suppose $r \geq 1$ and $k \geq 1$. Let ω be a smooth closed r-form defined near $[s]$. Suppose τ is a smooth $(r-1)$-form defined near $[s^{k-1}]$ such that $d\tau = \omega$ near $[s^{k-1}]$. If $k = r$, assume further that $\int_{\partial \langle s \rangle} \tau = \int_{\langle s \rangle} \omega$. Then there exists a smooth $(r-1)$-form τ' defined near $[s]$ such that $\tau' = \tau$ near $[s^{k-1}]$, and $d\tau' = \omega$ near $[s]$.*

Remark. That the integral conditions are necessary in (a$_r$) and (b$_r$) is a consequence of Stokes's theorem. For in (a$_r$), if τ exists, then

$$\int_{\partial \langle s \rangle} \omega = \int_{\partial \langle s \rangle} \tau = \int_{\langle s \rangle} d\tau = \int_{\langle s \rangle} 0 = 0;$$

and in (b$_r$), if τ' exists, then

$$\int_{\langle s \rangle} \omega = \int_{\langle s \rangle} d\tau' = \int_{\partial \langle s \rangle} \tau' = \int_{\partial \langle s \rangle} \tau.$$

PROOF OF LEMMA 3. PROOF BY INDUCTION. We shall first verify (a$_0$) and then establish that

$$(a_0) \Rightarrow (b_1) \Rightarrow (a_1) \Rightarrow (b_2) \Rightarrow \cdots.$$

(a_0): ω is a smooth 0-form; that is, a smooth function defined near $[s^{k-1}]$; and $d\omega = 0$. Hence ω is constant on the components of its domain. If $k > 1$, $[s^{k-1}]$ is connected, so ω is a constant function c in a neighborhood of $[s^{k-1}]$. Set $\tau = c$ in a neighborhood of $[s]$. If $k = 1$, then $\langle s \rangle = \langle v_0, v_1 \rangle$ for some pair of vertices v_0, v_1, and

$$0 = \int_{\partial\langle s \rangle} \omega = \omega(v_1) - \omega(v_0).$$

Thus the constant value of ω near v_1 equals the constant value of ω near v_0; that is, ω is constant near $[s^{k-1}]$ as before. Once again, set τ equal to this constant function on a neighborhood of $[s]$.

(a_{r-1}) $\Rightarrow$ (b_r): ω is a closed r-form ($r \geq 1$) defined on an open set containing $[s]$. By Poincaré's lemma (Section 5.2), ω is exact near $[s]$; that is, there exists a smooth $(r-1)$-form τ_1 defined near $[s]$ such that $d\tau_1 = \omega$ near $[s]$. (To see that Poincaré's lemma applies here, we need only note that any open set containing $[s]$ must contain another open set about $[s]$ which is diffeomorphic (smoothly homeomorphic with a smooth inverse) to an open ball. In fact, we can choose a star-shaped region containing $[s]$.) Now in general, τ_1 will not be equal to τ near $[s^{k-1}]$. Consider the difference $\tau_1 - \tau$ near $[s^{k-1}]$. It is closed since, near $[s^{k-1}]$,

$$d(\tau_1 - \tau) = \omega - \omega = 0.$$

Furthermore, if $k = (r - 1) + 1 = r$, then

$$\int_{\partial\langle s \rangle} (\tau_1 - \tau) = \int_{\partial\langle s \rangle} \tau_1 - \int_{\partial\langle s \rangle} \tau$$

$$= \int_{\langle s \rangle} d\tau_1 - \int_{\partial\langle s \rangle} \tau$$

$$= \int_{\langle s \rangle} \omega - \int_{\partial\langle s \rangle} \tau$$

$$= 0 \qquad \text{(by hypothesis).}$$

Thus we can apply (a_{r-1}) to the form $\tau_1 - \tau$. There exists a smooth closed $(r-1)$-form μ defined near $[s]$ such that $\mu = \tau_1 - \tau$ near $[s^{k-1}]$. Let $\tau' = \tau_1 - \mu$. Then τ' is a smooth $(r-1)$-form defined near $[s]$ such that $\tau' = \tau_1 - \mu = \tau$ near $[s^{k-1}]$, and $d\tau' = d\tau_1 - d\mu = \omega - 0 = \omega$ near $[s]$.

(b_r) $\Rightarrow$ (a_r): $\langle s \rangle = \langle v_0, \ldots, v_k \rangle$ for some choice of vertices $v_0, \ldots, v_k$; let $\langle t \rangle = \langle v_1, \ldots, v_k \rangle$. Let $F = [s^{k-1}] - (t)$. Since ω is closed, Poincaré's lemma asserts the existence of a smooth $(r-1)$-form μ defined near F such that $d\mu = \omega$ near F. (F is star-shaped; hence any open set containing F contains a star-shaped neighborhood U of F) (see Figure 6.7). In particular, $d\mu = \omega$ near $[t^{k-2}]$.

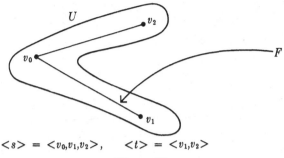

$$\langle s \rangle = \langle v_0, v_1, v_2 \rangle, \qquad \langle t \rangle = \langle v_1, v_2 \rangle$$

Figure 6.7

If $k > 1$, we would like to apply (b_r) to the forms ω and μ and the $(k-1)$-simplex t. In order to do this we must check if $k - 1 = r$, then $\int_{\langle t \rangle} \omega - \int_{\partial \langle t \rangle} \mu = 0$. But, letting $c = \partial \langle s \rangle - \langle t \rangle$ so that $\partial c = -\partial \langle t \rangle$,

$$
\begin{aligned}
\int_{\langle t \rangle} \omega - \int_{\partial \langle t \rangle} \mu &= \int_{\langle t \rangle} \omega + \int_{\partial c} \mu \\
&= \int_{\langle t \rangle} \omega + \int_c d\mu \\
&= \int_{\langle t \rangle} \omega + \int_c \omega \qquad \text{(since each simplex of c is contained} \\
&\qquad\qquad\qquad\qquad\quad \text{in F and $d\mu = \omega$ near F)} \\
&= \int_{\partial \langle s \rangle} \omega \\
&= 0 \qquad\qquad\qquad\qquad \text{(by hypothesis).}
\end{aligned}
$$

Applying (b_r), there exists a form μ' defined near $[t]$ such that $\mu' = \mu$ near $[t^{k-2}]$ and $d\mu' = \omega$ near $[t]$. Let μ_2 be the form defined near $[s^{k-1}]$ by glueing together μ and μ' along their common domain, an open set where they agree (Figure 6.8). Then $d\mu_2 = \omega$ near $[s^{k-1}]$, since both μ and μ' have this property in their domains of definition.

If $k = 1$, then $[s^{k-1}]$ consists of two vertices v_0, v_1. Since ω is closed, Poincaré's lemma guarantees the existence of smooth $(r-1)$-forms μ_i near v_i

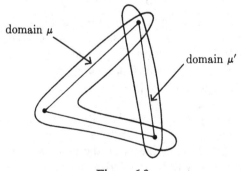

domain μ

domain μ'

Figure 6.8

($i = 0, 1$), with $d\mu_i = \omega$. By shrinking domains, we can assume (domain μ_0) and (domain μ_1) are disjoint (Figure 6.9). This defines μ_2 near $[s^{k-1}]$, with $d\mu_2 = \omega$ near $[s^{k-1}]$ as before.

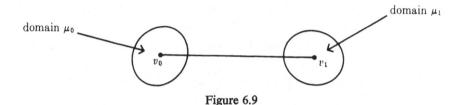

Figure 6.9

Finally, let f be a smooth function which is identically 1 in a small neighborhood of $[s^{k-1}]$, and identically zero outside the domain of μ_2. Then $f\mu_2$ is a smooth $(r-1)$-form defined near $[s]$. Let $\tau = d(f\mu_2)$. Then τ is a closed r-form defined near $[s]$, and we have near $[s^{k-1}]$

$$\tau = d(f\mu_2) = df \wedge \mu_2 + f \, d\mu_2 = d\mu_2 = \omega,$$

since $f \equiv 1$ and $df \equiv 0$ near $[s^{k-1}]$. $\qquad\square$

PROOF OF LEMMA 2. We shall construct inductively a sequence

$$\tau_0, \tau_1, \ldots, \tau_n = \tau \qquad (n = \dim X)$$

of $(l-1)$-forms such that

(1) τ_k is defined in a neighborhood of the k-skeleton $[K^k]$ of K,
(2) $d\tau_k = \omega$ near $[K^k]$,
(3) $\tau_k = \tau_{k-1}$ near $[K^{k-1}]$, and
(4) $\int_{l-1} (\tau_{l-1}) = c$.

Note that this will prove the lemma because (4) implies that for each oriented $(l-1)$-simplex $\langle s \rangle$ of $[K]$ and each $k \geq l - 1$,

$$\int_{l-1} (\tau_k)(\langle s \rangle) = \int_{\langle s \rangle} \tau_k = \int_{\langle s \rangle} \tau_{l-1} = \int_{l-1} (\tau_{l-1})(\langle s \rangle) = c(\langle s \rangle).$$

To construct τ_0, cover K^0 by a collection of mutually disjoint balls. Since ω is closed, ω is exact in each of these balls by Poincaré's lemma. Hence there exists a smooth $(l-1)$-form τ_0', defined on the union of these balls, such that $d\tau_0' = \omega$ there. If $l - 1 \neq 0$, take $\tau_0 = \tau_0'$. If $l - 1 = 0$, we want $\int_0 (\tau_0) = c$. But for v_j a vertex of $[K]$,

$$\int_0 (\tau_0')(\langle v_j \rangle) = \int_{\langle v_j \rangle} \tau_0' = \tau_0'(v_j).$$

Let $a_j = c(v_j) - \tau_0'(v_j)$, and define τ_0 on the ball about v_j by $\tau_0 = \tau_0' + a_j$. Then $d\tau_0 = d\tau_0' = \omega$ near $[K^0]$, and $\int_0 (\tau_0) = c$ as required.

172

Now assume that τ_{k-1} has been constructed with Properties (1)–(4). To construct τ_k, note that if we can find, for each k-simplex s, a smooth $(l-1)$-form $\tau_k(s)$ defined in a neighborhood of $[s]$ such that $d(\tau_k(s)) = \omega$ near $[s]$, and $\tau_k(s) = \tau_{k-1}$ near $[s^{k-1}]$, then glueing will yield a smooth $(l-1)$-form τ_k' satisfying (1)–(3).

To construct $\tau_k(s)$, we shall apply (b_l) of Lemma 3. Note that ω is a smooth closed l-form defined near $[s]$ and that τ_{k-1} is a smooth $(l-1)$-form defined near $[s^{k-1}]$ such that $d\tau_{k-1} = \omega$ near $[s^{k-1}]$. Furthermore, if $k = l$, then

$$\int_{\langle s \rangle} \omega = \int_l (\omega)(\langle s \rangle) \qquad (\langle s \rangle = s \text{ together with either orientation})$$

$$= \partial^* c(\langle s \rangle) \qquad \text{(by hypothesis)}$$

$$= c(\partial \langle s \rangle)$$

$$= \int_{k-1} (\tau_{k-1})(\partial \langle s \rangle) \qquad \text{(by (4) since } k = l)$$

$$= \int_{\partial \langle s \rangle} \tau_{k-1}.$$

Hence we can apply (b_l). There exists a smooth $(l-1)$-form $\tau_k(s)$ near $[s]$ such that $\tau_k(s) = \tau_{k-1}$ near $[s^{k-1}]$ and $d(\tau_k(s)) = \omega$ near $[s]$.

This constructs τ_k' satisfying (1)–(3). If $k \neq l-1$, set $\tau_k = \tau_k'$. If $k = l-1$, we have τ_{l-1}' satisfying (1)–(3), and we want τ_{l-1} such that $\int_{l-1}(\tau_{l-1}) = c$. Let $c_1 = c - \int_{l-1}(\tau_{l-1}')$, and define τ_{l-1} in a neighborhood of $[K^{l-1}]$ by

$$\tau_{l-1} = \tau_{l-1}' + \alpha_{l-1}(c_1),$$

where α_{l-1} is the linear map $C^{l-1}(K) \to C^\infty(X, \Lambda^{l-1}(X))$ defined in Lemma 1.

For each r and each oriented r-simplex $\langle s \rangle$, note that $\alpha_r(\varphi_{\langle s \rangle})$ is identically zero on a neighborhood of $X - \text{St}(s)$. In particular, $\alpha_r(\varphi_{\langle s \rangle})$ is identically zero near $[K^{r-1}]$. Since each $c \in C^r(K)$ is a linear combination of such $\varphi_{\langle s \rangle}$, we have $\alpha_r(c)$ identically zero near $[K^{r-1}]$ for each r-cochain c.

Applying this first with $r = l$, then with $r = l-1$, we find

$$d\tau_{l-1} = d\tau_{l-1}' + d \circ \alpha_{l-1}(c_1) = d\tau_{l-1}' + \alpha_l \circ \partial^*(c_1) = d\tau_{l-1}' = \omega$$

near $[K^{l-1}]$ and

$$\tau_{l-1} = \tau_{l-1}' + \alpha_{l-1}(c_1) = \tau_{l-1}' = \tau_{l-2}$$

near $[K^{l-2}]$. Thus τ_{l-1} satisfies (1)–(3) with $k = l-1$. Property (4) is also satisfied:

$$\int_{l-1}(\tau_{l-1}) = \int_{l-1}(\tau_{l-1}') + \int_{l-1} \circ \alpha_{l-1}(c_1)$$

$$= (c - c_1) + c_1$$

$$= c. \qquad \square$$

Remark 1. De Rham's theorem shows that the simplicial cohomology groups (with coefficients in R) of a smoothly triangulated manifold (X, K, h)

are isomorphic to the De Rham cohomology groups of X. In particular, these groups are independent of the triangulation (K, h) of X. Since the cohomology groups are dual to the homology groups, the groups $H_i(K, R)$, for $[K]$ a smooth manifold, also depend on $[K]$ only, not on the particular simplicial subdivision K.

Remark 2. The direct sum $\sum_{l=0}^{\dim X} \oplus H^l(X, d)$ can be given the structure of an associative algebra as follows. Recall that $\sum_{l=0}^{\dim X} \oplus C^\infty(X, \Lambda^l(X))$ is an associative algebra under exterior multiplication $\wedge$. $\sum \oplus Z^l(X, d)$ is a subalgebra, for if $d\omega = 0$ and $d\tau = 0$, then

$$d(\omega \wedge \tau) = (d\omega) \wedge \tau \pm \omega \wedge (d\tau) = 0.$$

$\sum \oplus B^l(X, d)$ is an ideal in $\sum \oplus Z^l(X, d)$, for if $\omega = d\mu$ and $d\tau = 0$, then $\omega \wedge \tau = d(\mu \wedge \tau)$. Hence

$$\sum_l \oplus H^l(X, d) = \sum_l \oplus (Z^l(X, d)/B^l(X, d)) \cong \sum_l \oplus \left(Z^l(X, d) \bigg/ \sum_l \oplus B^l(X, d) \right)$$

is also an associative algebra. In particular, $\sum_l \oplus H^l(X, d)$ is a ring, called the *De Rham cohomology ring* of X.

It is also possible to define a product, called the *cup product*, of simplicial cochains in such a way that $\sum_l \oplus H^l(K)$ becomes an algebra. It can be shown that the isomorphism $\int : \sum_l \oplus H^l(X, d) \to \sum_l \oplus H^l(K)$ is then an algebra isomorphism.

Remark 3. Lemma 3 contains in disguise a proof that

$$H^l(S^n, d) = \begin{cases} 0 & (\text{if } 0 < l < n), \\ R & (\text{if } l = 0, n). \end{cases}$$

For if ω is a closed l-form $(0 < l < n)$ defined on a neighborhood of the n-skeleton $[s^n]$ of an $(n + 1)$-simplex s, then it was shown that ω extends to a closed (and hence exact) l-form on $[s]$. This implies that $H^l(S^n, d) = 0$ for $0 < l < n$. It was shown for $l = n$ that any closed n-form ω, defined near $[s^n]$ such that $\int_{\partial\langle s \rangle} \omega = 0$, is also exact. The map $Z^n(S^n, d) \to R$ defined by $\omega \to \int_{\partial\langle s \rangle} \omega$ is then a homomorphism with kernel $B^n(S^n, d)$. Hence $H^n(S^n, d) = R$. Also $H^0(S^n, d) = R$ because S^n is connected.

We have tacitly assumed here that any closed l-form ω on S^n can be extended to a closed l-form defined in a neighborhood of S^n. $\Psi^*\omega$ is such an extension, where $\Psi: R^{n+1} - \{0\} \to S^n$ is radial projection.

Intrinsic Riemannian geometry of surfaces

7

7.1 Parallel translation and connections

Definition. Let M be an oriented Riemannian manifold of dimension 2. Let $T(M)$ denote the tangent bundle of M. Let

$$S(M) = [(m, v) \in T(M); \langle v, v \rangle = 1].$$

$S(M)$ is called the *sphere bundle*, or *circle bundle*, of M.

The notation (m, v) for a point of $T(M)$ {or $S(M)$} is redundant since $v \in T(M, m)$. Nevertheless, we use it to emphasize that v is a tangent vector at m.

Remarks

(1) $S(M)$ is a smooth manifold of dimension 3. The function $f: T(M) \to R^1$, given by $f(m, v) = \langle v, v \rangle$, is smooth, and $df \neq 0$ whenever $f = 1$, so the implicit function theorem applies.

(2) Note that the circle $S^1 = [z \in C; |z| = 1]$ is a group under (complex) multiplication. Since $e^{i\theta_1} \cdot e^{i\theta_2} = e^{i(\theta_1 + \theta_2)}$, the group S^1 is just the group of rotations of the oriented plane R^2. This group *acts* on $S(M)$: there exists a smooth map

$$A: S^1 \times S(M) \to S(M)$$

given by

$$A(g, (m, v)) = (m, gv) \qquad (g \in S^1; (m, v) \in S(M)),$$

where gv is the image of the vector v under rotation by g in the oriented plane $T(M, m)$ (Figure 7.1). So, if $g = e^{i\theta}$, and $\{v_1, v_2\}$ is any oriented orthonormal basis for $T(M, m)$, then $v = c_1 v_1 + c_2 v_2$ for some $c_1, c_2 \in R^1$, and

$$gv = (c_1 \cos \theta - c_2 \sin \theta)v_1 + (c_1 \sin \theta + c_2 \cos \theta)v_2.$$

We shall often denote $A(g, (m, v))$ by $g(m, v)$. Then $g: S(M) \to S(M)$ is a smooth map for each $g \in S^1$.

gv

θ

v

Figure 7.1

(3) If $\pi: S(M) \to M$ denotes projection, then $\pi^{-1}(m)$ is just the unit circle in $T(M, m)$. Moreover, if (m, v_1) and (m, v_2) are any two elements of $\pi^{-1}(m)$, then there exists a unique $g \in S^1$ such that $(m, v_2) = g(m, v_1)$. (Take $g = e^{i\theta}$, where θ is the positive angle of rotation from v_1 to v_2.)

(4) $S(M)$ is locally a product space. For let U be a coordinate neighborhood in M, with coordinate functions (x_1, x_2). Let e_1 be the vector field $(\partial/\partial x_1)/\|\partial/\partial x_1\|$, where $\|\partial/\partial x_1\| = \langle(\partial/\partial x_1), (\partial/\partial x_1)\rangle^{1/2}$. Then e_1 is a smooth vector field on U, which is everywhere of length 1. Thus e_1 defines a smooth map

$$c: U \to \pi^{-1}(U) \quad \text{by} \quad c(m) = (m, e_1(m)).$$

Clearly $\pi \circ c = i_U$. Now define $B: U \times S^1 \to \pi^{-1}(U)$ by

$$B(m, g) = gc(m) = (m, ge_1(m)) = A(g, (m, e_1(m))).$$

Then it is easy to verify that B is smooth, injective, and surjective; and that dB is everywhere nonsingular so that B^{-1} is also smooth.

(5) It is not true that $S(M)$ is globally a product of S^1 with M. If there exists a smooth nonzero vector field on M, then the above argument shows that $S(M)$ is diffeomorphic with $M \times S^1$. However, there do not exist such nonzero vector fields in general. (For example, $M = S^2$.)

For $M = R^2$, the notion of translating a tangent vector parallel to itself is clear. We now propose to generalize it and introduce the concept of *parallel translation* of tangent vectors on arbitrary 2-dimensional oriented Riemannian manifolds. It will turn out that we will be able to parallel translate vectors along curves from one point to another, but that the result will depend on the curve. In particular, if we parallel translate around a closed curve, we may not get back to our original vector. The new vector will differ from the original vector by a rotation; i.e., by an element of S^1. For $M = R^2$, a "flat" space, this rotation is zero. For arbitrary M, this rotation (or, more precisely, the limit of it as the curve shrinks to a point m) will measure the "curvature" of M at m.

We develop this notion of parallel translation in order to obtain an intrinsic meaning for curvature of the Riemannian manifold M; that is, a meaning independent of any ambient space in which M may lie. In Chapter 8 we will interpret this curvature differently when M is a submanifold of R^3.

We shall require that parallel translation be an isometry. Thus, parallel translation of a unit vector along a curve $\alpha: [a, b] \to M$ will determine a unit tangent vector $\tilde{\alpha}(t) \in T(M, \alpha(t))$ for each $t \in [a, b]$. If $v \in \pi^{-1}(\alpha(a))$, then parallel translation of v will determine a curve $\tilde{\alpha}: [a, b] \to S(M)$ such that $\pi \circ \tilde{\alpha} = \alpha$. Moreover, if

$$v_1 \in \pi^{-1}(\alpha(a)) \quad \text{and} \quad v_1 = gv \quad (g \in S^1),$$

then the curve $\tilde{\alpha}_1: [a, b] \to S(M)$, determined by parallel translating v_1, will be given by

$$\tilde{\alpha}_1(t) = g\tilde{\alpha}(t) \quad (t \in [a, b]).$$

Conversely, if, corresponding to each curve $\alpha: [a, b] \to M$ and each unit tangent vector v at $\alpha(a)$, there existed a unique "lift" $\tilde{\alpha}: [a, b] \to S(M)$, with the above properties, then a notion of parallel translation is defined (see Figure 7.2).

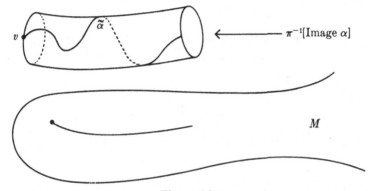

Figure 7.2

Recall that in the theory of covering spaces, each curve had a unique lift because the fibers $p^{-1}(x)$ were discrete. However, here the fibers $\pi^{-1}(m)$ are not discrete; they are circles. Hence lifts are not unique. In fact, we do not even know in which direction to start moving. (There is a whole line of vectors $\tilde{v} \in T(S(M), (m, v))$ such that $d\pi(\tilde{v}) = \dot{\alpha}(a)$; each of these is a candidate for $\dot{\tilde{\alpha}}(a)$.) So given $m \in M$ and $v \in T(M, m)$, we need a way of determining, for each curve α through m, an initial direction for $\tilde{\alpha}$; that is, we need a way of choosing, for each $\dot{\alpha}(a) \in T(M, m)$, a vector $\dot{\tilde{\alpha}}(a) \in T(S(M), (m, v))$ such that $d\pi(\dot{\tilde{\alpha}}(a)) = \dot{\alpha}(a)$. Choosing the vector $\dot{\tilde{\alpha}}(a)$ is more primitive than finding the lift $\tilde{\alpha}$, but it will turn out that when the choice is made at every point of $\pi^{-1}(\alpha([a, b]))$, the lift—hence the parallel translate—is determined.

A natural way of uniquely determining such a vector $\dot{\tilde{\alpha}}(a)$ would be to require that it lie in a given two-dimensional subspace of $T(S(M), (m, v))$ that is mapped isomorphically onto $T(M, m)$ by $d\pi$. Such a subspace will be complementary to the vertical space

$$d\pi^{-1}(0) = [t \in T(S(M), (m, v)); d\pi(t) = 0].$$

Definition. A *connection* on $S(M)$ is a choice of a two-dimensional subspace $\mathcal{H}(m, v)$ of $T(S(M), (m, v))$ at each point $(m, v) \in S(M)$ such that the following hold.

(1) $T(S(M), (m, v)) = \mathcal{H}(m, v) \oplus d\pi^{-1}(0)$; that is, the subspace $\mathcal{H}(m, v)$ is complementary to the vertical space at (m, v).
(2) $dg(\mathcal{H}(m, v)) = \mathcal{H}(m, gv)$ for each $g \in S^1$.
(3) The choice of $\mathcal{H}$ is smooth; that is, for each point $(m, v) \in S(M)$, there exists an open set U about (m, v) and smooth vector fields X and Y defined on U such that $\{X, Y\}$ spans $\mathcal{H}$ at each point of U.

Remark. There exists a smooth vector field V on $S(M)$ such that V spans the vertical space $d\pi^{-1}(0)$ at each point of $S(M)$. It is constructed as follows. Let $\partial/\partial\theta$ denote the usual unit tangent vector field on S^1. Then $\partial/\partial\theta$ is invariant under the action of $g \in S^1$ on S^1; that is,

$$dg\left(\frac{\partial}{\partial\theta}\right)\bigg|_h = \frac{\partial}{\partial\theta}\bigg|_{gh} \qquad \text{for each } h \in S^1.$$

For $(m, v) \in S(M)$, consider the smooth map $A^1 : S^1 \to S(M)$ defined by

$$A^1(g) = g(m, v) = (m, gv).$$

Define (see Figure 7.3)

$$V(m, v) = dA^1\left(\frac{\partial}{\partial\theta}\bigg|_1\right).$$

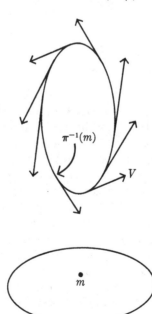

Figure 7.3

In terms of a local coordinate neighborhood U of m in M and of the corresponding direct sum representation

$$T(S(M), (m, v)) = T(M, m) \oplus T(S^1, g) \qquad ((m, v) \in \pi^{-1}(U)),$$

where g is such that $v = ge_1$, the vector field V is given by

$$V(m, v) = \left(0, \frac{\partial}{\partial \theta}\Big|_g\right).$$

In particular, note that V is smooth and never zero, that $d\pi(V) = 0$, and that $dh(V) = V$ for each $h \in S^1$.

Definition. Let $\mathscr{H}$ be a connection on $S(M)$. The 1-*form of* $\mathscr{H}$, or the *connection* 1-*form*, is the 1-form φ on $S(M)$ defined as follows. Let

$$q: T(S(M), (m, v)) = \mathscr{H}(m, v) \oplus d\pi^{-1}(0) \to d\pi^{-1}(0)$$

be the projection map. For $t \in T(S(M), (m, v))$, set $\varphi(t) = \lambda$, where λ is the real number such that $q(t) = \lambda V(m, v)$.

Local description of φ. Let X and Y be smooth vector fields defined in an open set U of $S(M)$ such that $\{X(m, v), Y(m, v)\}$ spans $\mathscr{H}(m, v)$ for each $(m, v) \in U$. Then $\{V(m, v), X(m, v), Y(m, v)\}$ is a basis for $T(S(M), (m, v))$ at each $(m, v) \in U$. Let $\{\varphi_1(m, v), \varphi_2(m, v), \varphi_3(m, v)\}$ be the dual basis for $T^*(S(M), (m, v))$. Then $\varphi_1, \varphi_2, \varphi_3$ are smooth 1-forms on U, and $\varphi = \varphi_1$. In particular,

(1) φ is smooth, since φ_1 is smooth.
(2) $\varphi(V) \equiv 1$.
(3) $g^*\varphi = \varphi$ for each $g \in S^1$. For if $t \in T(S(M), (m, v))$, then

$$t = \lambda V + t_1 \qquad (\lambda \in R; t_1 \in \mathscr{H}),$$

and

$$
\begin{aligned}
(g^*\varphi)(t) &= \varphi \circ dg(t) \\
&= \varphi(\lambda \, dg(V) + dg(t_1)) \\
&= \varphi(\lambda V) \qquad \text{(since } dg(\mathscr{H}) \subset \mathscr{H}) \\
&= \lambda \\
&= \varphi(t).
\end{aligned}
$$

Lemma. *Suppose ψ is any smooth 1-form on $S(M)$ such that $\psi(V) \equiv 1$ and $g^*\psi = \psi$. Then $\mathscr{H} = \psi^{-1}(0)$ is a connection on $S(M)$ with the property that its connection 1-form is ψ.*

PROOF. For each $(m, v) \in S(M)$, $\psi(m, v): T(S(M), (m, v)) \to R$ is a linear functional. Since $\dim T(S(M), (m, v)) = 3$, $\psi^{-1}(0)$ has dimension 2. $V \notin \psi^{-1}(0)$, so $\psi^{-1}(0)$ is a complement to the vertical space. $dg(\psi^{-1}(0)) = \psi^{-1}(0)$ because $g^*\psi = \psi$. $\qquad \square$

Remark. Let U be a coordinate neighborhood in M. We now exhibit a connection on $\pi^{-1}(U) = S(U) = U \times S^1$. Recall that, given coordinates (x_1, x_2) in U, a smooth map $c\colon U \to \pi^{-1}(U)$ is defined by

$$c(m) = (m, (\partial/\partial x_1)/\|\partial/\partial x_1\|).$$

For $m \in M$, let

$$\mathscr{H}_1(c(m)) = dc(T(U, m)).$$

Then $\mathscr{H}_1(c(m))$ is complementary to the vertical. For

$$d\pi(\mathscr{H}_1(c(m))) = d\pi \circ dc(T(U, m)) = d(\pi \circ c)(T(U, m)) = T(U, m)$$

so that $\mathscr{H}_1(c(m))$ is two-dimensional and $d\pi|_{\mathscr{H}_1(c(m))}$ is an isomorphism. Furthermore, $V \notin \mathscr{H}_1$ since $d\pi(V) = 0$.

Now set $\mathscr{H}_1(gc(m)) = dg(\mathscr{H}_1(c(m)))$.

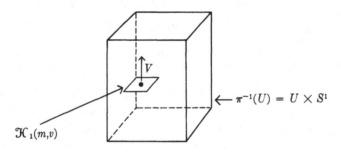

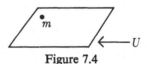

Figure 7.4

In terms of the product representation $\pi^{-1}(U) = U \times S^1$ given by c, $\mathscr{H}_1(m, v)$ is just the tangent space at (m, v) to the submanifold $U \times \{v\}$ (Figure 7.4). More precisely, letting $B\colon U \times S^1 \to \pi^{-1}(U)$ be the isomorphism defined by $B(m, g) = gc(m) = (m, ge_1(m))$,

$$\mathscr{H}_1(m, v) = dB(T(U \times \{g\}, (m, g))),$$

where $g \in S^1$ is such that $ge_1(m) = v$. The 1-form φ_1 of this connection is

$$\varphi_1 = (B^{-1})^*(\widetilde{d\theta}),$$

where $p\colon U \times S^1 \to S^1$ is projection, $d\theta$ is the 1-form on S^1 dual to $\partial/\partial\theta$, and

$$\widetilde{d\theta} = p^*(d\theta).$$

Note that $d\varphi_1 = 0$ for this special connection, for

$$\begin{aligned}
d\varphi_1 &= d[(B^{-1})^* \circ p^*(d\theta)] = d[(p \circ B^{-1})^*(d\theta)] \\
&= (p \circ B^{-1})^*(d(d\theta)) = 0.
\end{aligned}$$

Warning. $d(d\theta) = 0$, not because $d\theta$ is the differential of a 0-form (it is not), but because there are no nonzero 2-forms on S^1.

Our definition of a connection was motivated by a desire to construct a notion of parallel translation. We now prove that given a connection on $S(M)$, parallel translation is indeed defined.

Theorem. *Let $\mathcal{H}$ be a connection on $S(M)$ with 1-form φ. Let $\alpha: [a, b] \to M$ be a broken C^∞ curve in M. Let $v \in T(M, \alpha(a))$ with $\|v\| = 1$. Then there exists a unique broken C^∞ curve $\tilde{\alpha}: [a, b] \to S(M)$, called the* horizontal lift *of α, through $(\alpha(a), v)$, such that*

(1) $\pi \circ \tilde{\alpha} = \alpha$.
(2) $\dot{\tilde{\alpha}}(t) \in \mathcal{H}(\tilde{\alpha}(t))$; *that is,* $\varphi(\dot{\tilde{\alpha}}(t)) = 0$ *for all $t \in [a, b]$.*
(3) $\tilde{\alpha}(a) = (\alpha(a), v)$.

The vector $\tilde{\alpha}(b) \in T(M, \alpha(b))$ is the parallel translate *of v along α to $\alpha(b)$.*

The proof of this theorem requires two preliminary lemmas.

Lemma 1. *Let $\mathcal{H}_1$ and $\mathcal{H}_2$ be two connections on $S(M)$ with connection 1-forms φ_1 and φ_2. Then*

(1) $(\varphi_2 - \varphi_1)(V) \equiv 0$.
(2) $g^*(\varphi_2 - \varphi_1) = \varphi_2 - \varphi_1$ *(for all $g \in S^1$).*
(3) $\varphi_2 - \varphi_1 = \pi^*(\tau)$ *(for some smooth 1-form τ on M).*

PROOF. (1) and (2) are clear. We shall show that (1) and (2) imply (3). If ψ is any smooth 1-form on $S(M)$ with $\psi(V) \equiv 0$ and $g^*\psi = \psi$ for all $g \in S^1$, then $\psi = \pi^*(\tau)$ for some τ. To define τ on $v \in T(M, m)$, let $(m, v_1) \in \pi^{-1}(m)$, and let $w \in T(S(M), (m, v_1))$ be such that $d\pi(w) = v$. Set $\tau(v) = \psi(w)$. $\tau(v)$ is independent of the w chosen in $d\pi^{-1}(v)$ since $d\pi(w_1) = v$ implies that $d\pi(w_1 - w) = 0$, so that $w_1 - w = \lambda V$ for some λ. Thus

$$\psi(w_1) = \psi(w + \lambda V) = \psi(w) + \lambda\psi(V) = \psi(w).$$

Also, $\tau(v)$ is independent of the point (m, v_1) chosen in $\pi^{-1}(m)$ because if

$$(m, v_2) \in \pi^{-1}(m),$$

then $v_2 = gv_1$ for some $g \in S^1$. Moreover, if $w \in T(S(M), (m, v_1))$ satisfies $d\pi(w) = v$, then $dg(w) \in T(S(M), (m, v_2))$ satisfies $d\pi(dg(w)) = v$, and

$$\psi|_{(m, v_2)}(dg(w)) = \psi|_{g(m, v_1)}(dg(w))$$
$$= g^*\psi|_{(m, v_1)}(w) = \psi|_{(m, v_1)}(w).$$

τ is smooth because in a coordinate neighborhood U, $\tau(v) = \psi \circ dc(v)$, where

$$c: U \to \pi^{-1}(U)$$

is defined by $c(m) = (m, e_1(m))$. $\qquad\square$

Lemma 2. *Let $\alpha: [a, b] \to M$ be a smooth curve in M. Let $\tilde{\alpha}: [a, b] \to S(M)$ and $\tilde{\beta}: [a, b] \to S(M)$ be smooth curves such that $\pi \circ \tilde{\alpha} = \alpha$ and $\pi \circ \tilde{\beta} = \alpha$*

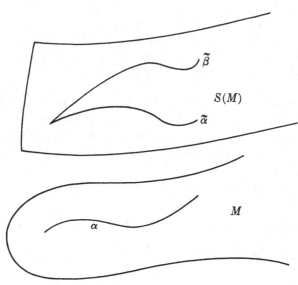

Figure 7.5

(see Figure 7.5). Suppose $\tilde{\alpha}$ is horizontal relative to some connection $\mathscr{H}$ on $S(M)$ with connection 1-form φ; that is, suppose $\varphi(\dot{\tilde{\alpha}}(t)) \equiv 0$. Then there exists a smooth function $\theta: [a, b] \to R$ such that

(1) $\tilde{\beta}(t) = e^{i\theta(t)}\tilde{\alpha}(t) \quad (t \in [a, b])$ *and*
(2) $\varphi(\dot{\tilde{\beta}}(t)) = (d\theta/dt)(t) \quad (t \in [a, b])$.

Furthermore, if $\tilde{\alpha}(a) = \tilde{\beta}(a)$, then θ can be chosen such that $\theta(a) = 0$.

PROOF. Let $\hat{g}: [a, b] \to S^1$ be defined by

$$\tilde{\beta}(t) = \hat{g}(t)\tilde{\alpha}(t) \qquad (t \in [a, b]).$$

It is easy to verify that $\hat{g}$ is a smooth curve. Since R is a covering space of S^1, and $[a, b]$ is simply connected, there exists a lift $\theta: [a, b] \to R$ of $\hat{g}$ (see Figure 7.6). Furthermore, if $\tilde{\alpha}(a) = \tilde{\beta}(a)$, then $\hat{g}(a) = 1$, and there exists a unique such lift with $\theta(a) = 0$.

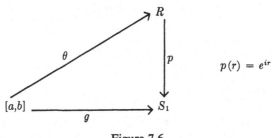

Figure 7.6

Since p is smooth and has a smooth inverse locally, θ is smooth. Furthermore,

$$\tilde{\beta}(t) = \hat{g}(t)\tilde{\alpha}(t) = p \circ \theta(t)\tilde{\alpha}(t) = e^{i\theta(t)}\tilde{\alpha}(t)$$

so (1) is satisfied.

To verify (2), first note that the tangent vector to the curve

$$\hat{g}: [a, b] \to S^1 \qquad [\hat{g}(t) = e^{i\theta(t)}]$$

is given by

$$\dot{\hat{g}}(t) = (p \circ \theta)\dot{}(t)$$
$$= d(p \circ \theta)\left(\frac{d}{dt}\right) = dp\left(d\theta\left(\frac{d}{dt}\right)\right) = dp\left(\frac{d\theta}{dt}\frac{d}{dt}\right) = \frac{d\theta}{dt}\,dp\left(\frac{d}{dt}\right) = \frac{d\theta}{dt}\frac{\partial}{\partial\theta}.$$

Restricting attention to a coordinate neighborhood $U \subset M$ and the corresponding product representation $\pi^{-1}(U) \cong U \times S^1$,

$$\tilde{\alpha}(t) = h(t)c(\alpha(t)) = (\alpha(t), h(t))$$

for some $h(t) = e^{i\psi(t)} \in S^1$, and

$$\tilde{\beta}(t) = (\alpha(t), \hat{g}(t)h(t)) = (\alpha(t), e^{i(\theta(t) + \psi(t))}).$$

The tangent vector to $\tilde{\alpha}$ at $\tilde{\alpha}(t)$ is then $(\dot{\alpha}(t), (d\psi/dt)(\partial/\partial\theta))$, whereas the tangent vector to $\tilde{\beta}$ at $\tilde{\beta}(t)$ is $(\dot{\alpha}(t), [(d\theta/dt) + (d\psi/dt)](\partial/\partial\theta))$; that is,

$$\dot{\tilde{\beta}}(t) = d(\hat{g}(t))(\dot{\tilde{\alpha}}(t)) + \left(0, \frac{d\theta}{dt}\frac{\partial}{\partial\theta}\right) = d(\hat{g}(t))(\dot{\tilde{\alpha}}(t)) + \frac{d\theta}{dt}\,V,$$

where $d(\hat{g}(t))$ is the differential of the map $\hat{g}(t): S(M) \to S(M)$. Since $\dot{\tilde{\alpha}}(t)$ is horizontal, and $d(\hat{g}(t))(\mathcal{H}) \subset \mathcal{H}$,

$$\varphi(\dot{\tilde{\beta}}(t)) = \frac{d\theta}{dt}\,\varphi(V) = \frac{d\theta}{dt}. \qquad \square$$

PROOF OF THE THEOREM. Note that it suffices to prove the theorem for a smooth curve α. For then we can uniquely lift each smooth portion of any broken curve.

Local existence. Let U be a coordinate neighborhood in M. We shall show the existence of unique horizontal lifts in $\pi^{-1}(U) = S(U)$. Let $c: U \to S(U)$ be as usual: $c(m) = (m, e_1(m))$ for $m \in U$. We shall first show that if $\mathcal{H}$ is the special connection $\mathcal{H}_1$ on $\pi^{-1}(U)$, constructed via the product structure, then α has a unique horizontal lift $\tilde{\alpha}_1$ such that $\tilde{\alpha}_1(a) = c(\alpha(a))$. Indeed, let $\tilde{\alpha}_1: [a, b] \to \pi^{-1}(U)$ be defined by $\tilde{\alpha}_1 = c \circ \alpha$. Then $\pi \circ \tilde{\alpha}_1 = \pi \circ c \circ \alpha = \alpha$ and $\dot{\tilde{\alpha}}_1(t) = dc(\dot{\alpha}(t)) \in \mathcal{H}_1(c(t))$, so $\tilde{\alpha}_1$ is a horizontal lift. Moreover, $\tilde{\alpha}_1$ is the unique $\mathcal{H}_1$-horizontal lift such that

$$\tilde{\alpha}_1(a) = c(\alpha(a)).$$

For if $\tilde{\alpha}_2$ were another such lift, then, by Lemma 2,

$$\tilde{\alpha}_2(t) = e^{i\theta(t)}\tilde{\alpha}_1(t)$$

for some smooth function θ with $\theta(a) = 0$; and $\varphi_1(\dot{\tilde{\alpha}}_2(t)) = d\theta/dt$, where φ_1 is the connection 1-form of $\mathcal{H}_1$. Now $\tilde{\alpha}_2$ is $\mathcal{H}_1$-horizontal if and only if

$\varphi_1(\overset{\circ}{\alpha}_2(t)) \equiv 0$; that is, $d\theta/dt \equiv 0$. Hence $\theta(t)$ must be constant. Since $\theta(a) = 0$, $\theta(t) \equiv 0$; that is, $\tilde{\alpha}_2(t) = \tilde{\alpha}_1(t)$ for all t; that is, $\tilde{\alpha}_2 = \tilde{\alpha}_1$.

Thus α admits a unique $\mathscr{H}_1$-horizontal lift $\tilde{\alpha}_1$ with $\tilde{\alpha}_1(a) = c(\alpha(a))$. Now consider our original connection with connection 1-form φ. Then, by Lemma 1,

$$\varphi_1 - \varphi = \pi^*\tau$$

for some smooth 1-form τ on U. Let $\tilde{\alpha}$ be any curve in $\pi^{-1}(U)$ such that $\pi \circ \tilde{\alpha} = \alpha$. Then, by Lemma 2, $\tilde{\alpha}(t) = e^{i\theta(t)}\tilde{\alpha}_1(t)$ and $\varphi_1(\overset{\circ}{\tilde{\alpha}}(t)) = d\theta/dt$. Thus $\tilde{\alpha}$ is an $\mathscr{H}$-horizontal lift of α if and only if $\varphi(\overset{\circ}{\tilde{\alpha}}(t)) \equiv 0$; that is, if and only if

$$(\varphi_1 - \varphi)(\overset{\circ}{\tilde{\alpha}}(t)) \equiv \varphi_1(\overset{\circ}{\tilde{\alpha}}(t)) = \frac{d\theta}{dt}.$$

But on the other hand,

$$(\varphi_1 - \varphi)(\overset{\circ}{\tilde{\alpha}}(t)) = (\pi^*\tau)(\overset{\circ}{\tilde{\alpha}}(t)) = \tau(d\pi\overset{\circ}{\tilde{\alpha}}(t)) = \tau(\overset{\circ}{\alpha}(t)).$$

Thus α is $\mathscr{H}$-horizontal if and only if $d\theta/dt = \tau(\overset{\circ}{\alpha}(t))$; that is, $\theta = \int_0^t \tau(\overset{\circ}{\alpha}(t))\,dt + \theta_0$ for some constant θ_0. Hence each $\mathscr{H}$-horizontal lift $\tilde{\alpha}$ of α is of the form

$$\tilde{\alpha}(t) = \hat{g}(t)(c \circ \alpha(t)),$$

where

$$\hat{g}(t) = e^{i\theta_0}e^{i\int_0^t \tau(\overset{\circ}{\alpha}(t))dt}$$

For each unit vector v in $T(U, \alpha(a))$, there is precisely one θ_0 with $0 \leq \theta_0 < 2\pi$ and $(\alpha(a), v) = e^{i\theta_0}(\alpha(a), e_1)$. The above formula, with this value of θ_0, then gives the unique $\mathscr{H}$-horizontal lift $\tilde{\alpha}$ with $\tilde{\alpha}(a) = (\alpha(a), v)$.

Global existence. To establish global existence, let $\alpha: [a, b] \to M$ and let

$$t_0 = \sup\,[t \in [a, b]; \ \alpha|_{[a,b]} \text{ has a (unique) lift } \tilde{\alpha}].$$

We shall show that $t_0 = b$. Suppose $t_0 \neq b$. Then consider the restriction of α to the interval $[t_0 - \varepsilon, t_0 + \varepsilon]$. By local existence, this has a unique lift $\tilde{\tilde{\alpha}}$ for some sufficiently small $\varepsilon > 0$, say with $\tilde{\tilde{\alpha}}(t_0) = (\alpha(t_0), w) \in S(M)$. Then $\tilde{\alpha}(t_0 - \varepsilon) = g\tilde{\tilde{\alpha}}(t_0 - \varepsilon)$ for some $g \in S^1$, and $g\tilde{\tilde{\alpha}}$ is a horizontal lift with $g\tilde{\tilde{\alpha}}(t_0 - \varepsilon) = \tilde{\alpha}(t_0 - \varepsilon)$. By uniqueness, $g\tilde{\tilde{\alpha}} = \tilde{\alpha}$ on the interval $[t_0 - \varepsilon, t_0)$. Hence $g\tilde{\tilde{\alpha}}$ extends $\tilde{\alpha}$ beyond t_0, contradicting the definition of t_0. $\square$

Remark. Note that, relative to the special connection $\mathscr{H}_1$ on $\pi^{-1}(U)$, parallel translation is independent of the curve. In fact, the vector field $e_1 = (\partial/\partial x_1)/\|\partial/\partial x_1\|$ is parallel along every curve in U.

7.2 Structural equations and curvature

Definition. Consider the circle bundle $S(M)$ of a smooth oriented Riemannian 2-manifold M. Two smooth 1-forms ω_1 and ω_2 are defined on $S(M)$ as follows. For $t \in T(S(M), (m, v))$,

$$\omega_1(t) = \langle d\pi(t), v \rangle,$$
$$\omega_2(t) = \langle d\pi(t), iv \rangle,$$

where $iv = e^{i\pi/2}v$ is the image of v under rotation through an angle of $\pi/2$ in $T(M, m)$. (We shall show below that these 1-forms are indeed smooth.)

Remark 1. Thus $\omega_1(t)$ and $\omega_2(t)$ are the components of $d\pi(t)$ relative to the orthonormal basis $\{v, iv\}$ for $T(M, m)$; that is,

$$d\pi(t) = \omega_1(t)v + \omega_2(t)(iv).$$

Remark 2. Suppose $\mathcal{H}$ is a connection on $S(M)$ with connection 1-form φ. Then $\{\varphi, \omega_1, \omega_2\}$ is a basis for $T^*(S(M), (m, v))$ for each $(m, v) \in S(M)$.

PROOF. Since $\dim T^*(S(M), (m, v)) = 3$, it suffices to show that there exists no nonzero $t \in T(S(M), (m, v))$ that is simultaneously annihilated by these three forms—for then these forms are linearly independent. But if $\omega_1(t) = \omega_2(t) = 0$, then $d\pi(t) = 0$, so t is vertical; that is, $t = \lambda V$ for some λ. Furthermore, if $\varphi(t) = 0$, then $\lambda = \varphi(\lambda V) = \varphi(t) = 0$, so $t = 0$. $\qquad\square$

Remark 3. Let $g = e^{i\theta} \in S^1$. Then

$$g^*\omega_1 = (\cos \theta)\omega_1 + (\sin \theta)\omega_2,$$
$$g^*\omega_2 = -(\sin \theta)\omega_1 + (\cos \theta)\omega_2.$$

PROOF. $gv = (\cos \theta)v + (\sin \theta)(iv)$. Hence, for $t \in T(S(M), (m, v))$,

$$
\begin{aligned}
g^*\omega_1|_{(m,v)}(t) &= \omega_1|_{(m,gv)}(dg(t)) \\
&= \langle d\pi \circ dg(t), gv \rangle \\
&= \langle d\pi(t), gv \rangle \\
&= \langle d\pi(t), (\cos \theta)v + (\sin \theta)iv \rangle \\
&= (\cos \theta)\omega_1(t) + (\sin \theta)\omega_2(t).
\end{aligned}
$$

Similarly,

$$g^*\omega_2(t) = -(\sin \theta)\omega_1(t) + (\cos \theta)\omega_2(t). \qquad\square$$

Remark 4. $g^*(\omega_1 \wedge \omega_2) = \omega_1 \wedge \omega_2$ for all $g = e^{i\theta} \in S^1$. For,

$$g^*(\omega_1 \wedge \omega_2) = g^*\omega_1 \wedge g^*\omega_2 = (\cos^2 \theta + \sin^2 \theta)\omega_1 \wedge \omega_2 = \omega_1 \wedge \omega_2.$$

Furthermore, $\omega_1 \wedge \omega_2(t_1, t_2) = 0$ if either t_1 or t_2 is vertical. Hence, as in the proof that $\varphi - \varphi_1 = \pi^*\tau$ for some τ (Section 7.1), the 2-form $\omega_1 \wedge \omega_2$ is the image under π^* of a (unique) form on M.

Definition. The *volume element* of a smooth oriented Riemannian 2-manifold M is the smooth 2-form, vol, on M such that

$$\pi^*(\text{vol}) = \omega_1 \wedge \omega_2;$$

that is, for $v_1, v_2 \in T(M, m)$, vol $(v_1, v_2) = \omega_1 \wedge \omega_2|_{(m,v)}(v_1', v_2')$ for any

$$(m, v) \in \pi^{-1}(m) \subset S(M) \quad \text{and} \quad v_1', v_2' \in T(S(M), (m, v))$$

such that $d\pi(v_i') = v_i$ $(i = 1, 2)$.

Remark 5. Suppose U is a coordinate neighborhood in M with coordinate functions (x_1, x_2). Let $e_1 = (\partial/\partial x_1)/\|\partial/\partial x_1\|$ and let ω_1', ω_2' be the smooth 1-forms on U that at each $m \in U$ form the basis for $T^*(M, m)$ dual to $\{e_1(m), ie_1(m)\}$. Let

$$c: U \to \pi^{-1}(U) \subset S(M)$$

be given by $c(m) = (m, e_1(m))$. Then, for $v \in T(M, m)$,

$$(c^*\omega_1)(v) = \omega_1(dc(v)) = \langle d\pi \circ dc(v), e_1 \rangle = \langle v, e_1 \rangle = \omega_1'(v),$$

so $\omega_1' = c^*\omega_1$. Similarly, $\omega_2' = c^*\omega_2$. In particular,

$$\omega_1' \wedge \omega_2' = c^*\omega_1 \wedge c^*\omega_2 = c^*(\omega_1 \wedge \omega_2) = c^* \circ \pi^*(\text{vol}) = (\pi \circ c)^*(\text{vol});$$

so, since $\pi \circ c = i_U$,

$$\boxed{\text{vol}|_U = \omega_1' \wedge \omega_2'.}$$

Now let $\tilde\omega_i = \pi^*\omega_i'$ $(i = 1, 2)$. Then $\tilde\omega_1$ and $\tilde\omega_2$ are smooth 1-forms on $\pi^{-1}(U) \subset S(M)$ and

$$\tilde\omega_1 \wedge \tilde\omega_2 = \pi^*\omega_1' \wedge \pi^*\omega_2' = \pi^*(\omega_1' \wedge \omega_2') = \pi^*(\text{vol}) = \omega_1 \wedge \omega_2.$$

Moreover, at each point $(m, e_1(m))$ of $c(U)$, $\omega_i = \tilde\omega_i$. For, if

$$t \in T(S(M), (m, e_1)),$$

then

$$\begin{aligned}
\omega_1(t)e_1 + \omega_2(t)(ie_1) &= d\pi(t) \\
&= \omega_1'(d\pi(t))e_1 + \omega_2'(d\pi(t))(ie_1) \\
&= \tilde\omega_1(t)e_1 + \tilde\omega_2(t)(ie_1).
\end{aligned}$$

Note further that, for $g = e^{i\theta} \in S^1$,

$$g^*\tilde\omega_i = g^* \circ \pi^*\omega_i' = (\pi \circ g)^*\omega_i' = \pi^*\omega_i' = \tilde\omega_i.$$

Thus, from Remark 3 above,

$$\begin{aligned}
(g^*\omega_1)|_{(m,e_1)} &= (\cos\theta)\omega_1 + (\sin\theta)\omega_2|_{(m,e_1)} \\
&= (\cos\theta)\tilde\omega_1 + (\sin\theta)\tilde\omega_2|_{(m,e_1)} \\
&= (\cos\theta)g^*\tilde\omega_1 + (\sin\theta)g^*\tilde\omega_2|_{(m,e_1)}.
\end{aligned}$$

Applying $(g^{-1})^*$, the forms ω_1 and $\tilde\omega_i$ at (m, ge_1) are related by

$$\omega_1 = (\cos\theta)\tilde\omega_1 + (\sin\theta)\tilde\omega_2.$$

Similarly,

$$\omega_2 = -(\sin\theta)\tilde\omega_1 + (\cos\theta)\tilde\omega_2.$$

In particular, the above formulae show that ω_1 and ω_2 are smooth.

Remark 6. For higher dimensional Riemannian manifolds, the volume element is obtained similarly. If U is a coordinate neighborhood in the oriented Riemannian manifold M, with coordinate functions $(x_1, \ldots, x_n)$

such that $dx_1 \wedge \cdots \wedge dx_n$ gives the orientation of U, consider the vector fields $\partial/\partial x_1, \ldots, \partial/\partial x_n$. Using the Gram–Schmidt orthogonalization process, we obtain smooth vector fields $e_1, \ldots, e_n$ on U which form an orthonormal basis for the tangent space at each point. Let $\omega_1', \ldots, \omega_n'$ be the dual 1-forms. Then the n-form $\text{vol}|_U = \omega_1' \wedge \cdots \wedge \omega_n'$ is independent of the (oriented) coordinate system on U and thus defines a global nonzero n-form vol.

Given an oriented Riemannian 2-manifold M and a connection on $S(M)$ with connection 1-form φ, the 1-forms φ, ω_1, ω_2 form a basis for the cotangent space at each point of $S(M)$. Hence the 2-forms $\omega_1 \wedge \omega_2$, $\omega_1 \wedge \varphi$, $\omega_2 \wedge \varphi$ form a basis for the 2-forms at each point of $S(M)$. Hence $d\varphi$, $d\omega_1$, $d\omega_2$ can be expressed in terms of this basis. The resulting formulae are called the *Cartan structural equations*. We now derive them, beginning with the second structural equation.

Second structural equation. On $\pi^{-1}(U)$, for a coordinate neighborhood U, let φ_1 denote the connection 1-form of the special connection $\mathcal{H}_1$. Then $d\varphi_1 = 0$ so that

$$d\varphi = d\varphi - d\varphi_1 = d(\varphi - \varphi_1) = d(\pi^* \tau) = \pi^*(d\tau)$$

for some smooth 1-form τ on U. Now $d\tau$ is a 2-form on U, hence is a multiple of the volume element; that is, $d\tau = -K \, \text{vol}$ for some smooth function K on U. Thus

$$d\varphi = \pi^*(-K \, \text{vol}) = \pi^*(-K)\pi^*(\text{vol})$$

or

$$\boxed{d\varphi = -(K \circ \pi)\omega_1 \wedge \omega_2.}$$

The smooth function K is independent of the coordinates used, since it is determined by this last formula. Thus K is a smooth function on M, called the *curvature* of the connection φ.

First structural equation. On $\pi^{-1}(U)$, for a coordinate neighborhood U, we have seen that at $e^{i\theta}c(m)$,

$$\omega_1 = (\cos \theta)\tilde{\omega}_1 + (\sin \theta)\tilde{\omega}_2,$$
$$\omega_2 = -(\sin \theta)\tilde{\omega}_1 + (\cos \theta)\tilde{\omega}_2.$$

Now

$$d\tilde{\omega}_i = d(\pi^* \omega_i') = \pi^*(d\omega_i') = \pi^*(a_i \, \text{vol}) = (a_i \circ \pi)\omega_1 \wedge \omega_2$$

for some smooth function a_i on U. Thus setting $\tilde{a}_i = a_i \circ \pi$,

$$dw_1 = -(\sin \theta) \, d\theta \wedge \tilde{\omega}_1 + (\cos \theta)\tilde{a}_1\omega_1 \wedge \omega_2$$
$$+ \cos \theta \, d\theta \wedge \tilde{\omega}_2 + (\sin \theta)\tilde{a}_2\omega_1 \wedge \omega_2$$
$$= d\theta \wedge \omega_2 + (\tilde{a}_1 \cos \theta + \tilde{a}_2 \sin \theta)\omega_1 \wedge \omega_2.$$

If $\mathcal{H}$ is the special connection $\mathcal{H}_1$ on $\pi^{-1}(U)$, then $\varphi_1 = d\theta$, thus for this special connection,

$$(*) \qquad d\omega_1 = \varphi_1 \wedge \omega_2 + b_1\omega_1 \wedge \omega_2$$

for some smooth function b_1 on $\pi^{-1}(U)$. Similarly,

$$d\omega_2 = -\varphi_1 \wedge \omega_1 + b_2\omega_1 \wedge \omega_2.$$

For an arbitrary connection form φ, $\varphi_1 - \varphi = \pi^*\tau$ for some smooth 1-form $\tau = c_1\omega_1' + c_2\omega_2'$ on U. Hence

$$
\begin{aligned}
\varphi_1 - \varphi &= \pi^*(c_1\omega_1' + c_2\omega_2') \\
&= (c_1 \circ \pi)\tilde{\omega}_1 + (c_2 \circ \pi)\tilde{\omega}_2 \\
&= f_1\omega_1 + f_2\omega_2
\end{aligned}
$$

for some smooth functions f_1, f_2 on $\pi^{-1}(U)$, since $\tilde{\omega}_1, \tilde{\omega}_2$ span the same space at each point as ω_1, ω_2. Thus

$$\varphi_1 = \varphi + f_1\omega_1 + f_2\omega_2,$$

and, by substituting into (*),

$$
\begin{aligned}
d\omega_1 &= \varphi \wedge \omega_2 + f_1\omega_1 \wedge \omega_2 + b_1\omega_1 \wedge \omega_2 \\
&= \varphi \wedge \omega_2 + (f_1 + b_1)\omega_1 \wedge \omega_2.
\end{aligned}
$$

This, together with the corresponding equation for $d\omega_2$, gives the first structural equations as follows:

$$
\boxed{
\begin{aligned}
d\omega_1 &= \varphi \wedge \omega_2 + h_1\omega_1 \wedge \omega_2, \\
d\omega_2 &= -\varphi \wedge \omega_1 + h_2\omega_1 \wedge \omega_2,
\end{aligned}
}
$$

where h_1, h_2 are smooth functions on $S(M)$. Note that although these equations were derived over a coordinate neighborhood, they are independent of coordinates. Thus they are valid globally.

Although one might expect that by choosing an appropriate connection φ on $S(M)$, the coefficients of $d\omega_i$ relative to the basis $\{\varphi \wedge \omega_1, \varphi \wedge \omega_2, \omega_1 \wedge \omega_2\}$ could be prescribed fairly arbitrarily, this is not the case. In fact, $d\omega_1$ never has a component in the $\varphi \wedge \omega_1$ direction, and $d\omega_2$ never has a component in the $\varphi \wedge \omega_2$ direction. Moreover, the components of $d\omega_1$ and $d\omega_2$ in the $\varphi \wedge \omega_2$ and $\varphi \wedge \omega_1$ directions, respectively, must always be $+1$ and -1.

It is natural to ask whether the first structural equations can be made simpler by an appropriate choice of connection on $S(M)$. In particular, can φ be chosen such that $h_1 \equiv 0$ and $h_2 \equiv 0$? The answer is yes, and the choice is unique.

Theorem. *Let M be an oriented Riemannian 2-manifold. Then there exists a unique connection ψ on $S(M)$ such that*

$$
\boxed{
\begin{aligned}
d\omega_1 &= \psi \wedge \omega_2, \\
d\omega_2 &= -\psi \wedge \omega_1.
\end{aligned}
}
$$

This connection is called the **Riemannian connection.**

PROOF. Let φ be any connection on $S(M)$. If ψ is any other connection on $S(M)$, then, as above,

$$\varphi - \psi = x_1\omega_1 + x_2\omega_2$$

for some x_1 and x_2. Solving for φ and substituting in the first structural equations for φ, we obtain

$$d\omega_1 = \psi \wedge \omega_2 + (h_1 + x_1)\omega_1 \wedge \omega_2,$$
$$d\omega_2 = -\psi \wedge \omega_1 + (h_2 + x_2)\omega_1 \wedge \omega_2.$$

Thus

$$d\omega_1 = \psi \wedge \omega_2,$$
$$d\omega_2 = -\psi \wedge \omega_1$$

if and only if $x_1 = -h_1$, $x_2 = -h_2$. This gives both existence and uniqueness. $\quad\square$

The Cartan structural equations have a dual formulation in terms of vector fields. Let V, E_1, E_2 be the smooth vector fields on $S(M)$ that form the dual basis to $\varphi, \omega_1, \omega_2$. Then E_1 and E_2 are horizontal at each point since $\varphi(E_1) = \varphi(E_2) = 0$. Moreover,

$$d\pi(E_1(m, v)) = \omega_1(E_1)v + \omega_2(E_1)(iv) = v,$$

so $E_1(m, v)$ is the unique horizontal vector at (m, v) whose image under $d\pi$ is v. Similarly, $d\pi(E_2(m, v)) = iv$. The structural equations then become

$$\boxed{\begin{aligned} [V, E_1] &= E_2, \\ [V, E_2] &= -E_1, \\ [E_1, E_2] &= (K \circ \pi)V - h_1E_1 - h_2E_2. \end{aligned}}$$

If $\varphi = \psi$, the 1-form of the Riemannian connection, the last boxed equation reduces to

$$\boxed{[E_1, E_2] = (K \circ \pi)V.}$$

To verify these equations, apply the formula

$$d\tau(V_1, V_2) = \tfrac{1}{2}\{V_1\tau(V_2) - V_2\tau(V_1) - \tau([V_1, V_2])\}$$

nine times, as τ runs through the set $\{\varphi, \omega_1, \omega_2\}$, and V_1, V_2 runs through the set $\{V, E_1, E_2\}$.

Remark. If K is constant, these formulae show that $\{V, E_1, E_2\}$ spans a finite-dimensional Lie algebra.

From now on, for an oriented Riemannian 2-manifold M, let the connection chosen be the Riemannian connection, and let K be the curvature function for that connection.

7.3 Interpretation of curvature

We now show that the curvature K of M measures the amount of rotation obtained in parallel translating vectors around small closed curves in M. The intuitive reason is this. On $S(M)$ we have the vector fields E_1, E_2, and V, and we know that for the Riemannian connection,

$$[E_1, E_2] = (K \circ \pi)V.$$

But $[E_1, E_2](m, v)$ is just the tangent vector to the curve through (m, v) obtained by following the integral curves of E_1 and E_2 forward and then backward through parameter distances of $\sqrt{s}$ (Figure 7.7; see Section 5.3).

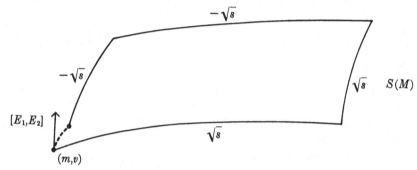

Figure 7.7

Projecting this figure down to M, we obtain a rectangular-shaped figure which is "nearly" closed; that is, the curve obtained through m has zero tangent vector at m because it is the projection of $[E_1, E_2](m, v)$, which is vertical (see Figure 7.8).

Now the integral curves in $S(M)$ are the horizontal lifts of the curves in M; that is, these curves are obtained by parallel translating v around the curves in M. The endpoints of the curve through (m, v)—dotted in Figure 7.9—essentially differ by an element of S^1, namely the rotation $g = e^{i\theta}$, which sends v into its parallel translate around the rectangle in M.

Since the area of the rectangle in M is approximately $\sqrt{s} \cdot \sqrt{s} = s$, the

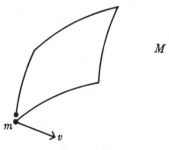

Figure 7.8

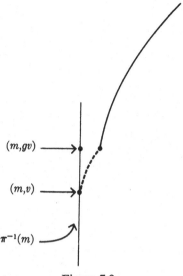

(m, gv) ⟶

(m, v) ⟶

$\pi^{-1}(m)$ ⟶

Figure 7.9

limit as $s \to 0$ of the angle of rotation θ divided by the area of the rectangle is equal to the coefficient of V, namely $K(m)$.

Stated precisely and in somewhat greater generality, the result we have been discussing is as follows.

Theorem 1. *Let M be an oriented Riemannian 2-manifold. Let $\langle s \rangle$ be an oriented 2-simplex in R^2, and let $h: [s] \to M$ be a map which has a smooth extension mapping a neighborhood of $[s]$ into M. Let α be the closed broken C^∞ curve in M obtained by restricting h to $\partial \langle s \rangle$. Then the rotation obtained by parallel translation around the closed curve α is*

$$e^{i \int_{\langle s \rangle} h^*(K \, \mathrm{vol})}$$

so that the angle of rotation is $\int_{\langle s \rangle} h^*(K \, \mathrm{vol})$.

Remark. Note that this result contains the result discussed above. To obtain $K(m)$, take the limit of $\int_{\langle s \rangle} h^*(K \, \mathrm{vol}) / \int_{\langle s \rangle} h^*(\mathrm{vol})$ as $\langle s \rangle$ shrinks to zero and $h(\langle s \rangle)$ shrinks to m. However, the theorem says more. For example, it is possible to have $K > 0$ on $h([s])$ and still get a trivial rotation upon parallel translating around α, namely when the total angle of rotation $\int_{\langle s \rangle} h^*(K \, \mathrm{vol})$ is an integer multiple of 2π.

PROOF OF THEOREM 1. Let $\langle s \rangle = \langle v_0, v_1, v_2 \rangle$ for some vertices v_0, v_1, v_2, and let $w_0 \in T(M, h(v_0))$ be a unit vector. The lines in $[s]$ through v_1 cover $[s]$; their images under h are curves in M which cover $h([s])$. Let $\tilde{h}: [s] \to S(M)$ be obtained by mapping each of these curves into its horizontal lift in $S(M)$ through

$$(h(v_1), w_1) \in S(M),$$

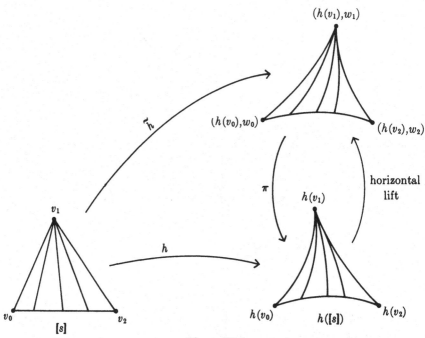

Figure 7.10

where w_1 is the parallel translate of w_0 along the curve $\alpha|_{\langle v_0, v_1 \rangle}$ to $h(v_1)$ (Figure 7.10).

By construction, $\pi \circ \tilde{h} = h$. Moreover, $\tilde{h}$ has a smooth extension mapping a neighborhood of $[s]$ into $S(M)$. This may be checked via local coordinates; we omit the computation.

Now

$$\int_{\langle s \rangle} h^*(K \,\text{vol}) = \int_{\langle s \rangle} (\pi \circ \tilde{h})^*(K \,\text{vol})$$

$$= \int_{\langle s \rangle} \tilde{h}^*[\pi^*(K \,\text{vol})]$$

$$= \int_{\langle s \rangle} \tilde{h}^*[(K \circ \pi)\omega_1 \wedge \omega_2]$$

$$= -\int_{\langle s \rangle} \tilde{h}^*(d\varphi) \qquad \text{(second structural equation)}$$

$$= -\int_{\langle s \rangle} d(\tilde{h}^*\varphi)$$

$$= -\int_{\partial \langle s \rangle} \tilde{h}^*\varphi \qquad \text{(Stokes's theorem)}$$

$$= -\int_{\partial \langle s \rangle} \varphi\left(d\tilde{h}\left(\frac{d}{dt}\right)\right) dt$$

$$= -\int_{\partial \langle s \rangle} \varphi\left(d\tilde{\beta}\left(\frac{d}{dt}\right)\right) dt,$$

where $\tilde{\beta} = \tilde{h}|_{\partial\langle s\rangle}$. Let $\tilde{\alpha}$ denote the horizontal lift of α through $\tilde{h}(v_0) = (\alpha(v_0), w_0)$. Since $\tilde{\beta}|_{\langle v_0, v_1\rangle} = \tilde{h}|_{\langle v_0, v_1\rangle}$ and $\tilde{\beta}|_{\langle v_1, v_2\rangle} = \tilde{h}|_{\langle v_1, v_2\rangle}$ are horizontal

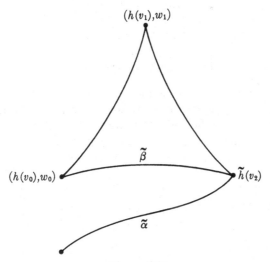

$(h(v_1), w_1)$

$(h(v_0), w_0)$

$\tilde{\beta}$

$\tilde{h}(v_2)$

$\tilde{\alpha}$

Figure 7.11

by construction of $\tilde{h}$, we have $\tilde{\beta}|_{\langle v_0, v_1\rangle} = \tilde{\alpha}|_{\langle v_0, v_1\rangle}$ and $\tilde{\beta}|_{\langle v_1, v_2\rangle} = \tilde{\alpha}|_{\langle v_1, v_2\rangle}$. By Lemma 2, Section 7.1, there exists a function $f: \langle v_2, v_0\rangle \to R$ with $f(v_2) = 0$ such that

$$\tilde{\beta}(t) = e^{if(t)}\tilde{\alpha}(t).$$

But $\tilde{\alpha}$ is the horizontal lift of α, so that $\tilde{\alpha}(v_0)$ is the parallel translate of w_0 around α. On the other hand, $\tilde{\beta}(v_0) = w_0$. Hence $e^{if(v_0)}$ is just the rotation mapping the parallel translate of w_0 around α into w_0; that is, $e^{-if(v_0)}$ rotates w_0 into its parallel translate around α. By the second statement of Lemma 2, Section 7.1,

$$\varphi\left(d\tilde{\beta}\left(\frac{d}{dt}\right)\right) = \frac{df}{dt}$$

for $t \in \langle v_2, v_0\rangle$. Moreover, $\varphi(d\tilde{\beta}(d/dt)) = 0$ on $\langle v_0, v_1\rangle$ and $\langle v_1, v_2\rangle$ since β is horizontal there. Thus since $\partial\langle s\rangle = \langle v_0, v_1\rangle + \langle v_1, v_2\rangle + \langle v_2, v_0\rangle$,

$$\int_{\langle s\rangle} h^*(K \text{ vol}) = -\int_{\langle v_2, v_0\rangle} \varphi\left(d\tilde{\beta}\left(\frac{d}{dt}\right)\right) dt$$

$$= -\int_{\langle v_2, v_0\rangle} \frac{df}{dt} dt$$

$$= -f(v_0)$$

$$= \text{the angle of rotation from } w_0 \text{ to its parallel translate around } \alpha. \qquad \square$$

193

Definitions. Let $\alpha: [a, b] \to M$ be a smooth curve. The *length* $l(\alpha)$ of α is the real number

$$l(\alpha) = \int_a^b \|\dot{\alpha}(t)\| \, dt.$$

The *arc length* along α is the function $s: [a, b] \to R$ given by

$$s(t) = \int_a^t \|\dot{\alpha}(\tau)\| \, d\tau.$$

Remark. l and s are defined because $t \to \|\dot{\alpha}(t)\|$ is continuous. Note that the function s is of class C^1, but it is not necessarily smooth because $t \to \|\dot{\alpha}(t)\|$ is not necessarily differentiable where $\dot{\alpha}(t) = 0$. If, however, $\|\dot{\alpha}(t)\| \neq 0$ for all t, then s is smooth and monotonically increasing.

Definition. A curve $\alpha: [a, b] \to M$ is said to be *parameterized by arc length* if $\|\dot{\alpha}(t)\| = 1$ for all $t \in [a, b]$. In this case, $s(t) = t - a$ for all $t \in [a, b]$.

Remark. Given any curve $\alpha: [a, b] \to M$ with $\|\dot{\alpha}(t)\| \neq 0$ for all $t \in [a, b]$, a new curve $\alpha_1: [0, l(\alpha)] \to M$, parameterized by arc length, is obtained by setting

$$\alpha_1 = \alpha \circ s^{-1}.$$

Then $\operatorname{Im} \alpha_1 = \operatorname{Im} \alpha$, and $l(\alpha_1) = l(\alpha)$.

Remark. The concept of arc length extends to broken C^∞ curves α since $\|\dot{\alpha}(t)\|$ is defined at all but a finite number of points.

Definition. Given a smooth curve $\alpha: [a, b] \to M$ parameterized by arc length, a smooth curve $\alpha': [a, b] \to S(M)$ is defined by

$$\alpha'(t) = (\alpha(t), \dot{\alpha}(t)) \qquad (t \in [a, b]).$$

α is said to be a *geodesic* in M if α' is horizontal; that is, if α' is the horizontal lift of α through $(\alpha(a), \dot{\alpha}(a)) \in S(M)$. Note that if α is a geodesic, the parallel translate of $\dot{\alpha}(0)$ along α to $\alpha(t)$ is just $\dot{\alpha}(t)$; that is, the tangent to α translates into itself, and α is a "straight line" of the surface.

To measure how far a curve α is from being "straight," we measure how far α' is from being horizontal. Suppose, then, α is parameterized by arc length so that $\alpha': [a, b] \to S(M)$ is a curve in $S(M)$.

Definition. The *geodesic curvature* $k_\alpha(t)$ of α at $t \in [a, b]$ is $\psi(d\alpha'(d/dt))$, where ψ is the 1-form of the Riemannian connection.

Notation. If $\alpha: [a, b] \to M$ is a broken C^∞ curve with $\dot{\alpha}(t) \neq 0$ for all $t \in [a, b]$, let $\tau(\alpha) = \int_0^{l(\alpha)} k_{\alpha_1}(t) \, dt$, where α_1 is the new curve obtained from α by parameterizing by arc length.

If M is a smoothly triangulated manifold, then τ can be considered as a 1-cochain (relative to the triangulation).

Lemma (the Gauss–Bonnet theorem for 2-simplices). *Let M be an oriented Riemannian 2-manifold. Let $\langle s \rangle$ be an oriented 2-simplex in R^2, and let $h: [s] \to M$ be a map which has a smooth nonsingular extension mapping a neighborhood of $[s]$ into M. Let α be the closed broken C^∞ curve in M obtained by restricting h to $\partial \langle s \rangle$. Then*

$$\int_{\langle s \rangle} h^* K(\text{vol}) = -\tau(\alpha) + \sum (\text{interior angles of } h[s]) - \pi.$$

PROOF. From Theorem 1 above, $e^{i \int_{\langle s \rangle} h^*(K \, \text{vol})}$ is the rotation obtained by parallel translation around the closed curve α. Suppose α is broken up into its three smooth curves α_0, α_1, and α_2 so that $\tau(\alpha) = \sum \tau(\alpha_i)$ and $\alpha_i: [a_i, a_{i+1}] \to M$ with $a_0 = a$ and $a_3 = b$. By Lemma 2, Section 7.1, $e^{i\tau(\alpha_i)}$ is the rotation from the parallel translate of $\dot{\alpha}_i(a_i)$ to $\dot{\alpha}_i(a_{i+1})$. Hence, from the picture in M (Figure 7.12), we get that parallel translation around the closed curve α is given by $e^{i(-\tau(\alpha) - \sum \text{exterior angles})}$. Hence, by taking logarithms, we get

$$\int_{\langle s \rangle} h^*(K \, \text{vol}) = -\tau(\alpha) - \sum (\text{exterior angles}) + 2\pi l,$$

where l is an integer.

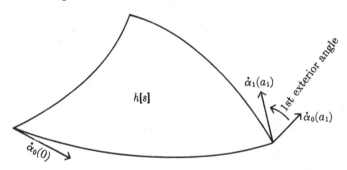

Figure 7.12

We use a continuity argument to show $l = 1$. Suppose ρ_0 is a flat Riemannian metric in a neighborhood of $h[s]$ (say transferred from R^2 via h). Then $K = 0$, $\tau(\alpha) = 0$, and $\sum (\text{exterior angles})$ is 2π. Hence for the flat Riemannian metric, $l = 1$. Suppose ρ is our given Riemannian metric, and let $\rho_t = t\rho_0 + (1 - t)\rho$ be a family of metrics, $t \in [0, 1]$. Let K_t, $\tau_t(\alpha)$, exterior angles$_t$ be the usual entities for ρ_t. These are continuous functions of t. Hence l is a continuous function of t. Since it is an integer for all t and equal to 1 for $t = 0$, we obtain $l = 1$. (You can also obtain this result by checking that $l = 1$ for small triangles and taking barycentric subdivisions.)

Since interior angle + exterior angle = π, the lemma is proved. $\square$

Definition. Let M be an oriented, connected, smoothly triangulated 2-manifold. For each 2-simplex s in M, let $\langle s \rangle$ denote this simplex oriented consistently with M. That is, if $h: K \to M$ is the triangulation and ω is a

2-form on M giving its orientation, let the orientation of s be given by the 2-form $h_s^*\omega$. Let

$$c = \sum_s \langle s \rangle.$$

Then c is a cycle called the *fundamental cycle* of M. Given any 2-form μ on M, the *integral* of μ over M is defined by

$$\int_M \mu = \int_c h^*\mu.$$

Exercise. Prove that c is a cycle.

Remark. The integral can be defined without use of a triangulation. Let M be a compact oriented n-manifold, and let μ be an n-form on M. Let $\{U_j, f_j\}$ be a smooth partition of unity on M, where $\{U_j\}$ is a finite covering of M by coordinate neighborhoods. Then integration of n-forms is defined on each U_j by pulling the forms back to R^n through the coordinate systems. The integral of μ over M is then given by

$$\int_M \mu = \sum_j \int_{U_j} f_j \mu.$$

This is independent of the partition of unity used, for if $\{V_k, g_k\}$ is another such partition, then

$$\sum_j \int_{U_j} f_j \mu = \sum_{j,k} \int_{U_j \cap V_k} f_j g_k \mu = \sum_k \int_{V_k} g_k \mu.$$

Theorem 2 (Gauss–Bonnet theorem). *Let M be an oriented, connected, smoothly triangulated, Riemannian 2-manifold. Then*

$$\frac{1}{2\pi} \int_M K \,\mathrm{vol} = \chi(M) = \beta_0 - \beta_1 + \beta_2.$$

($\chi(M)$ is the Euler characteristic of M.)

PROOF. Note that each 1-simplex t of M is an edge of precisely two 2-simplices of M. For given any point $m \in (t)$, there exists, by the implicit function theorem, a coordinate ball U about m such that $(t) \cap U$ is mapped into a straight line in R^2. By choosing U small enough, t must divide U into precisely two pieces. These pieces must lie in distinct 2-simplices, and, since open simplices are disjoint, there can be no other 2-simplex with t as an edge.

Thus, since each 2-simplex has three 1-simplices as edges, the total number n_1 of 1-simplices of M is given by $n_1 = 3n_2/2$, where n_2 is the number of 2-simplices of M. Letting n_0 denote the number of vertices in M, the Euler characteristic (Section 6.1) is given by

$$\begin{aligned}
\chi &= n_0 - n_1 + n_2 \\
&= n_0 - (3n_2/2) + n_2 \\
&= n_0 - (n_2/2).
\end{aligned}$$

Now we apply the previous lemma, and

$$\frac{1}{2\pi}\int_M K \, \mathrm{vol} = \frac{1}{2\pi}\int_c h^*(K\,\mathrm{vol})$$

$$= \frac{1}{2\pi}\sum_s \int_{\langle s\rangle} h^*(K\,\mathrm{vol})$$

$$= \frac{1}{2\pi}\sum_s \left(-\tau(\partial\langle s\rangle) + \sum (\text{interior angles of } h[s]) - \pi\right)$$

$$= \frac{1}{2\pi}\left(-\tau(\partial c) + \sum_s \left(\sum \text{interior angles of } h[s]\right) - n_2\pi\right).$$

But $\partial c = 0$, and $\sum_s (\sum \text{interior angles of } h[s])$ equals the sum over all vertices v in M of the sum of the interior angles at v of all 2-simplices with v as a vertex. Taking a coordinate neighborhood of V contained in St (v), we see that for each v, the sum of these interior angles at v is exactly 2π (Figure 7.13). Hence

$$\frac{1}{2\pi}\int_M K(\mathrm{vol}) = \frac{1}{2\pi}(2\pi n_0 - n_2\pi)$$

$$= n_0 - \frac{n_2}{2} = \chi(M). \qquad \square$$

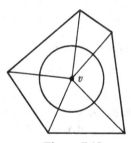

Figure 7.13

Remark. Note that this theorem holds for any connection on $S(M)$, since only the second structural equation was used in the proof.

Corollary 1. *Let M be any Riemannian 2-manifold homeomorphic with the sphere S^2. Then*

$$\int_M K \, \mathrm{vol} = 4\pi.$$

Corollary 2. *Let M be any Riemannian manifold homeomorphic with the torus $S^1 \times S^1$. Then*

$$\int_M K \, \mathrm{vol} = 0.$$

Corollary 3. *Let M be as in the theorem. Suppose there exists on M a smooth vector field which is never zero. Then $\chi(M) = 0$. In particular, there exists no nonzero vector field on S^2.*

PROOF. Let X be such a vector field and let $Y = X/\|X\|$. Then Y is smooth, and $\|Y(m)\| = 1$ for all $m \in M$, so Y is a smooth map $M \to S(M)$. On $S(M)$,

$$d\psi = -(K \circ \pi)\omega_1 \wedge \omega_2 = -\pi^*(K \text{ vol}).$$

Hence

$$\begin{aligned} d(Y^*\psi) = Y^*(d\psi) &= -Y^* \circ \pi^*(K \text{ vol}) \\ &= -(\pi \circ Y)^*(K \text{ vol}) \\ &= -K \text{ vol}, \end{aligned}$$

since $\pi \circ Y = i_M$. Thus K vol is exact and

$$2\pi\chi = \int_c K \text{ vol} = -\int_c d(Y^*\varphi) = -\int_{\partial c} Y^*\varphi = 0. \qquad \square$$

7.4 Geodesic coordinate systems

Let M be an oriented Riemannian 2-manifold, and consider the vector field E_1 on $S(M)$. Let U_t denote the local 1-parameter group of transformations on $S(M)$ associated to the vector field E_1. Then, for each $m_0 \in M$ and $z \in \pi^{-1}(m_0)$, there exists an $\varepsilon > 0$ and an open set W_z about z in $S(M)$ such that the map $[-\varepsilon, \varepsilon] \times W_z \to S(M)$ given by $(t, w) \to U_t(w)$ is smooth. Since $\pi^{-1}(m_0) \cong S^1$ is compact, $\pi^{-1}(m_0)$ can be covered by a finite number of such sets W_z. Taking ε to be the minimum of the corresponding numbers ε, the map $\mu: [0, \varepsilon] \times S^1 \to S(M)$, given by

$$\mu(t, g) = U_t(m_0, gv_0) \qquad (v_0 \text{ fixed} \in T(M, m_0); \|v_0\| = 1),$$

is well defined and is smooth. Since for each $g \in S^1$, the curve $t \to \mu(t, g)$ in $S(M)$ is horizontal with tangent vector E_1, the curve $t \to \pi \circ \mu(t, g)$ is a geodesic starting at m_0 with tangent vector $d\pi(E_1(m_0, gv_0)) = gv_0$ at m_0 (see Figure 7.14).

Let D_ε denote the open disc of radius ε about the origin in R^2, and let $P: [0, \varepsilon) \times S^1 \to D_\varepsilon$ (P for polar map) be given by

$$P(t, g) = tg.$$

Then P is a smooth map. Moreover, the restriction of P to $(0, \varepsilon) \times S^1$ is injective and maps $(0, \varepsilon) \times S^1$ onto $D_\varepsilon - \{0\}$. Since, morever, $\pi \circ \mu(0, g) = m_0$ for all $g \in S^1$, a map $G: D_\varepsilon \to M$ is defined by $G = \pi \circ \mu \circ P^{-1}$.

$$\begin{array}{ccc} [0, \varepsilon) \times S^1 & \overset{\mu}{\longrightarrow} & S(M) \\ \downarrow{\scriptstyle P} & & \downarrow{\scriptstyle \pi} \\ D_\varepsilon & \overset{G}{\longrightarrow} & M \end{array}$$

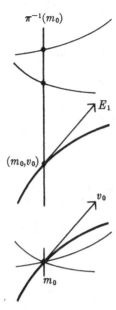

Figure 7.14

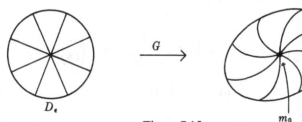

Figure 7.15

Exercise. Verify that this map G is smooth and that $dG(0)$ is an isomorphism, so that G maps D_ε diffeomorphically onto an open set of M for ε sufficiently small. (Note that $G(0) = m_0$ and that G maps radial straight lines in D_ε into geodesics through m_0; see Figure 7.15.)

Definition. The map $G: D_\varepsilon \to M$, where $\varepsilon > 0$ is sufficiently small that G maps D_ε diffeomorphically onto its image, is called a *polar*, or *geodesic*, *coordinate system* about m_0.

Remark. Let $\partial/\partial r$ and $\partial/\partial\theta$ denote the natural vector fields on $(0, \varepsilon) \times S^1$. Then $dP(\partial/\partial r)$ and $dP(\partial/\partial\theta)$ are the tangent vectors to the polar coordinate curves in $D_\varepsilon - \{0\}$. Note that $dP(\partial/\partial\theta)$ can be smoothly extended to D_ε by setting $dP(\partial/\partial\theta)|_0 = 0$. The following lemma asserts that these orthogonal vector fields $dP(\partial/\partial r)$ and $dP(\partial/\partial\theta)$ are mapped by geodesic coordinate systems into orthogonal vector fields; that is, that the orthogonal curves $r = $ constant

199

and $\theta = $ constant are mapped into orthogonal curves in M by geodesic coordinate systems.

Gauss's Lemma. $\langle dG \circ dP(\partial/\partial r), dG \circ dP(\partial/\partial \theta) \rangle = 0.$

PROOF. $dG \circ dP = d(G \circ P) = d(\pi \circ \mu) = d\pi \circ d\mu.$ Moreover, $d\mu(\partial/\partial r) = E_1,$ hence for $(m, v) \in$ Image μ,

$$d\pi\left[d\mu\left(\frac{\partial}{\partial r}\right)\Big|_{(m,v)}\right] = d\pi(E_1(m, v)) = v.$$

Thus

$$\left\langle dG \circ dP\left(\frac{\partial}{\partial r}\right), dG \circ dP\left(\frac{\partial}{\partial \theta}\right)\right\rangle = \left\langle d\pi \circ d\mu\left(\frac{\partial}{\partial r}\right), d\pi \circ d\mu\left(\frac{\partial}{\partial \theta}\right)\right\rangle$$

$$= \omega_1\left(d\mu\left(\frac{\partial}{\partial \theta}\right)\right)$$

$$= (\mu^*\omega_1)\left(\frac{\partial}{\partial \theta}\right).$$

To see that this is zero, consider

$$d(\mu^*\omega_1) = \mu^*(d\omega_1) = \mu^*(\psi \wedge \omega_2) \qquad \text{(by first structural equation)}$$
$$= \mu^*\psi \wedge \mu^*\omega_2.$$

Then

$$d(\mu^*\omega_1)\left(\frac{\partial}{\partial r}, \frac{\partial}{\partial \theta}\right) = \frac{1}{2}\left\{(\mu^*\psi)\left(\frac{\partial}{\partial r}\right)(\mu^*\omega_2)\left(\frac{\partial}{\partial \theta}\right) - (\mu^*\psi)\left(\frac{\partial}{\partial \theta}\right)(\mu^*\omega_2)\left(\frac{\partial}{\partial r}\right)\right\}$$

$$= \frac{1}{2}\left\{\psi\left(d\mu\left(\frac{\partial}{\partial r}\right)\right)\omega_2\left(d\mu\left(\frac{\partial}{\partial \theta}\right)\right) - \psi\left(d\mu\left(\frac{\partial}{\partial \theta}\right)\right)\omega_2\left(d\mu\left(\frac{\partial}{\partial r}\right)\right)\right\}.$$

But $d\mu(\partial/\partial r) = E_1,$ so that $\psi(d\mu(\partial/\partial r)) = 0$ since E_1 is horizontal, and $\omega_2(d\mu(\partial/\partial r)) = 0.$ Hence

$$0 = d(\mu^*\omega_1)\left(\frac{\partial}{\partial r}, \frac{\partial}{\partial \theta}\right)$$

$$= \frac{1}{2}\left\{\frac{\partial}{\partial r}\left[(\mu^*\omega_1)\left(\frac{\partial}{\partial \theta}\right)\right] - \frac{\partial}{\partial \theta}\left[(\mu^*\omega_1)\left(\frac{\partial}{\partial r}\right)\right] - (\mu^*\omega_1)\left(\left[\frac{\partial}{\partial r}, \frac{\partial}{\partial \theta}\right]\right)\right\}.$$

But

$$(\mu^*\omega_1)\left(\frac{\partial}{\partial r}\right) = \omega_1\left(d\mu\left(\frac{\partial}{\partial r}\right)\right) = \omega_1(E_1) \equiv 1,$$

so the second term is zero. Moreover, the bracket $[(\partial/\partial r), (\partial/\partial \theta)] = 0$ by equality of mixed partial derivatives, so the third term is zero also. Thus

$$\frac{\partial}{\partial r}\left[(\mu^*\omega_1)\left(\frac{\partial}{\partial \theta}\right)\right] \equiv 0,$$

and $(\mu^*\omega_1)(\partial/\partial\theta)$ is independent of r; that is,

$$\left\langle dG \circ dP\left(\frac{\partial}{\partial r}\right), dG \circ dP\left(\frac{\partial}{\partial\theta}\right)\right\rangle$$

is independent of r. But as $r \to 0$, $dP(\partial/\partial\theta) \to 0$, so that this inner product $\to 0$. Since it is independent of r, it must therefore be identically zero. $\quad\square$

Remark. By Gauss's lemma, $dG(dP(\partial/\partial\theta))$ is always orthogonal to the radial geodesics of our geodesic coordinate system.

Next we study the behavior of the length of $dG(dP(\partial/\partial\theta))$ as we move along a radial geodesic. Since $\omega_1(d\mu(\partial/\partial\theta))$ and $\omega_2(d\mu(\partial/\partial\theta))$ are the components of $d\pi(d\mu(\partial/\partial\theta))$ relative to an orthonormal basis, and since

$$\omega_1(d\mu(\partial/\partial\theta)) = (\mu^*\omega_1)(\partial/\partial\theta) = 0$$

from the proof of Gauss's lemma, $(\mu^*\omega_2)(\partial/\partial\theta) = \omega_2(d\mu(\partial/\partial\theta))$ is, at least up to sign, equal to the length of $d\pi \circ d\mu(\partial/\partial\theta) = dG \circ dP(\partial/\partial\theta)$.

Now from the first structural equation,

$$d(\mu^*\omega_2) = \mu^*(d\omega_2) = -\mu^*(\psi \wedge \omega_1) = -\mu^*\psi \wedge \mu^*\omega_1.$$

Thus

$$d(\mu^*\omega_2)\left(\frac{\partial}{\partial r}, \frac{\partial}{\partial\theta}\right) = -\frac{1}{2}\left[(\mu^*\psi)\left(\frac{\partial}{\partial r}\right)(\mu^*\omega_1)\left(\frac{\partial}{\partial\theta}\right) - (\mu^*\psi)\left(\frac{\partial}{\partial\theta}\right)(\mu^*\omega_1)\left(\frac{\partial}{\partial r}\right)\right].$$

But $(\mu^*\omega_1)(\partial/\partial\theta) = 0$ by Gauss's lemma, and

$$(\mu^*\omega_1)\left(\frac{\partial}{\partial r}\right) = \omega_1\left(d\mu\left(\frac{\partial}{\partial r}\right)\right) = \omega_1(E_1) \equiv 1,$$

so that

$$d(\mu^*\omega_2)\left(\frac{\partial}{\partial r}, \frac{\partial}{\partial\theta}\right) = \frac{1}{2}(\mu^*\psi)\left(\frac{\partial}{\partial\theta}\right).$$

On the other hand,

$$d(\mu^*\omega_2)\left(\frac{\partial}{\partial r}, \frac{\partial}{\partial\theta}\right)$$
$$= \frac{1}{2}\left\{\frac{\partial}{\partial r}\left[(\mu^*\omega_2)\left(\frac{\partial}{\partial\theta}\right)\right] - \frac{\partial}{\partial\theta}\left[(\mu^*\omega_2)\left(\frac{\partial}{\partial r}\right)\right] - (\mu^*\omega_2)\left(\left[\frac{\partial}{\partial r}, \frac{\partial}{\partial\theta}\right]\right)\right\}$$
$$= \frac{1}{2}\frac{\partial}{\partial r}\left[(\mu^*\omega_2)\left(\frac{\partial}{\partial\theta}\right)\right]$$

since

$$(\mu^*\omega_2)\left(\frac{\partial}{\partial r}\right) = \omega_2\left(d\mu\left(\frac{\partial}{\partial r}\right)\right) = \omega_2(E_1) = 0$$

and $[(\partial/\partial r), (\partial/\partial\theta)] = 0$. Thus

$$\boxed{\frac{\partial}{\partial r}\left[(\mu^*\omega_2)\left(\frac{\partial}{\partial\theta}\right)\right] = (\mu^*\psi)\left(\frac{\partial}{\partial\theta}\right).}$$

Now applying the second structural equation,

$$d(\mu^*\psi) = \mu^*(d\psi) = \mu^*(-(K\circ\pi)\omega_1 \wedge \omega_2) = -(K\circ\pi\circ\mu)\mu^*\omega_1 \wedge \mu^*\omega_2.$$

Thus,

$$d(\mu^*\psi)\left(\frac{\partial}{\partial r}, \frac{\partial}{\partial\theta}\right)$$

$$= -\frac{1}{2}(K\circ\pi\circ\mu)\left[(\mu^*\omega_1)\left(\frac{\partial}{\partial r}\right)(\mu^*\omega_2)\left(\frac{\partial}{\partial\theta}\right) - (\mu^*\omega_1)\left(\frac{\partial}{\partial\theta}\right)(\mu^*\omega_2)\left(\frac{\partial}{\partial r}\right)\right]$$

$$= -\frac{1}{2}(K\circ G\circ P)(\mu^*\omega_2)\left(\frac{\partial}{\partial\theta}\right),$$

as before. Finally,

$$d(\mu^*\psi)\left(\frac{\partial}{\partial r}, \frac{\partial}{\partial\theta}\right)$$

$$= \frac{1}{2}\left\{\frac{\partial}{\partial r}\left[(\mu^*\psi)\left(\frac{\partial}{\partial\theta}\right)\right] - \frac{\partial}{\partial\theta}\left[(\mu^*\psi)\left(\frac{\partial}{\partial r}\right)\right] - (\mu^*\psi)\left(\left[\frac{\partial}{\partial r}, \frac{\partial}{\partial\theta}\right]\right)\right\}$$

$$= \frac{1}{2}\frac{\partial}{\partial r}\left[(\mu^*\psi)\left(\frac{\partial}{\partial\theta}\right)\right]$$

since

$$(\mu^*\psi)\left(\frac{\partial}{\partial r}\right) = \psi\left(d\mu\left(\frac{\partial}{\partial r}\right)\right) = \psi(E_1) = 0 \quad\text{and}\quad \left[\frac{\partial}{\partial r}, \frac{\partial}{\partial\theta}\right] = 0.$$

Thus

$$\boxed{\frac{\partial}{\partial r}\left[(\mu^*\psi)\left(\frac{\partial}{\partial\theta}\right)\right] = -(K\circ G\circ P)(\mu^*\omega_2)\left(\frac{\partial}{\partial\theta}\right).}$$

Differentiating the first boxed equation above and substituting into the second, we obtain

$$\frac{\partial^2}{\partial r^2}\left[(\mu^*\omega_2)\left(\frac{\partial}{\partial\theta}\right)\right] + (K\circ G\circ P)\left[(\mu^*\omega_2)\left(\frac{\partial}{\partial\theta}\right)\right] = 0.$$

Moreover,

$$(\mu^*\omega_2)\left(\frac{\partial}{\partial\theta}\right)\Big|_{r=0} = \omega_2\left(d\mu\left(\frac{\partial}{\partial\theta}\right)\right)\Big|_{r=0} = \omega_2(V) = 0,$$

and, from the first boxed equation above,

$$\frac{\partial}{\partial r}\left[(\mu^*\omega_2)\left(\frac{\partial}{\partial\theta}\right)\right]\Big|_{r=0} = (\mu^*\psi)\left(\frac{\partial}{\partial\theta}\right)\Big|_{r=0} = \psi\left(d\mu\left(\frac{\partial}{\partial\theta}\right)\right)\Big|_{r=0} = \psi(V) = 1.$$

In particular, $(\mu^*\omega_2)(\partial/\partial\theta) \geq 0$ (at least for small values of r), and so $(\mu^*\omega_2)(\partial/\partial\theta)$ is indeed equal to the length of $dG \circ dP(\partial/\partial\theta)$.

Thus we have shown the following theorem.

Theorem 1. *Let $G: D_\varepsilon \to M$ be a geodesic coordinate system about m_0. For $g \in S^1$, consider the geodesic $r \to G \circ P(r, g)$ through m_0. Let $f(r)$ denote the length of the vector field $dG \circ dP(\partial/\partial\theta)$ along this geodesic. Then f satisfies the differential equation*

$$\frac{d^2f}{dr^2} + (K \circ G \circ P)f = 0 \qquad \text{(Jacobi's equation)}$$

as well as the initial conditions $f(0) = 0$ and $f'(0) = 1$.

Remark. Note that if K is constant, this differential equation can be solved explicitly. Namely, if $K > 0$ then $f(r) = (1/\sqrt{K})\sin(\sqrt{K}r)$. If $K = 0$, then $f(r) = r$; and if $K < 0$, then $f(r) = (1/\sqrt{K})\sinh(\sqrt{-K}r)$.

Theorem 2. *Geodesics minimize arc length locally; that is, if α is a geodesic in M starting at m_0, then there exists an $\varepsilon > 0$ such that $l(\alpha|_{[0,\varepsilon]}) \leq l(\beta)$ for all broken C^∞ curves β in M from m_0 to $\alpha(\varepsilon)$.*

PROOF. Let $G: D_\varepsilon \to M$ be a geodesic coordinate system about m_0 with $\varepsilon < l(\alpha)$. Then since α is a geodesic,

$$\alpha(t) = G \circ P(t, g_0) = G(tg_0) \qquad (0 \leq t < \varepsilon)$$

for some $g_0 \in S^1$. Let $p': D_\varepsilon - \{0\} \to D_\varepsilon$ be the map which rotates each point of D_ε onto the radial line through g_0; that is, p' is defined by $p'(rg) = rg_0$.

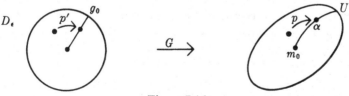

Figure 7.16

Let $U = G(D_\varepsilon)$, and let $p: U - \{m_0\} \to U$ be the corresponding map in U; that is, $p = G \circ p' \circ G^{-1}$ (Figure 7.16). Then $dp: T(U, m) \to T(U, p(m))$ decreases lengths; that is,

$$\|dp(v)\| \leq \|v\| \quad \text{for all } v \in T(U, m), \, m \in U - \{m_0\}.$$

For by Gauss's lemma, $dG \circ dP(\partial/\partial r)$ and $dG \circ dP(\partial/\partial\theta)$ are orthogonal vectors in $T(U, m)$. Moreover,

$$\left\|dp\left(dG \circ dP\left(\frac{\partial}{\partial r}\right)\right)\right\| = \left\|dG\left(dp' \circ dP\left(\frac{\partial}{\partial r}\right)\right)\right\| = \left\|dG \circ dP\left(\frac{\partial}{\partial r}\right)\right\|,$$

$$\left\|dp\left(dG \circ dP\left(\frac{\partial}{\partial\theta}\right)\right)\right\| = \left\|dG\left(dp' \circ dP\left(\frac{\partial}{\partial\theta}\right)\right)\right\| = \|dG(0)\| = 0.$$

It follows that the magnitude of $dp(v)$ equals the magnitude of the orthogonal projection of v onto $dG \circ dP(\partial/\partial r)$, which is less than or equal to $\|v\|$.

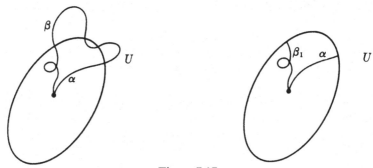

Figure 7.17

Now let $\beta: [a, b] \to M$ be any broken C^∞ curve from m_0 to $\alpha(\varepsilon)$. Let

$$c = \inf [t \in [a, b]; \beta(r) \notin U].$$

Then $\beta(c) \notin U$, but $\beta(t) \in U$ for all $t < c$. Let $\beta_1 = \beta|_{[a,c]}$ (Figure 7.17). Then $l(\beta_1) \leq l(\beta)$. Now consider the curve $p \circ \beta_1: [a, c] \to M$. Then

$$\|p \circ \beta_1(t)\| = \|dp(\dot\beta_1(t))\| \leq \|\dot\beta_1(t)\|$$

since dp decreases lengths, hence

$$l(p \circ \beta_1) \leq l(\beta_1) \leq l(\beta).$$

But Image $p \circ \beta_1 = $ Image α, so, since α is one-to-one, we have $p \circ \beta_1(t) = \alpha \circ f(t)$ for $t \in [a, c)$ where $f = \alpha^{-1} \circ p \circ \beta_1$. Thus

$$l(\beta) \geq l(p \circ \beta_1) = \int_a^c \|p \circ \beta_1(t)\| \, dt$$

$$= \int_a^c \|\alpha \circ f(t)\| \, dt$$

$$= \int_a^c \|\dot\alpha(f(t))\| |f'(t)| \, dt$$

$$\geq \int_a^c \|\dot\alpha(f(t))\| f'(t) \, dt = \int_0^\varepsilon \|\dot\alpha(\tau)\| \, d\tau$$

$$= l(\alpha|_{[0,\varepsilon]}). \qquad \square$$

Remark. The proof above also shows that if $G: D_\varepsilon \to M$ is geodesic coordinate system about m_0 and β is any broken C^∞ curve starting at m_0 and ending at some point outside the geodesic neighborhood $G(D_\varepsilon)$, then $l(\beta) \geq \varepsilon$.

Theorem 3. *Let* $\rho: M \times M \to R$ *be defined by*

$$\rho(m_1, m_2) = \inf [l(\alpha); \alpha \text{ a broken } C^\infty \text{ curve in } M \text{ starting at}$$
$$m_1 \text{ and ending at } m_2].$$

Then ρ *is a metric on* M, *and the metric topology on* M *is the same as the manifold topology on* M.

PROOF. Clearly $\rho(m_1, m_2) = \rho(m_2, m_1)$. Also,

$$\rho(m_1, m_3) \leq \rho(m_1, m_2) + \rho(m_2, m_3)$$

because, given any $\varepsilon > 0$, let α and β be curves from m_1 to m_2 and from m_2 to m_3, respectively, such that

$$l(\alpha) < \rho(m_1, m_2) + \frac{\varepsilon}{2}, \qquad l(\beta) < \rho(m_2, m_3) + \frac{\varepsilon}{2}.$$

Then

$$\rho(m_1, m_3) \leq l(\alpha) + l(\beta) < \rho(m_1, m_2) + \rho(m_2, m_3) + \varepsilon.$$

The remaining requirement $\rho(m_1, m_2) = 0 \Leftrightarrow m_1 = m_2$ for a metric is also satisfied because if $m_1 \neq m_2$, let $G: D_\varepsilon \to M$ be a geodesic coordinate system about m_1 not containing m_2. Then every curve from m_1 to m_2 has length at least ε, so $\rho(m_1, m_2) \geq \varepsilon$.

The two topologies on M agree because for each $m_0 \in M$, there exists $\varepsilon_0(m_0) > 0$ and a geodesic coordinate system $G: D_{\varepsilon_0(m_0)} \to M$ about m_0. Then $G(D_\varepsilon) = B_{m_0}(\varepsilon)$ for all $\varepsilon \leq \varepsilon_0(m_0)$. $\qquad\square$

Theorem 4. *Let* M_1 *and* M_2 *be oriented Riemannian 2-manifolds. Let* $m_1 \in M_1$ *and* $m_2 \in M_2$. *Suppose* $G_1: D_\varepsilon \to M_1$ *and* $G_2: D_\varepsilon \to M_2$ *are geodesic coordinate systems about* m_1 *and* m_2, *respectively (see Figure 7.18). If* $K_2 \circ G_2 \circ G_1^{-1} = K_1$, *where* K_i *is the curvature of* M_i ($i = 1, 2$), *then* $G_2 \circ G_1^{-1}$ *is an isometry.*

PROOF. We must show that $d(G_2 \circ G_1^{-1}) = dG_2 \circ dG_1^{-1}$ is an isometry at each point. Since $G_2 \circ G_1^{-1}$ maps radial geodesics into radial geodesics, $dG_2 \circ dG_1^{-1}$ preserves lengths in the radial direction. By Gauss's lemma, $dG_2 \circ dG_1^{-1}$ preserves orthogonality. Thus we need only verify that lengths in the θ-direction are preserved; that is, we need only show that

$$\left\| dG_2 \circ dP\left(\frac{\partial}{\partial\theta}\right) \right\| = \left\| dG_1 \circ dP\left(\frac{\partial}{\partial\theta}\right) \right\|.$$

But, if we fix $g \in S^1$ and let

$$f_j(r) = \left\| dG_j \circ dP\left(\frac{\partial}{\partial\theta}\bigg|_{(r,g)}\right) \right\| \qquad (j = 1, 2),$$

205

then Jacobi's equation says that

$$\frac{d^2 f_j}{dr^2} + (K_j \circ G_j \circ P) f_j = 0,$$

where $f_j(0) = 0$ and $f_j'(0) = 1$ $(j = 1, 2)$. By uniqueness of solutions, $f_1(r) = f_2(r)$ for all r; that is,

$$\left\| dG_2 \circ dP \left(\frac{\partial}{\partial \theta} \right)_{(r,g)} \right\| = \left\| dG_1 \circ dP \left(\frac{\partial}{\partial \theta} \Big|_{(r,g)} \right) \right\|$$

for all $(r, g) \in (0, \varepsilon) \times S^1$.　　　　　　　　　　　　□

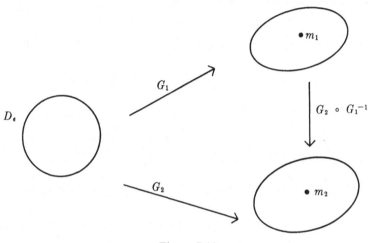

Figure 7.18

Corollary. *If M has constant curvature K, then given any pair m_1, m_2 of points in M and v_1, v_2 of unit tangent vectors $(v_i \in T(M, m_i))$, there exists an isometry Φ mapping a neighborhood of m_1 onto a neighborhood of m_2 such that $\Phi(m_1) = m_2$ and $d\Phi(v_1) = v_2$.*

PROOF. Let $G_i: D_\varepsilon \to M_i$ be a geodesic coordinate system about m_i $(i = 1, 2)$. Let $g \in S^1$ be the rotation of D_ε which maps $dG_1^{-1}(v_1)$ onto $dG_2^{-1}(v_2)$. Then $G_2 \circ g: D_\varepsilon \to M$ is another geodesic coordinate system about m_2. Set $\Phi = G_2 \circ g \circ G_1^{-1}$.　　　　　　　　　　　□

　　Remark. Thus, if $K \equiv 0$, M is locally isometric with R^2 (with its usual Riemannian structure). However, it is not true that M is globally isometric with R^2 (consider, for example, the torus $S^1 \times S^1$ with its product structure). However, it can be shown that if M is simply connected and $K \equiv 0$, then M is isometrically the same as R^2.

　　Remark. Jacobi's equation also gives some information in the case of non-constant curvature. For example, suppose $K \geq C$ for some constant C. Then if f is the solution of Jacobi's equation $f'' + Kf = 0$, and f_1 is the solution of $f_1'' + Cf_1 = 0$, then the theory of ordinary differential equations (Sturm–Liouville systems) tells us that $f \le f_1$; that is, the geodesics in M spread more slowly than do geodesics in a space of constant curvature C; that is, geodesics emanating from m_0 will come back together faster in M than in a space of constant curvature C.

7.5 Isometries and spaces of constant curvature

Let M_1, M_2 be smooth oriented Riemannian 2-manifolds. Suppose $f: M_1 \to M_2$ is an isometry that preserves the orientation. Thus f is smooth, injective, surjective, and it has a smooth inverse. Moreover, df preserves the inner product on tangent spaces,

$$\langle df(v_1), df(v_2) \rangle = \langle v_1, v_2 \rangle \qquad (v_1, v_2 \in T(M_1, m_1); \ m_1 \in M_1),$$

and if τ is any 2-form giving the orientation of M_2, then $f^*\tau$ gives the orientation of M_1. Note that since f preserves the Riemannian structure and the orientation, f preserves everything defined in terms of these; for example, the curvature, the Riemannian connection, and the parallel translation. Five explicit statements follow.

(1) f induces a map $\tilde{f}: S(M_1) \to S(M_2)$, defined by

$$\tilde{f}(m_1, v) = (f(m_1), df(v)) \qquad ((m_1, v) \in S(M_1)).$$

This map is smooth and has a smooth inverse, namely $\tilde{f}^{-1} = \widetilde{f^{-1}}$. Note that

$$\pi_2 \circ \tilde{f} = f \circ \pi_1.$$

$$
\begin{array}{ccc}
S(M_1) & \xrightarrow{\tilde{f}} & S(M_2) \\
\downarrow{\scriptstyle \pi_1} & & \downarrow{\scriptstyle \pi_2} \\
M_1 & \xrightarrow{f} & M_2
\end{array}
$$

Moreover, since df is an orientation preserving isometry at each point,

$$\tilde{f}(g(m_1, v)) = g\tilde{f}(m_1, v)$$

for all $(m_1, v) \in S(M_1)$ and $g \in S^1$; that is, $\tilde{f} \circ g = g \circ \tilde{f}$ for each $g \in S^1$.

(2) $\tilde{f}$ preserves the forms ω_1, ω_2 and ψ (ψ the 1-form of the Riemannian connection). For, if $t \in T(S(M_1), (m_1, v))$, then

$$
\begin{aligned}
(\tilde{f}^*\omega_1^{M_2})(t) &= \omega_1^{M_2}(d\tilde{f}(t)) \\
&= \langle df(v), d\pi_2(d\tilde{f}(t)) \rangle \\
&= \langle df(v), df(d\pi_1(t)) \rangle \\
&= \langle v, d\pi_1(t) \rangle \\
&= \omega_1^{M_1}(t);
\end{aligned}
$$

that is,

$$\tilde{f}^*\omega_1{}^{M_2} = \omega_1{}^{M_1}.$$

Similarly,

$$\tilde{f}^*\omega_2{}^{M_2} = \omega_2{}^{M_1}.$$

To check that the connection is preserved, we use the uniqueness of the Riemannian connection. Note that $\tilde{f}^*\psi^{M_2}$ is a connection form on $S(M_1)$ since

$$(\tilde{f}^*\psi^{M_2})(V) = \psi^{M_2}(d\tilde{f}(V)) = \psi^{M_2}(V) = 1,$$

and

$$g^*(\tilde{f}^*\psi^{M_2}) = g^* \circ \tilde{f}^*(\psi^{M_2}) = (\tilde{f} \circ g)^*(\psi^{M_2}) = (g \circ \tilde{f})^*\psi^{M_2}$$
$$= \tilde{f}^*(g^*\psi^{M_2}) = \tilde{f}^*\psi^{M_2}$$

for each $g \in S^1$. Moreover,

$$d\omega_1{}^{M_2} = \psi^{M_2} \wedge \omega_2{}^{M_2},$$
$$d\omega_2{}^{M_2} = -\psi^{M_2} \wedge \omega_1{}^{M_2},$$

so that

$$d\omega_1{}^{M_1} = d(\tilde{f}^*\omega_1{}^{M_2}) = \tilde{f}^*(d\omega_1{}^{M_2}) = (\tilde{f}^*\psi^{M_2}) \wedge (\tilde{f}^*\omega_2{}^{M_2})$$
$$= (\tilde{f}^*\psi^{M_2}) \wedge \omega_2{}^{M_1},$$
$$d\omega_2{}^{M_1} = d(\tilde{f}^*\omega_2{}^{M_2}) = \tilde{f}^*(d\omega_2{}^{M_2}) = -(\tilde{f}^*\psi^{M_2}) \wedge (\tilde{f}^*\omega_1{}^{M_2})$$
$$= -(\tilde{f}^*\psi^{M_2}) \wedge \omega_1{}^{M_1}.$$

Thus the connection form $\tilde{f}^*\psi^{M_2}$ satisfies the simplified form of the structural equations. By uniqueness,

$$\tilde{f}^*\psi^{M_2} = \psi^{M_1}.$$

(3) f preserves curvature; that is, $f^*K^{M_2} = K^{M_1}$. For

$$d\psi^{M_i} = -(K^{M_i} \circ \pi_i)\omega_1{}^{M_i} \wedge \omega_2{}^{M_i} \qquad (i = 1, 2),$$

so that

$$-(K^{M_1} \circ \pi_1)\omega_1{}^{M_1} \wedge \omega_2{}^{M_1} = d\psi^{M_1}$$
$$= d(\tilde{f}^*\psi^{M_2})$$
$$= \tilde{f}^*(d\psi^{M_2})$$
$$= \tilde{f}^*((-K^{M_2} \circ \pi_2)\omega_1{}^{M_2} \wedge \omega_2{}^{M_2})$$
$$= -(K^{M_2} \circ \pi_2 \circ \tilde{f})(\tilde{f}^*\omega_1{}^{M_2}) \wedge (\tilde{f}^*\omega_2{}^{M_2})$$
$$= -(K^{M_2} \circ f \circ \pi_1)\omega_1{}^{M_1} \wedge \omega_2{}^{M_1},$$

hence

$$K^{M_1} \circ \pi_1 = K^{M_2} \circ f \circ \pi_1;$$

that is,

$$K^{M_1} = K^{M_2} \circ f = f^*K^{M_2}.$$

(4) f preserves parallel translation; that is, if $\alpha: [a, b] \to M_1$, then the horizontal lift $\widetilde{f \circ \alpha}$ of $f \circ \alpha$ through $(f \circ \alpha(a), df(v))$ is given by

$$\widetilde{f \circ \alpha} = \tilde{f} \circ \tilde{\alpha},$$

where $\tilde{\alpha}$ is the horizontal lift of α through $(\alpha(a), v)$. The reason is that

$$\tilde{f}^* \psi^{M_2} = \psi^{M_1} \quad \text{and} \quad d\tilde{f}(\mathscr{H}^{M_1}) = \mathscr{H}^{M_2}.$$

(5) f maps geodesics into geodesics, for α is a geodesic if and only if $\tilde{\alpha}$ is parallel along α.

Remark. Note that, given M, the set of all orientation-preserving isometries $h: M \to M$ forms a group $\mathscr{H}$ under composition. We call this group $\mathscr{H}$ the *group of isometries* of M.

Definition. $\mathscr{H}$ is *transitive* if for each $m_1, m_2 \in M$, there exists $h \in \mathscr{H}$ such that $h(m_1) = m_2$.

Theorem 1. *Let M be a smooth oriented Riemannian 2-manifold. Suppose the group of isometries of M is transitive. Then M has constant curvature.*

PROOF. Let $m_0 \in M$. Then for any $m \in M$, there exists $h \in \mathscr{H}$ such that $h(m_0) = m$. Thus $K(m_0) = (h^*K)(m_0) = K(h(m_0)) = K(m)$. ∎

Definition. Let $\mathscr{H}$ denote the group of isometries of M. For $m_0 \in M$, let $\mathscr{K}_{m_0}$ denote the subgroup of $\mathscr{H}$ leaving m_0 fixed; that is,

$$\mathscr{K}_{m_0} = [h \in \mathscr{K}; h(m_0) = m_0].$$

$\mathscr{K}_{m_0}$ is called the *isotropy group* of M at m_0.

Remark. Note that for $k \in \mathscr{K}_{m_0}$, $dk: T(M, m_0) \to T(M, m_0)$ is an orientation-preserving isometry; that is, dk is a rotation of $T(M, m_0)$. Thus for each $k \in \mathscr{K}_{m_0}$, there exists $g \in S^1$ such that $dk(v) = gv$ for all $v \in T(M, m_0)$. Moreover, since $d(k_1 \circ k_2) = dk_1 \circ dk_2$, the map $\Phi: \mathscr{K}_{m_0} \to S^1$ defined by $k \to dk(m_0)$ is a homomorphism.

Lemma. *If M is connected, the homomorphism $\Phi: \mathscr{K}_{m_0} \to S^1$ is injective.*

PROOF. Let $G: D_\varepsilon \to M$ be a geodesic coordinate system about m_0. Note that for $k \in \mathscr{K}_{m_0}$, $k: G(D_\varepsilon) \to G(D_\varepsilon)$ since k is an isometry. Moreover, $G^{-1} \circ k \circ G$ is an orientation-preserving isometry of D_ε leaving 0 fixed; that is, $G^{-1} \circ k \circ G = g$ for some rotation $g \in S^1$. Since $dg(0)$ is also rotation by g,

$$dk(m_0) = d(G \circ g \circ G^{-1})(m_0) = dG \circ g \circ dG^{-1}(m_0).$$

Thus

$$k \in \text{kernel } \Phi \Leftrightarrow dk(m_0) = \text{identity}$$
$$\Leftrightarrow g = \text{identity}$$
$$\Leftrightarrow k|_{G(D_\varepsilon)} = G \circ g \circ G^{-1} = \text{identity}.$$

Thus if $k \in$ kernel Φ, the set

$$N = [m \in M; k(m) = m \text{ and } dk: T(M, m) \to T(M, m) = \text{identity}]$$

is an open set in M. On the other hand, since k and dk are continuous, N is closed in M. Since M is connected, $N = M$; that is, $k =$ identity on M. Thus

$$\text{kernel } \Phi = \text{(identity)}$$

and Φ is injective. $\square$

Theorem 2. *Let M be a connected oriented Riemannian 2-manifold. Suppose $\mathscr{H}^1$ is a subgroup of the group $\mathscr{H}$ of isometries of M such that*

(1) *$\mathscr{H}^1$ is transitive, and*
(2) *for some $m_0 \in M$, the homomorphism*

$$\Phi: \mathscr{K}_{m_0}^{\ 1} \to S^1$$

is surjective, where $\mathscr{K}_{m_0}^{\ 1} = \mathscr{K}_{m_0} \cap \mathscr{H}^1$. Then $\mathscr{H} = \mathscr{H}^1$.

PROOF. Suppose $h \in \mathscr{H}$. By transitivity of $\mathscr{H}^1$, there exists $h^1 \in \mathscr{H}^1$ such that

$$h^1(h(m_0)) = m_0.$$

Then $h^1 h \in \mathscr{K}_{m_0}$. Since $\Phi | \mathscr{K}_{m_0}^{\ 1}$ is surjective, there exists $k^1 \in \mathscr{K}_{m_0}^{\ 1}$ such that

$$\Phi(k^1) = \Phi(h^1 h).$$

Since Φ is injective (by the lemma), $k^1 = h^1 h$; that is, $h = k^1(h^1)^{-1} \in \mathscr{H}^1$. Thus $\mathscr{H} \subset \mathscr{H}^1$; that is, $\mathscr{H} = \mathscr{H}^1$. $\square$

Definition. For each $m \in M$, the set

$$[km = k(m); k \in \mathscr{K}_{m_0}]$$

is called the *orbit* of $\mathscr{K}_{m_0}$ through m in M.

Theorem 3. *Let M be an oriented Riemannian 2-manifold. Suppose, for some $m_0 \in M$, $\Phi: \mathscr{K}_{m_0} \to S^1$ is surjective. Then, in a geodesic coordinate neighborhood of m_0, the geodesics through m_0 are the orthogonal trajectories (reparameterized by arc length) of the orbits of $\mathscr{K}_{m_0}$ (see Figure 7.19).*

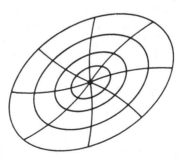

Figure 7.19

PROOF. If $G: D_\varepsilon \to M$ is a geodesic coordinate neighborhood of m_0, then the orbits of $\mathscr{K}_{m_0}$ are the images under G of the concentric circles about the origin in D_ε (see the proof of the lemma above). These orbits are orthogonal to the radial geodesics by Gauss's lemma. By the uniqueness of orthogonal trajectories, these trajectories are the radial geodesics. □

EXAMPLE 1. Consider the plane R^2 with the usual Riemannian structure. Since $\|\partial/\partial r_1\| = 1$, the map $c: R^2 \to S(R^2)$ given by $c(m) = (m, e_1) = (m, (\partial/\partial r_1))$ is globally defined, so the special connection $\mathscr{H}_1$ is defined on all of $S(R^2)$. Moreover, $e_1 = \partial/\partial r_1$ and $ie_1 = \partial/\partial r_2$, so $\omega_1' = dr_1$ and $\omega_2' = dr_2$. Thus, by Remark 5, Section 7.2,

$$\omega_1 = (\cos \theta)\tilde{\omega}_1 + (\sin \theta)\tilde{\omega}_2$$
$$= (\cos \theta)\pi^* \, dr_1 + (\sin \theta)\pi^* \, dr_2.$$

Similarly,

$$\omega_2 = -(\sin \theta)\pi^* \, dr_1 + (\cos \theta)\pi^* \, dr_2.$$

Since $\varphi_1 = d\theta$, the first structural equations become

$$d\omega_1 = -(\sin \theta) \, d\theta \wedge (\pi^* \, dr_1) + (\cos \theta) \, d\theta \wedge (\pi^* \, dr_2)$$
$$= \varphi_1 \wedge \omega_2,$$

and, similarly,

$$d\omega_2 = -\varphi_1 \wedge \omega_1.$$

Thus the structural equations are in the simplified form, and, by uniqueness, the special connection on $S(R^2)$ is the Riemannian connection. Moreover, as we have seen for the special connection, $d\varphi_1 = 0$, so the curvature K of R^2 is identically zero.

Clearly, each translation and each rotation of R^2 is an isometry. Let $\mathscr{H}^1$ be the group generated by rotations and translations of R^2. Then $\mathscr{H}^1$ is transitive (in fact, the group of translations is transitive), and $\Phi: \mathscr{K}_0^1 \to S^1$ is surjective ($\mathscr{K}_0^1$ the group of rotations about 0 in R^2); so, by Theorem 2, $\mathscr{H}^1$ is the full group of isometries of R^2.

For $m_0 \in R^2$, $\mathscr{K}_{m_0}$ is the group of rotations of R^2 about m_0. Since the orbits of $\mathscr{K}_{m_0}$ are circles about m_0, the geodesics through m_0 are straight lines by Theorem 3.

EXAMPLE 2. Let S^2 denote the sphere of radius r about the origin in R^3, with the induced Riemannian structure. Note that the group $\mathscr{H}$ of rotations about the origin in R^3 is a group of isometries of S^2. It is the full group of isometries of S^2 by Theorem 2. Since $\mathscr{H}$ is transitive on S^2, the curvature K of S^2 is constant by Theorem 1. By the Gauss–Bonnet theorem,

$$\frac{1}{2\pi} \int_{S^2} K \, \mathrm{vol} = \chi(S^2) = 2;$$

that is,

$$K \text{ vol } (S^2) = 4\pi, \qquad K = \frac{4\pi}{\text{vol } (S^2)} = \frac{4\pi}{4\pi r^2} = \frac{1}{r^2}.$$

The geodesics through m in S^2 are the orthogonal trajectories of the orbits of the group of rotations of R^3 that leave m fixed. These orbits are circles about m, so the geodesics through m are the great circles through m (Figure 7.20).

Figure 7.20

EXAMPLE 3. Let D denote the open disc of radius 1 in R^2. Regarding D as a subset of the complex plane, consider the Riemannian structure (called the *Poincaré*, or *hyperbolic*, *metric*) on D defined by

$$\langle v_1, v_2 \rangle = \frac{v_1 \cdot v_2}{(1 - p\bar{p})^2} \qquad (v_1, v_2 \in T(D, p); \, p \in D),$$

where $v_1 \cdot v_2$ denotes the usual inner product (dot product) on $T(D, p)$ and $\bar{p}$ is the complex conjugate of p, so that $p\bar{p} = r^2$, where r is the Euclidean distance from p to the origin. Thus,

$$\left\| \frac{\partial}{\partial r_1} \right\|_p = \left\| \frac{\partial}{\partial r_2} \right\|_p = \frac{1}{1 - p\bar{p}}$$

and $\langle (\partial/\partial r_1), (\partial/\partial r_2) \rangle = 0$. Note that the radial lines in D through the origin have infinite length. For, if $\alpha: [0, 1) \to D$ is given by $\alpha(t) = te^{i\theta}$ for some θ, then

$$l(\alpha) = \int_0^1 \|\dot{\alpha}(t)\| \, dt = \int_0^1 \frac{1}{1 - t^2} \, dt = \infty.$$

Let

$$\mathscr{C} = \left[f: D \to D; f(z) = e^{i\theta} \frac{z - p}{1 - z\bar{p}} \text{ for some } p \in D, 0 \le \theta < 2\pi \right].$$

From elementary complex variable theory, $\mathscr{C}$ is a group (the conformal group) of transformations of D onto itself. $\mathscr{C}$ is, in fact, a group of isometries of D.

For let $f \in \mathscr{C}$. If we identify $T(D, z)$ with $T(D, f(z))$ by identifying both with R^2 in the usual way, then, since conformal maps preserve angles, $df(z) = \lambda(z)g(z)$ for some real number $\lambda(z) > 0$ and some rotation $g(z) \in S^1$. The magnification factor $\lambda(z)$ is given by $\lambda(z) = |(df/dz)(z)|$. But an elementary computation shows that

$$\left| \frac{df}{dz}(z) \right| (1 - z\bar{z}) = 1 - f(z)\overline{f(z)}.$$

Since vectors at z which are unit vectors in the Poincaré metric have Euclidean length $1 - z\bar{z}$, this implies that $df(z)$ is a rotation which maps unit vectors (in the Poincaré metric) into unit vectors; that is, $df(z)$ is an isometry for each z, so $f: D \to D$ is an isometry.

Thus $\mathscr{C}$ is a group of isometries of D. $\mathscr{C}$ is transitive because, for $p \in D$,

$$f_p: z \to \frac{(z - p)}{(1 - z\bar{p})}$$

maps p onto the origin, hence for each pair $p_1, p_2 \in D$, $f_{p_2}^{-1} \circ f_{p_1}$ maps p_1 onto p_2. Moreover, the isotropy subgroup $\mathscr{K}_0$ of $\mathscr{C}$ at the origin is given by

$$\mathscr{K}_0 = [f: D \to D; f(z) = e^{i\theta}z];$$

that is, $\mathscr{K}_0$ is the group of rotations in R^2. Hence $\Phi: \mathscr{K}_0 \to S^1$ is surjective, so, by Theorem 2, $\mathscr{C}$ is the full group of isometries of D. In particular, since $\mathscr{C}$ is transitive, the curvature K of D is constant by Theorem 1.

Since the isotropy group $\mathscr{K}_0$ at 0 is the rotation group, whose orbits are circles about 0, Theorem 3 tells us that the geodesics through 0 are the radial lines (suitably parameterized). Since isometries map geodesics into geodesics, and since the image of a radial straight line under a conformal transformation of D is either a radial straight line or a circle which meets the boundary of D at right angles, we see (Figure 7.21) that such lines and circles are the geodesics

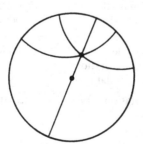

Figure 7.21

in D. In particular, the sum of the interior angles of a geodesic triangle in D is less than π, so that, by the Gauss–Bonnet theorem, the constant value of K is negative (see Figure 7.22).

Exercise. Find the constant value of K. [*Hint*: K can be found either by computing the area of a geodesic triangle and applying the Gauss–Bonnet theorem or by use of Jacobi's equation.]

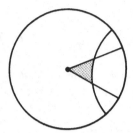

Figure 7.22

Remark. Although we have exhibited the hyperbolic space D as a submanifold of R^2, note that the inclusion map is not an isometry. In fact, it can be shown that the space D cannot even be imbedded as a Riemannian submanifold of R^3.

Remark. We have exhibited three Riemannian manifolds R^2, S^2, and D of constant curvature $K = 0$, $K > 0$, and $K < 0$, respectively. It turns out that any two simply connected complete Riemannian 2-manifolds of the same constant curvature K are isometric. So these three examples are essentially the only examples of simply connected complete 2-manifolds of constant curvature. These three examples are important in several different mathematical disciplines, as illustrated in Table 7.1.

Table 7.1

Space	Riemannian geometry	Group theory (group of isometries)	Complex variables	Elementary geometry
R^2	$K = 0$	Euclidean motions in R^2 (rotations + translations)	Complex plane	Euclidean geometry
S^2	$K > 0$	Rotation group in R^3	Riemann sphere	Spherical geometry (elliptic geometry on P^2)
D	$K < 0$	Conformal group	Disc (hyperbolic plane, or upper half-plane)	Hyperbolic geometry

Relative to the last column in particular, these three examples are the only "elementary" geometries. Under the replacements *points → points, straight lines → geodesics, length → arc length, angles → angles,* and *congruence → isometry,* most of the axioms for Euclidean geometry are satisfied by these examples. The most noteworthy exception is that Euclid's fifth postulate, the parallel postulate, fails to hold on D: given a "straight line" l and a point p not on l, there are infinitely many "straight lines" through p which never meet l (see Figure 7.23).

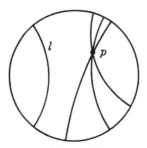

Figure 7.23

In the case of elliptic geometry, it is customary to use P^2 as model rather than S^2, so that any pair of "straight lines" will meet in at most one point. In this elliptic geometry a "straight line" fails to divide the remaining points into two disconnected parts.

In the case of complex variable theory, R^2, S^2, and D are the only simply connected complex 1-manifolds, a strong form of Riemann's mapping theorem.

8 Imbedded manifolds in R^3

In the previous chapter we studied the intrinsic geometry of a surface. Now we look at the properties of a surface as it lies in R^3.

Let (M, f) be a submanifold of R^3, where M is an oriented 2-manifold. Let M be given the Riemannian structure induced from f: for $v_1, v_2 \in T(M, m)$,

$$\langle v_1, v_2 \rangle_m = \langle df(v_1), df(v_2) \rangle_{f(m)}.$$

Definition. Given (M, f), the *spherical map* on M is the map $s: M \to S^2$ defined as follows. For $m \in M$, let $s(m) \in T(R^3, f(m))$ be the unit vector perpendicular to $df(T(M, m))$ that is consistent with the orientations of $T(M, m)$ and $T(R^3, f(m))$; that is, if v is a unit vector in $T(M, m)$, then $s(m)$ is the (unique) unit vector in $T(R^3, f(m))$ such that $\{df(v), df(iv), s(m)\}$ is an oriented orthonormal basis for $T(R^3, f(m))$. Identifying $T(R^3, f(m))$ with R^3 in the usual way, we may regard $s(m)$ as a point in S^2. (Figure 8.1).

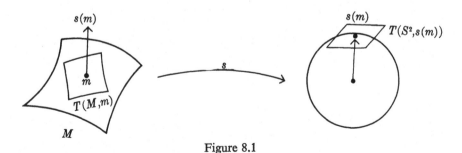

Figure 8.1

We shall find it convenient to identify $T(M, m)$ with $df(T(M, m))$. Furthermore, since $df(T(M, m))$ and $T(S^2, s(m))$ are the same subspace of R^3 under

216

the usual identification of the tangent spaces to R^3 with R^3, we shall identify $T(M, m)$ and $T(S^2, s(m))$. Then

$$ds(m): T(M, m) \rightarrow T(S^2, s(m)) = T(M, m),$$

so $ds(m)$ is a linear transformation from $T(M, m)$ into itself. The main result of this chapter is the theorem of Gauss: The curvature $K(m)$ of M at m is equal to the determinant of $ds(m)$. In fact, Gauss originally defined K (for imbedded manifolds) to be this determinant, and he was surprised to discover that K was in fact intrinsic; that is, K depends only on the Riemannian structure on M and not on the particular imbedding f.

Let us first note what manifolds in R^3 look like in local coordinates in terms of the spherical map. For $f: M \rightarrow R^3$, let $f_j = r_j \circ f \, (j = 1, 2, 3)$, so that

$$f(m) = (f_1(m), f_2(m), f_3(m))$$

for all $m \in M$. Here r_j are the coordinate functions on R^3. Let $\mu: U \rightarrow M$ be a smooth injective map from an open set $U \subset R^2$ onto an open set in M such that μ^{-1} is smooth, and such that if τ is a 2-form on M which gives the orientation of M, then $\mu^*\tau$ is a positive multiple of $dr_1 \wedge dr_2$. Thus μ^{-1} is an oriented coordinate system on $\mu(U)$. Let $x_i = r_i \circ \mu^{-1} \, (i = 1, 2)$ denote the coordinate functions on $\mu(U)$. Then $\partial/\partial x_i = d\mu(\partial/\partial r_i)$ and

$$df\left(\frac{\partial}{\partial x_i}\right) = d(f \circ \mu)\left(\frac{\partial}{\partial r_i}\right) = \sum_{k=1}^{3} \frac{\partial(f_k \circ \mu)}{\partial r_i} \frac{\partial}{\partial r_k} \qquad (i = 1, 2),$$

so that the matrix for df, relative to the bases $\{\partial/\partial x_i\}$ for $T(M, m)$ and $\{\partial/\partial r_j\}$ for $T(R^3, f(m))$, is

$$df \sim \begin{pmatrix} \dfrac{\partial(f_1 \circ \mu)}{\partial r_1} & \dfrac{\partial(f_1 \circ \mu)}{\partial r_2} \\[2ex] \dfrac{\partial(f_2 \circ \mu)}{\partial r_1} & \dfrac{\partial(f_2 \circ \mu)}{\partial r_2} \\[2ex] \dfrac{\partial(f_3 \circ \mu)}{\partial r_1} & \dfrac{\partial(f_3 \circ \mu)}{\partial r_2} \end{pmatrix}.$$

The Riemannian structure on $\mu(U)$ is then given by

$$\left\langle \frac{\partial}{\partial x_i}, \frac{\partial}{\partial x_j} \right\rangle = \sum_{k=1}^{3} \frac{\partial(f_k \circ \mu)}{\partial r_i} \frac{\partial(f_k \circ \mu)}{\partial r_j} \qquad (i, j \in \{1, 2\}).$$

Since $df(\partial/\partial x_1) \times df(\partial/\partial x_2)$—cross product in R^3—is perpendicular to $df(T(M, m))$ at each point $m \in U$ and is consistent with the orientations of $T(M, m)$ and $T(R^3, f(m))$, the spherical map on $\mu(U)$ is given by

$$s(m) = \frac{df(\partial/\partial x_1) \times df(\partial/\partial x_2)}{\|df(\partial/\partial x_1) \times df(\partial/\partial x_2)\|}$$

$$= \frac{1}{a} \left(\frac{\partial(f_1 \circ \mu)}{\partial r_1}, \frac{\partial(f_2 \circ \mu)}{\partial r_1}, \frac{\partial(f_3 \circ \mu)}{\partial r_1} \right) \times \left(\frac{\partial(f_1 \circ \mu)}{\partial r_2}, \frac{\partial(f_2 \circ \mu)}{\partial r_2}, \frac{\partial(f_3 \circ \mu)}{\partial r_2} \right),$$

217

where a is the magnitude of this cross product. In particular, we see that s is a smooth map.

Now consider a fixed point $m_0 \in M$. Altering f by a translation and rotation if necessary (this will leave the induced Riemannian structure on M unchanged), we may assume that $f(m_0) = 0$ and that $df(T(M, m_0))$ is tangent to the (r_1, r_2)-plane R^2 and is oriented consistently with the natural orientation of R^2; that is, $f^*(dr_1 \wedge dr_2)$ gives the orientation of $T(M, m_0)$ (Figure 8.2). Let $p: R^3 \to R^2$ be projection; that is,

$$p(u_1, u_2, u_3) = (u_1, u_2).$$

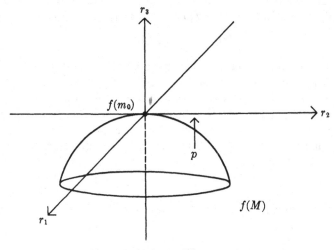

Figure 8.2

Then $p \circ f: M \to R^2$; $p \circ f(m_0) = 0$; and

$$d(p \circ f)(m_0) = dp \circ df(m_0) = df(m_0)$$

maps $T(M, m_0)$ isomorphically onto $T(R^2, 0)$. By the inverse function theorem, there exists a neighborhood U of 0 in R^2 such that $(p \circ f)^{-1}$ is defined and is smooth on U. Let $\mu = (p \circ f)^{-1}: U \to M$. Then μ^{-1} is a local coordinate system about m_0. Moreover, since $p \circ f \circ \mu = i_U$,

$$f \circ \mu(u_1, u_2) = (u_1, u_2, g(u_1, u_2)) \qquad ((u_1, u_2) \in U),$$

where $g = r_3 \circ f \circ \mu: U \to R^1$. Thus, if x_1, x_2 are the coordinate functions on $\mu(U)$, then the matrix for df relative to this coordinate system is

$$df \sim \begin{pmatrix} 1 & 0 \\ 0 & 1 \\ \dfrac{\partial g}{\partial r_1} & \dfrac{\partial g}{\partial r_2} \end{pmatrix},$$

and the spherical map on $\mu(U)$ is given by

$$s(\mu(u_1, u_2)) = \frac{1}{a(u_1, u_2)} \left(1, 0, \frac{\partial g}{\partial r_1}(u_1, u_2)\right) \times \left(0, 1, \frac{\partial g}{\partial r_2}(u_1, u_2)\right)$$

$$= \frac{1}{a(u_1, u_2)} \left(-\frac{\partial g}{\partial r_1}(u_1, u_2), -\frac{\partial g}{\partial r_2}(u_1, u_2), 1\right);$$

that is,

$$s \circ \mu = \frac{1}{a}\left(-\frac{\partial g}{\partial r_1}, -\frac{\partial g}{\partial r_2}, 1\right),$$

where $a = \sqrt{(\partial g/\partial r_1)^2 + (\partial g/\partial r_2)^2 + 1}$. Note that at $m_0 = \mu(0)$, $s(m_0) \perp df(T(M, m_0)) = R^2$ implies that $(\partial g/\partial r_1)(0) = (\partial g/\partial r_2)(0) = 0$ and $a(0) = 1$.

Now under our identification of $T(M, m_0)$ with $df(T(M, m_0))$, $(\partial/\partial x_i)(m_0)$ becomes identified with $(\partial/\partial r_i)(0)$ $(i = 1, 2)$, so that the entries b_{ij} for the matrix for

$$ds: T(M, m_0) \rightarrow T(M, m_0)$$

relative to the basis $\{\partial/\partial x_i\}$ are given by

$$b_{ji} = \left\langle ds\left(\frac{\partial}{\partial x_i}\right), \frac{\partial}{\partial x_j}\right\rangle_{m_0} = \left\langle d(s \circ \mu)\left(\frac{\partial}{\partial r_i}\right), \frac{\partial}{\partial r_j}\right\rangle_0$$

$$= dr_j\left(d(s \circ \mu)\left(\frac{\partial}{\partial r_i}\right)\right)\bigg|_0 = d(r_j \circ s \circ \mu)\left(\frac{\partial}{\partial r_i}\right)\bigg|_0$$

$$= \frac{\partial}{\partial r_i}(r_j \circ s \circ \mu)\bigg|_0 .$$

Thus

$$b_{11} = \frac{\partial}{\partial r_1}\left(-\frac{1}{a}\frac{\partial g}{\partial r_1}\right)\bigg|_0 = \frac{1}{a^2}\frac{\partial a}{\partial r_1}\frac{\partial g}{\partial r_1}\bigg|_0 - \frac{1}{a}\frac{\partial^2 g}{\partial r_1^2}\bigg|_0 = -\frac{\partial^2 g}{\partial r_1^2}\bigg|_0 ,$$

and, similarly,

$$b_{12} = -\frac{\partial^2 g}{\partial r_1 \partial r_2}\bigg|_0 = b_{21},$$

$$b_{22} = -\frac{\partial^2 g}{\partial r_2^2}\bigg|_0 ;$$

that is,

$$ds(m_0) \sim \begin{pmatrix} -\dfrac{\partial^2 g}{\partial r_1 \partial r_1} & -\dfrac{\partial^2 g}{\partial r_2 \partial r_1} \\ -\dfrac{\partial^2 g}{\partial r_1 \partial r_2} & -\dfrac{\partial^2 g}{\partial r_2 \partial r_2} \end{pmatrix}_0 .$$

In particular, assuming Gauss's theorem (which still remains to be proved) that K is the determinant of ds, we obtain a qualitative description of the behavior of $f(M)$ near the point $f(m_0)$ under various assumptions on the curvature $K(m_0)$. For from the critical point theory of functions of two variables, we know that the function g will have a maximum at the critical point 0, provided that the eigenvalues of the matrix

$$\begin{pmatrix} \dfrac{\partial^2 g}{\partial r_1{}^2} & \dfrac{\partial^2 g}{\partial r_1\, \partial r_2} \\[2ex] \dfrac{\partial^2 g}{\partial r_1\, \partial r_2} & \dfrac{\partial^2 g}{\partial r_2{}^2} \end{pmatrix}_0$$

are both negative and that g will have a minimum if the eigenvalues are both positive. Thus if $K(m_0) > 0$, then near 0, $f(M)$ lies either completely above or completely below the R^2-plane. On the other hand, if $K(m_0) < 0$, then $ds\,(m_0)$ has one positive and one negative eigenvalue, and g has a saddle point at 0. If $K(m_0) = 0$, the behavior of $f(M)$ near 0 is undetermined (see Figure 8.3).

In order to prove Gauss's theorem, we need an explicit description of the local geometry of imbedded manifolds.

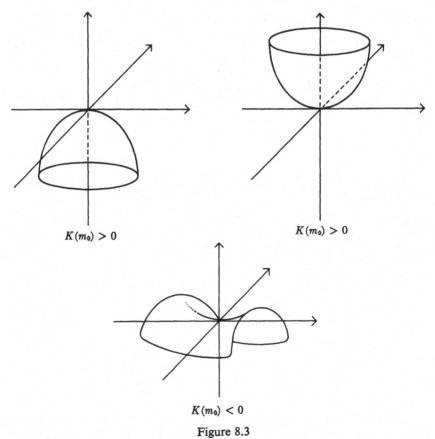

$K(m_0) > 0$

$K(m_0) > 0$

$K(m_0) < 0$

Figure 8.3

Theorem 1. *Let (M, f) be an oriented 2-manifold in R^3. Let $e_1, e_2 \colon S(M) \to R^3$ be defined by*

$$e_1(m, v) = df(v),$$
$$e_2(m, v) = df(iv).$$

Let ψ be the 1-form on $S(M)$ defined by

$$\psi = \langle de_1, e_2 \rangle;$$

that is, for $(m, v) \in S(M)$ and $t \in T(S(M), (m, v))$, $de_1(t) \in T(R^3, e_1(m, v)) = R^3$, and

$$\psi(t) = \langle de_1(t), e_2(m, v) \rangle.$$

Then ψ is the 1-form of the Riemannian connection on M.

PROOF

Invariance. For $g = e^{i\theta} \in S^1$,

$$g^*\psi = \psi \circ dg = \langle de_1 \circ dg, e_2 \circ g \rangle = \langle d(e_1 \circ g), e_2 \circ g \rangle.$$

But, for $(m, v) \in S(M)$,

$$e_1 \circ g(m, v) = e_1(m, \cos \theta \, v + \sin \theta \, (iv))$$
$$= \cos \theta \, df(v) + \sin \theta \, df(iv),$$
$$e_2 \circ g(m, v) = e_2(m, \cos \theta \, v + \sin \theta \, (iv))$$
$$= -\sin \theta \, df(v) + \cos \theta \, df(iv),$$

so

$$e_1 \circ g = \cos \theta \, e_1 + \sin \theta \, e_2,$$
$$e_2 \circ g = -\sin \theta \, e_1 + \cos \theta \, e_2.$$

Thus

$$d(e_1 \circ g) = \cos \theta \, de_1 + \sin \theta \, de_2.$$

Now since $\langle e_1, e_1 \rangle \equiv 1$, $\langle e_2, e_2 \rangle \equiv 1$, and $\langle e_1, e_2 \rangle \equiv 0$,

$$0 = d\langle e_1, e_1 \rangle = 2\langle de_1, e_1 \rangle,$$
$$0 = d\langle e_2, e_2 \rangle = 2\langle de_2, e_2 \rangle,$$
$$0 = d\langle e_1, e_2 \rangle = \langle de_1, e_2 \rangle + \langle e_1, de_2 \rangle;$$

so that $\langle de_1, e_1 \rangle = \langle de_2, e_2 \rangle = 0$, and $\langle e_1, de_2 \rangle = -\langle de_1, e_2 \rangle$. Hence

$$g^*\psi = \langle d(e_1 \circ g), e_2 \circ g \rangle = (\cos^2 \theta + \sin^2 \theta)\langle de_1, e_2 \rangle = \psi.$$

Normalization. For $(m, v) \in S(M)$, $V(m, v) = \dot\gamma(0)$, where $\gamma \colon R \to S(M)$ is given by

$$\gamma(\theta) = (m, \cos \theta \, v + \sin \theta \, (iv)).$$

Hence

$$\psi(V(m, v)) = \psi(\dot{\gamma}(0)) = \langle de_1(\dot{\gamma}(0)), e_2(\gamma(0)) \rangle$$
$$= \langle (e_1 \overset{\cdot}{\circ} \gamma)(0), df(iv) \rangle.$$

But

$$e_1 \circ \gamma(\theta) = \cos \theta \; df(v) + \sin \theta \; df(iv),$$

so that

$$(e_1 \overset{\cdot}{\circ} \gamma)(0) = -\sin \theta \; df(v) + \cos \theta \; df(iv)|_0 = df(iv),$$

and

$$\psi(V(m, v)) = 1.$$

Thus ψ is a connection form on M. To verify that it is the form of the Riemannian connection, it suffices to check that the simplified forms of the first structural equations are satisfied. To do this, note that when $(m, v) \in S(M)$ and $t \in T(S(M), (m, v))$,

$$\omega_1(t) = \langle d\pi(t), v \rangle = \langle df \circ d\pi(t), df(v) \rangle = \langle d(f \circ \pi)(t), e_1(m, v) \rangle,$$

so

$$\omega_1 = \langle d(f \circ \pi), e_1 \rangle.$$

Similarly,

$$\omega_2 = \langle d(f \circ \pi), e_2 \rangle.$$

Also note that the component of $d(f \circ \pi)$ in each r_j-direction is the exterior differential of a function, namely,

$$\left\langle d(f \circ \pi), \frac{\partial}{\partial r_j} \right\rangle = dr_j \circ d(f \circ \pi) = d(r_j \circ f \circ \pi),$$

so that $d(d(f \circ \pi)) = 0$, where the exterior derivative of a 3-tuple of forms is computed componentwise. Hence

$$d\omega_i = d\langle d(f \circ \pi), e_i \rangle$$
$$= \langle dd(f \circ \pi), e_i \rangle - \langle d(f \circ \pi), de_i \rangle.$$

(∗)
$$\boxed{d\omega_i = -\langle d(f \circ \pi), de_i \rangle.}$$

But

$$d(f \circ \pi) = \langle d(f \circ \pi), e_1 \rangle e_1 + \langle d(f \circ \pi), e_2 \rangle e_2,$$

since $e_1(m, v)$ and $e_2(m, v)$ form an orthonormal basis for $df(T(M, m))$ for each $(m, v) \in S(M)$. Thus

$$d(f \circ \pi) = \omega_1 e_1 + \omega_2 e_2,$$

hence, from (∗),

$$d\omega_1 = -\langle \omega_1 e_1 + \omega_2 e_2, de_1 \rangle = -\omega_2 \wedge \langle e_2, de_1 \rangle = -\omega_2 \wedge \psi = \psi \wedge \omega_2,$$

$$d\omega_2 = -\langle \omega_1 e_1 + \omega_2 e_2, de_2 \rangle = -\omega_1 \wedge \langle e_1, de_2 \rangle = \omega_1 \wedge \psi = -\psi \wedge \omega_1. \quad \square$$

Interpretation of parallel translation for imbedded manifolds

Let α be a curve in M, and $\tilde\alpha$ some lift to $S(M)$. Then $e_1 \circ \tilde\alpha$ is a curve in R^3 lying on the unit sphere, so that $d(e_1 \circ \tilde\alpha)(d/dt)_{t_0} = (d/dt)(e_1 \circ \tilde\alpha)_{t_0}$ is always perpendicular to $e_1 \circ \tilde\alpha(t_0)$. Now $\tilde\alpha$ is horizontal if and only if $\psi(\dot{\tilde\alpha}) = 0$. By Theorem 1, this is the case if and only if $d(e_1 \circ \tilde\alpha)(d/dt)_{t_0}$ is perpendicular to $e_2 \circ \tilde\alpha(t_0)$. Since e_1 and e_2 span the tangent space, $\tilde\alpha$ is horizontal if and only if $(d/dt)(e_1 \circ \tilde\alpha)_{t_0}$ is perpendicular to $df(T(M, \alpha(t_0)))$ for all t_0.

Thus is X is a unit vector field defined along α and $\tilde\alpha(t) = (\alpha(t), X(\alpha(t)))$, then X is parallel along $\alpha \Leftrightarrow \tilde\alpha$ is horizontal $\Leftrightarrow (d/dt)(e_1 \circ \tilde\alpha) = (d/dt)(X \circ \alpha)$ is always perpendicular to the manifold; that is, X is parallel along α if and only if $X \circ \alpha$ is "constant along the manifold": the tangential component of the derivative of $X \circ \alpha$ in R^3 is identically zero. For imbedded manifolds, this characterization can be used in place of the abstract notion of connection.

Theorem 2 (Gauss). *Let (M, f) be an oriented 2-manifold in R^3. Let $s: M \to S^2$ be the spherical map. Then for each $m \in M$, $\det ds(m) = K(m)$.*

PROOF. Let $\tilde s: S(M) \to S(S^2)$ be defined by

$$\tilde s(m, v) = (s(m), df(v)).$$

Let ω_1, ω_2 and ψ be the structural 1-forms on $S(M)$, and let $\bar\omega_1$, $\bar\omega_2$, and $\bar\psi$ be the corresponding forms on $S(S^2)$. Then

(I) $$\tilde s^* \bar\psi = \psi.$$

For

$$\tilde s^* \bar\psi = \bar\psi \circ d\tilde s = \langle d\bar e_1 \circ d\tilde s, \bar e_2 \circ \tilde s \rangle = \langle d(\bar e_1 \circ \tilde s), \bar e_2 \circ \tilde s \rangle.$$

But

$$\bar e_1 \circ \tilde s(m, v) = \bar e_1(s(m), df(v)) = df(v) = e_1(m, v),$$

$$\bar e_2 \circ \tilde s(m, v) = \bar e_2(s(m), df(v)) = i \, df(v) = s(m) \times df(v)$$
$$= df(iv) = e_2(m, v);$$

so that $\bar e_i \circ \tilde s = e_i$, and $\tilde s^* \bar\psi = \langle de_1, e_2 \rangle = \psi$.

(II) $$\begin{cases} \tilde s^* \bar\omega_1 = a_{11}\omega_1 + a_{12}\omega_2, \\ \tilde s^* \bar\omega_1 = a_{21}\omega_1 + a_{22}\omega_2, \end{cases}$$

where the a_{ij} are the smooth functions on $S(M)$ such that

$$\begin{pmatrix} a_{11} & a_{12} \\ a_{21} & a_{22} \end{pmatrix}_{(m, v)}$$

is the matrix for $ds(m)$ relative to the basis $\{v, iv\}$ for $T(M, m)$.

223

For if $(m, v) \in S(M)$ and $t \in T(S(M), (m, v))$, then

$$(\tilde{s}^*\bar{\omega}_1)(t) = \bar{\omega}_1 \circ d\tilde{s}(t)$$
$$= \langle d\bar{\pi}(d\tilde{s}(t)), df(v) \rangle \quad (\text{since } \tilde{s}(m, v) = (s(m), df(v)))$$
$$= \langle d(\bar{\pi} \circ \tilde{s})(t), df(v) \rangle$$
$$= \langle d(s \circ \pi)(t), df(v) \rangle$$
$$= \langle ds(d\pi(t)), df(v) \rangle$$
$$= \langle ds(\omega_1(t)v + \omega_2(t)(iv)), df(v) \rangle$$
$$= \langle ds(v), v \rangle \omega_1(t) + \langle ds(iv), v \rangle \omega_2(t),$$

since $df(v)$ is identified with v. Similarly,

$$(\tilde{s}^*\bar{\omega}_2)(t) = \langle ds(v), iv \rangle \omega_1(t) + \langle ds(iv), iv \rangle \omega_2(t),$$

completing the proof of (II).

(III) For each $m \in M$, $\det ds(m) = K(m)$.

For

$$-(K \circ \pi)\omega_1 \wedge \omega_2 = d\psi = d(\tilde{s}^*\bar{\psi}) = \tilde{s}^*(d\bar{\psi})$$
$$= \tilde{s}^*(-\bar{\omega}_1 \wedge \bar{\omega}_2) \quad (\text{since } S^2 \text{ has constant curvature 1})$$
$$= -\tilde{s}^*\bar{\omega}_1 \wedge \tilde{s}^*\bar{\omega}_2$$
$$= -(a_{11}\omega_1 + a_{12}\omega_2) \wedge (a_{21}\omega_1 + a_{22}\omega_2)$$
$$= -(a_{11}a_{22} - a_{12}a_{21})\omega_1 \wedge \omega_2,$$

so that

$$K(m) = \det \begin{pmatrix} a_{11} & a_{12} \\ a_{21} & a_{22} \end{pmatrix}_{(m,v)} = \det ds(m), \qquad \square$$

Remark. The linear transformation $ds(m)$ is called the *second fundamental form* of M at m. Note that it is a self-adjoint linear transformation, since its matrix relative to the special coordinates discussed above is the symmetric matrix

$$\begin{pmatrix} -\dfrac{\partial^2 g}{\partial r_1{}^2} & -\dfrac{\partial^2 g}{\partial r_2\,\partial r_1} \\[2ex] -\dfrac{\partial^2 g}{\partial r_1\,\partial r_2} & -\dfrac{\partial^2 g}{\partial r_2{}^2} \end{pmatrix}.$$

Theorem 3. *Let (M, f) be a compact oriented 2-manifold in R^3. Then there exists $m_0 \in M$ such that $K(m_0) > 0$.*

PROOF. Consider intuitively any sphere with M inside. Shrink this sphere until it just touches M at some point m_0. Then the curvature of M at m_0 is greater than the curvature of this sphere.

Consider more precisely the function $r^2 \circ f$ on M, where r is the distance from the origin in R^3. Let $m_0 \in M$ be a point where $r^2 \circ f$ assumes its maximum. By rotating about the origin in R^3 if necessary, we may assume that

$f(m_0)$ is on the r^3 axis; that is, that $f(m_0) = (0, 0, c)$ for some c. Then $r_3 \circ f$ also has an extremum at m_0 since $|r_3 \circ f|(m_0) = r \circ f(m_0) \geq r \circ f(m) \geq |r_3 \circ f|(m)$ for all $m \in M$. Hence for $v \in T(M, m_0)$,

$$\left\langle df(v), \frac{\partial}{\partial r^3} \right\rangle = dr_3(df(v)) = d(r_3 \circ f)(v) = 0;$$

that is, $df(T(M, m_0))$ is perpendicular to the r_3 axis. We may assume that the orientation of $df(T(M, m_0))$ agrees with the orientation on the r_1, r_2 plane. (Otherwise, a rotation through an angle of π about the r_1 axis will accomplish this.) We may further assume, by rotating about the r_3 axis if necessary, that $\partial/\partial r_1$ and $\partial/\partial r_2$ are eigenvectors of $ds(m_0)$. Then since $f(m_0) = (0, 0, c)$, the special coordinates are valid on the translate if $f(M)$ by $(0, 0, -c)$; that is, there exists $\mu: U \to M$, $U \subset R^2$ such that

$$f \circ \mu(u_1, u_2) = (u_1, u_2, c + g(u_1, u_2))$$

for some $g: U \to R$ with $g(0, 0) = 0$ and

$$\frac{\partial g}{\partial r_1}\bigg|_0 = \frac{\partial g}{\partial r_2}\bigg|_0 = 0.$$

Then, since $\partial/\partial r_1$ and $\partial/\partial r_2$ are eigenvectors of $ds(m_0)$, $ds(m_0)$ has matrix

$$\begin{pmatrix} \dfrac{\partial^2 g}{\partial r_1{}^2} & 0 \\[2ex] 0 & \dfrac{\partial^2 g}{\partial r_2{}^2} \end{pmatrix}_0.$$

But $r^2 \circ f \circ \mu$ has a maximum at $(0, 0)$. Hence since

$$r^2 \circ f \circ \mu(u_1, u_2) = u_1{}^2 + u_2{}^2 + [c + g(u_1, u_2)]^2,$$

then

$$0 \geq \frac{\partial^2}{\partial r_j{}^2}(r^2 \circ f \circ \mu)\bigg|_0 = \left[2 + 2\left(\frac{\partial g}{\partial r_j}\right)^2 + 2(c + g)\frac{\partial^2 g}{\partial r_j{}^2} \right]_0$$

$$= 2 + 2c \frac{\partial^2 g}{\partial r_j{}^2}\bigg|_0$$

for $j = 1, 2$. Thus

$$\frac{\partial^2 g}{\partial r_1{}^2}\bigg|_0 \leq -\frac{1}{c} \quad \text{and} \quad \frac{\partial^2 g}{\partial r_2{}^2}\bigg|_0 \leq -\frac{1}{c},$$

so that

$$K(m_0) = \det ds(m_0) = \frac{\partial^2 g}{\partial r_1{}^2}\frac{\partial^2 g}{\partial r_2{}^2}\bigg|_0 \geq \frac{1}{c^2} > 0. \qquad \square$$

Corollary. *The torus $S^1 \times S^1$ with its Riemannian product structure cannot be imbedded as a submanifold of R^3.*

PROOF. Since the covering map $R^2 \to S^1 \times S^1$ is a local isometry, $S^1 \times S^1$ has curvature identically zero. $\square$

Theorem 4. *Let (M, f) be a compact connected oriented 2-manifold in R^3. Suppose the curvature K of M is never zero. Then in fact $K > 0$ everywhere. Furthermore, the spherical map $s: M \to S^2$ is injective, surjective, and has a smooth inverse; that is, M and S^2 are diffeomorphic (isomorphic as manifolds).*

PROOF. The first statement is a consequence of the previous theorem. For the second statement, consider the map $\pi \circ s$ mapping M into the projective plane P^2, where $\pi: S^2 \to P^2$ is projection. We show that $\pi \circ s$ is a covering map. First, since $\det ds = K > 0$, ds is everywhere nonsingular, hence so is $d\pi \circ ds = d(\pi \circ s)$. By the inverse function theorem, $\pi \circ s$ is locally one-to-one and is an open map. Moreover, $\pi \circ s$ is onto. For if $p \in P^2$—say $p = \pi(\tilde{p})$, $\tilde{p} \in S^2$—consider the plane in R^3 perpendicular to $\tilde{p}$. Moving this plane out in the direction of $\tilde{p}$ toward infinity and then back until it just touches M at m_0, we find that $s(m_0) = \pm \tilde{p}$, so that $\pi \circ s(m_0) = p$. More precisely, let $m_0 \in M$ be such that the function $m \to \langle f(m), \tilde{p} \rangle$ assumes its maximum at m_0. Then $\tilde{p} \perp df(T(M, m_0))$ because, for $v \in T(M, m_0)$,

$$\langle df(v), \tilde{p} \rangle = d(\langle f, \tilde{p} \rangle)(v) = 0.$$

Hence $s(m_0) = \pm \tilde{p}$, and $\pi \circ s(m_0) = p$. Finally, since M is compact, $(\pi \circ s)^{-1}(p)$ is finite for each $p \in P^2$. (It is closed in M, hence compact; on the other hand, each point in $\pi^{-1}(p)$ is open in the relative topology, since π is locally one-to-one.) Hence for U a sufficiently small open set about p, $(\pi \circ s)^{-1}(U)$ is a union of disjoint open sets each mapped homeomorphically onto U by $\pi \circ s$. Thus $\pi \circ s$ is a covering map, and M is a covering space of P^2. But the only covering spaces of P^2 are S^2 and P^2 itself. Since P^2 is not orientable, M must be homeomorphic to S^2 and s must be a smooth homeomorphism with a smooth inverse. $\square$

Theorem 5. *Let (M_1, f_1) and (M_2, f_2) be oriented 2-manifolds in R^3. Suppose M_1 and M_2 are tangent along a curve α in R^3; that is, suppose there exist curves $\alpha_i: [a, b] \to M_i$ $(i = 1, 2)$ such that*

$$f_1 \circ \alpha_1 = f_2 \circ \alpha_2 = \alpha,$$

and

$$df_1(T(M_1, \alpha_1(t))) = df_2(T(M_2, \alpha_2(t)))$$

for all $t \in [a, b]$. Then parallel translation along α is the same in both manifolds; if $v_1 \in T(M_1, \alpha_1(a))$ and $v_2 \in T(M_2, \alpha_2(a))$ with $df_1(v_1) = df_2(v_2)$, then

df_1 (*parallel translate of v_1 along α_1*)
        $= df_2$ (*parallel translate of v_2 along α_2*).

PROOF. Let $s_i: M_i \to S^2$ denote the spherical map $(i = 1, 2)$, and let $\tilde{s}_i: S(M_i) \to S(S^2)$ denote the corresponding map on the circle bundles. Since

$$df_1(T(M_1, \alpha_1(t))) = df_2(T(M_2, \alpha_2(t)))$$

for all $t \in [a, b]$, we have $s_1 \circ \alpha = \pm s_2 \circ \alpha$. We may assume that $s_1 \circ \alpha = +s_2 \circ \alpha$, for otherwise we may reverse the orientation on M_2 so that this is the case. (Note that orientation reversal on M_2 has the effect of changing the sign of the connection form on $S(M_2)$ and hence does not affect parallelism.)

From the proof of Theorem 2,

$$\tilde{s}_i^* \bar{\psi} = \psi_i,$$

where ψ_i is the connection form on $S(M_i)$. Let $\tilde{\alpha}_i: [a, b] \to S(M_i)$ denote the horizontal lift of α_i through $(\alpha_i(a), v_i)$. Then

$$\bar{\psi}\left(d(\tilde{s}_i \circ \tilde{\alpha}_i)\left(\frac{d}{dt}\right)\right) = \bar{\psi}\left(d\tilde{s}_i\left(d\tilde{\alpha}_i\left(\frac{d}{dt}\right)\right)\right) = (\tilde{s}_i^* \bar{\psi})\left(d\tilde{\alpha}_i\left(\frac{d}{dt}\right)\right)$$

$$= \psi_i\left(d\tilde{\alpha}_i\left(\frac{d}{dt}\right)\right) = 0$$

since $\tilde{\alpha}_i$ is horizontal. Hence $\tilde{s}_i \circ \tilde{\alpha}_i: [a, b] \to S(S^2)$ $(i = 1, 2)$ is the horizontal lift of $s_1 \circ \alpha_1 = s_2 \circ \alpha_2$ through

$$\tilde{s}_1 \circ \tilde{\alpha}_1(a) = (s_1(\alpha_1(a)), df_1(v_1))$$
$$= (s_2(\alpha_2(a)), df_2(v_2))$$
$$= \tilde{s}_2 \circ \tilde{\alpha}_2(a).$$

In particular, $\tilde{s}_1 \circ \tilde{\alpha}_1 = \tilde{s}_2 \circ \tilde{\alpha}_2$. Hence, if $v_i(t)$ denotes the parallel translate of v_i along α_i to $\alpha_i(t)$, then $\tilde{\alpha}_i(t) = (\alpha_i(t), v_i(t))$ $(i = 1, 2)$, and

$$(s_1(\alpha_1(t)), df_1(v_1(t))) = \tilde{s}_1(\tilde{\alpha}_1(t))$$
$$= \tilde{s}_1 \circ \tilde{\alpha}_1(t)$$
$$= \tilde{s}_2 \circ \tilde{\alpha}_2(t)$$
$$= (s_2(\alpha_2(t)), df_2(v_2(t)));$$

that is, $df_1(v_1(t)) = df_2(v_2(t))$. $\square$

Remark. Theorem 5 gives us a way of seeing geometrically how parallel translation behaves for submanifolds in R^3. Let $\alpha: [a, b] \to M$ be a curve. Consider the family of tangent planes to $f(M)$ along the curve $f \circ \alpha$ in R^3. For $t_0, t \in [a, b]$, the intersection of the tangent plane at $f \circ \alpha(t_0)$ and the tangent plane at $f \circ \alpha(t)$ will generally be a line. The limit of this line, as $t \to t_0$, will be a line through $f \circ \alpha(t_0)$ in the plane $df(T(M, \alpha(t_0)))$. The collection of all these lines, for $t_0 \in [a, b]$, forms a surface D, called a developable surface. It turns out that such a surface D is flat $(K = 0)$ and is tangent to $f(M)$ along $f \circ \alpha$. Hence parallel translation in M along α is the same as parallel translation along $f \circ \alpha$ in D. But since D is flat, it is locally isometric to a piece of the plane; that is, in a neighborhood of α, D can be rolled out on the plane where parallel translation is ordinary translation. In particular, parallel translation along a circle on a sphere in R^3 is the same as parallel translation along that circle in the cone tangent to the sphere along that circle. But the cone can be rolled out on the plane where parallel translation is ordinary translation; then the cone can be rolled back around the sphere to find the parallel translate on the sphere (Figure 8.4).

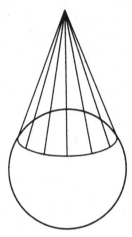

Figure 8.4

Theorem 6. *Let M be an oriented connected 2-manifold, and let $f, g: M \to R^3$ be two isometric imbeddings. Suppose that the second fundamental forms of (M, f) and (M, g) are the same. Then there exists a rigid motion (isometry) Φ of R^3 such that $f = \Phi \circ g$.*

SKETCH OF PROOF. Choose a point $(m_0, v_0) \in S(M)$ and a rigid motion Φ carrying $g(m_0)$ into $f(m_0)$, $dg(v_0)$ into $df(v_0)$, and $dg(T(M, m_0))$ onto $df(T(M, m_0))$. Let $f_1 = f$ and $f_2 = \Phi \circ g$; then $f_1(m_0) = f_2(m_0)$, $df_1(v_0) = df_2(v_0)$, and

$$df_1(T(M, m_0)) = df_2(T(M, m_0)).$$

Hence $\tilde{s}_1(m_0, v_0) = \tilde{s}_2(m_0, v_0)$. Furthermore, because Φ is an isometry, the second fundamental forms of (M, f_1) and (M, f_2) are equal.

Note that as a result of formulas (I) and (II) in the proof of Theorem 2, the fact that the second fundamental forms for (M, f_i), $i = 1, 2$, are equal implies

$$\tilde{s}_i: S(M) \to S(S^2)$$

has the following property: $\tilde{s}_1{}^*(\bar{\psi}) = \tilde{s}_2{}^*(\bar{\psi})$ and $\tilde{s}_1{}^*(\bar{\omega}_i) = \tilde{s}_2{}^*(\bar{\omega}_i)$, $i = 1, 2$.

Note also that $S(S^2)$ can be identified with the space of 3×3 orthogonal matrices of determinant one, for a point of $S(S^2)$ consists of a pair (u, v), where u is a unit vector in R^3, and v is a unit vector perpendicular to u. The matrix corresponding to (u, v) is the one whose columns are u, v, and $u \times v$.

Lemma. *Let α_1, α_2 be two curves $[0, 1] \to S(S^2)$ such that $\alpha_1{}^*(\bar{\psi}) = \alpha_2{}^*(\bar{\psi})$ and $\alpha_1{}^*(\bar{\omega}_i) = \alpha_2{}^*(\bar{\omega}_i)$ ($i = 1, 2$). Let β be the curve $\beta(t) = \alpha_2(t)^{-1} \cdot \alpha_1(t)$, where inverse is matrix inverse and the dot indicates matrix multiplication. Then $\beta^*(\bar{\psi}) = \beta^*(\bar{\omega}_i) = 0$.*

PROOF. We leave the proof of this lemma to the student. It involves computing $d\beta$ in terms of $d\alpha_1$ and $d\alpha_2$.

Corollary to the Lemma. *If in addition, $\alpha_1(0) = \alpha_2(0)$, then $\alpha_1 = \alpha_2$.*

PROOF. Since $\bar{\omega}_1$, $\bar{\omega}_2$, $\bar{\psi}$ span the cotangent space at each point of $S(S^2)$, the lemma implies that $d\beta$ is identically 0, so that β is a constant map. Since $\alpha_1(0) = \alpha_2(0)$, $\beta(0)$ is the identity matrix, and, therefore, so is $\beta(t)$ for all $t \in [0, 1]$. Hence $\alpha_1(t) = \alpha_2(t)$ for all $t \in [0, 1]$.

We are now in a position to show that $f_1 = f_2$. Let γ be a curve in $S(M)$ starting at (m_0, v_0), and let $\alpha_i = \tilde{s}_i \circ \gamma$. Then the curves α_i satisfy the hypothesis of the lemma, and $\alpha_1(0) = \alpha_2(0)$, so we conclude from the corollary that $\alpha_1 = \alpha_2$. In particular, $\tilde{s}_1(\gamma(t)) = \tilde{s}_2(\gamma(t))$ ($t \in [0, 1]$). Since any point of $S(M)$ is reachable by a smooth curve starting at (m_0, v_0), we have $\tilde{s}_1 = \tilde{s}_2$; that is, for every $(m, v) \in S(M)$, $s_1(m) = s_2(m)$ and $df_1(v) = df_2(v)$.

Consider the map $f_1 - f_2 \colon M \to R^3$, where $(f_1 - f_2)(m) = f_1(m) - f_2(m)$. Since $df_1(v) - df_2(v) = 0$ for any unit vector, we conclude $d(f_1 - f_2) = 0$; that is, $f_1 = f_2 + $ constant. Since $f_1(m_0) = f_2(m_0)$, we have $f_1 = f_2$. $\qquad\square$

Bibliography

Topology

1. Alexandroff, P. and Hopf, H. *Topologie.* Berlin: Springer-Verlag, 1935.
2. Bourbaki, N. *Topologie Générale.* Paris: Hermann, 1953.
3. Eilenberg, S. and Steenrod, N. E. *Foundations of Algebraic Topology.* Princeton, N.J.: Princeton University Press, 1952.
4. Greenberg, M., *Lectures on Algebraic Topology.* New York: Benjamin, 1967.
5. Hocking, J. G. and Young, G. S. *Topology.* Reading, Mass.: Addison-Wesley, 1961.
6. Hu, S. T. *Elements of General Topology.* San Francisco: Holden-Day, 1964.
7. Kelley, J. L. *General Topology.* New York: Van Nostrand, 1955; Springer-Verlag, 1975.
8. Lefschetz, S. *Introduction to Topology.* Princeton, N.J.: Princeton University Press, 1949.
9. Pontryagin, L. S. *Foundations of Combinatorial Topology.* Rochester, N.Y.: Graylock, 1952.
10. Seifert, H. and Threlfall, W. *Lehrbuch der Topologie.* Leipzig: Teubner, 1934.
11. Sierpinski, W. *Introduction to General Topology.* Toronto, Ontario: University of Toronto Press, 1934.
12. Spanier, R. H. *Algebraic Topology.* New York: McGraw-Hill, 1966; Springer-Verlag, 1981.
13. Wallace, A. H. *Introduction to Algebraic Topology.* New York: Pergamon, 1957.

Geometry

1. Auslander, L. and Mackenzie, R. E. *Introduction to Differentiable Manifolds.* New York: McGraw-Hill, 1963.
2. Bishop, R. L. and Crittenden, R. J. *Geometry of Manifolds.* New York: Academic, 1964.
3. De Rham, G. *Variétés Differentiables.* Paris: Hermann, 1960.
4. DoCarmo, M. *Differential Geometry of Curves and Surfaces.* Englewood Cliffs, N.J.: Prentice-Hall, 1976.
5. Flanders, H. *Differential Forms, with Applications to the Physical Sciences.* New York: Academic, 1963.
6. Guggenheimer, H. W. *Differential Geometry.* New York: McGraw-Hill, 1963.
7. Hicks, N. J. *Notes on Differential Geometry.* Princeton, N.J.: Van Nostrand, 1965.
8. Kobayashi, S. and Nomizu, K. *Foundations of Differential Geometry*, Vol. I. New York: Interscience (Wiley), 1963.
9. Laugwitz, D. *Differential and Riemannian Geometry.* New York: Academic, 1965.
10. O'Neill, B. *Elementary Differential Geometry.* New York: Academic, 1966.
11. Sternberg, S. *Lectures on Differential Geometry.* Englewood Cliffs, N.J.: Prentice-Hall, 1964.
12. Warner, F. *Foundations of Differentiable Manifolds and Lie Groups.* Glenview, Ill.: Scott, Foresman, 1971; Springer-Verlag, 1983.
13. Whitney, H. *Geometric Integration Theory.* Princeton, N.J.: Princeton University Press, 1957.

Index